Advances in Sustainable Utilization and Optimal Decision of Land Resources

Advances in Sustainable Utilization and Optimal Decision of Land Resources

Editors

Dianfeng Liu
Wenwu Tang
Jianxin Yang

MDPI • Basel • Beijing • Wuhan • Barcelona • Belgrade • Manchester • Tokyo • Cluj • Tianjin

Editors
Dianfeng Liu
Wuhan University
Wuhan, China

Wenwu Tang
University of North Carolina at Charlotte
Charlotte, NC, USA

Jianxin Yang
China University of Geosciences
Wuhan, China

Editorial Office
MDPI
St. Alban-Anlage 66
4052 Basel, Switzerland

This is a reprint of articles from the Special Issue published online in the open access journal *Sustainability* (ISSN 2071-1050) (available at: https://www.mdpi.com/journal/sustainability/special_issues/sustainable_utilization_and_optimal_decision).

For citation purposes, cite each article independently as indicated on the article page online and as indicated below:

LastName, A.A.; LastName, B.B.; LastName, C.C. Article Title. *Journal Name* **Year**, *Volume Number*, Page Range.

ISBN 978-3-0365-8344-0 (Hbk)
ISBN 978-3-0365-8345-7 (PDF)

Contents

About the Editors

Dianfeng Liu

Dr. Dianfeng Liu received his Ph.D. in land resource management from Wuhan University in 2011. He is currently a professor working at the School of Resource and Environmental Sciences at Wuhan University, China. His research focuses on sustainable land use and modelling, land use change and ecological effects, rural studies, and spatial planning. He has published over 70 papers and led three national natural science funds of China. He is a committee member of the big data branch of the China Society of Natural Resources and a member of the technology innovation team of the Ministry of Natural Resources, China.

Wenwu Tang

Dr Wenwu Tang holds a Ph.D. degree from the University of Iowa and was a Post-doc and Research Scientist at the U.S. National Center for Supercomputing Applications and the University of Illinois at Urbana-Champaign. His research interests include cyberinfrastructure and high-performance geocomputation, agent-based modeling, and land change modeling. Dr Tang has over 100 peer-reviewed publications, and his research has been supported by federal and state funding agencies, including USDA Forest Services, US CDC, US Fish and Wildlife, NCDOT, NC Forest Service, and Electric Power Research Institute.

Jianxin Yang

Dr. Jianxin Yang received his Ph.D. in land resource management from China University of Geosciences in 2019. He is currently an associate professor working at the School of Public Administration at China University of Geosciences, Wuhan. His research mainly focuses on modeling and analysis of land use change and its ecological effects, spatial planning, spatial-temporal statistics, multi-agent modeling, and complex network analysis. He has published over 35 papers and led/participated in 4 projects supported by national and provincial natural science funds of China.

Preface to "Advances in Sustainable Utilization and Optimal Decision of Land Resources"

Land resources are essential for human beings to survive. With rapid socioeconomic development and increasing urbanization, land resources have borne increasing pressure from multiple stakeholders, which accelerates land use competition and conflicts. In this context, the sustainable utilization of land resources has been regarded as one of the key indicators in the 2030 agenda for worldwide Sustainable Development Goals (SDGs), and has attracted a great deal of attention from countries around the world. The optimal utilization of land resources needs to take into account multiple aspects of land systems and interactions with other physical and socioeconomic systems (e.g., the ecosystem, climatic system, and human activities) to ensure that land use changes promote long-term ecological stability and human well-being.

This book explores various aspects of sustainable land use decisions, from urban to rural contexts, and the ecological effects of land use change. The book is divided into three sections: urban land use, rural land use, and land use change and its ecological effects.

The first section focuses on urban land use, discussing the challenges of urban growth and sprawl, urban green space, and industrial land transformation in China. This section presents new methodologies and models that use big data to understand resident behavior and evaluate the ecological benefits of urban green spaces. The section also explores the factors that influence industrial land transformation and its potential impacts on sustainable urban development.

The second section discusses rural land use, emphasizing the importance of a market-driven approach to rural construction in China. The section examines the intrinsic dynamics among local governments, market capital, and villagers in the market-driven pattern, highlighting its potential to fulfill villagers' interests and enhance sustainable rural development. Additionally, the section investigates the role of farmers as the primary decision-makers in revitalizing empty plots of land and promoting the reduction of chemical fertilizers to achieve green and sustainable agriculture.

The third section explores land use change and its ecological effects, discussing the fundamental role of land use in ecological civilization. The section presents results that guide land use changes to promote ecological stability and human well-being. The section also provides insights into the complexity of ecological systems.

The book's contributors come from diverse fields, including geography, environmental science, urban planning, and agriculture. They provide a comprehensive overview of the challenges and opportunities in sustainable land use decisions, presenting new methodologies and models that can help policymakers and practitioners make informed decisions.

In conclusion, this book aims to provide readers with an understanding of sustainable land use decisions, from urban to rural contexts, and the ecological effects of land use change. The book's interdisciplinary approach and diverse perspectives offer valuable insights into the complex issues of land use and its impacts on the environment and human well-being. We hope that this book will contribute to the ongoing efforts to promote sustainable land use decisions worldwide.

Dianfeng Liu, Wenwu Tang, and Jianxin Yang

Editors

sustainability

MDPI

Article

Urban Land Expansion Simulation Considering the Diffusional and Aggregated Growth Simultaneously: A Case Study of Luoyang City

Renyang Wang [1], Weishan He [2], Dang Wu [3], Lu Zhang [4,*] and Yujia Li [5,*]

1. Research Institute of New Economic, Ningbo University of Finance & Economics, Ningbo 315175, China; wangrenyang@nbufe.edu.cn
2. Anhui Transport Consulting & Design Institute CO., LTD., Hefei 230088, China; hwshan333@163.com
3. Wuhan Land Use and Urban Spatial Planning Research Center, Wuhan 430010, China; wudang@wlsp.org.cn
4. School of Public Administration, Central China Normal University, Wuhan 430079, China
5. College of Public Administration, Huazhong University of Science & Technology, Wuhan 430074, China

* Correspondence: zhanglu54522@mail.ccnu.edu.cn (L.Z.); yujialee@hust.edu.cn (Y.L.)

Abstract: Restricted by urban development stages, natural conditions, urban form and structure, diffusional growth occupies a large proportion of area in many cities. Traditional cellular automata (CA) has been widely applied in urban growth studies because it can simulate complex system evolution with simple rules. However, due to the limitation of neighborhood conditions, it is insufficient for simulating urban diffusional growth process. A maximum entropy mode was used to estimate three layers of probability spaces: the probability layer of cell transformation from non-urban status to urban status (PLCT), the probability layer for aggregated growth (PLAP), and the probability layer for diffusional growth (PLOP). At the same time, a maxent category selected CA model (MaxEnt-CSCA) was designed to simulate aggregated and diffusional urban expansion processes simultaneously. Luoyang City, with a large proportion of diffusional urban expansion (65.29% in 2009–2018), was used to test the effectiveness of MaxEnt-CSCA. The results showed that: (1) MaxEnt-CSCA accurately simulated aggregated growth of 47.40% and diffusional growth of 37.13% in Luoyang from 2009 to 2018, and the overall Kappa coefficient was 0.78; (2) The prediction results for 2035 showed that future urban expansion will mainly take place in Luolong District and the counties around the main urban area, and the distribution pattern of Luolong District will change from the relative diffusion state to the aggregation stage. This paper also discusses the applicable areas of MaxEnt-CSCA and illustrates the importance of selecting an appropriate urban expansion model in a region with a large amount of diffusional growth.

Keywords: diffusional urban expansion; maximum information entropy; cellular automata; Luoyang City

Citation: Wang, R.; He, W.; Wu, D.; Zhang, L.; Li, Y. Urban Land Expansion Simulation Considering the Diffusional and Aggregated Growth Simultaneously: A Case Study of Luoyang City. *Sustainability* **2021**, *13*, 9781. https://doi.org/10.3390/su13179781

Academic Editor: Jacques Teller

Received: 12 July 2021
Accepted: 27 August 2021
Published: 31 August 2021

1. Introduction

Cellular automata (CA) has been widely used in the field of urban expansion modeling because it can simulate complex macro-scale evolutions using simple micro-scale conversion rules [1–3]. Limited by the rules of neighboring cells, most existing urban expansion CA models only consider the situation where the urban land is added around existing urban land [4,5], and can actually simulate an urban plane extending in a specific direction. However, in the real world, urban built-up space is often developed piece by piece with a diffusion form and may not be alongside the original urban land patch [6]. As a result, simply describing the dynamic changes of urban expansion as a consecutive extended surface is very insufficient. Urban land dynamics are actually a collection of complex evolution forms containing internally connected dots, the expansion of planes and the integrated interactions between dots and planes [7,8]. The growth of "dots" occurs in different periods of urban development [6]. In the early stage of urban expansion, a large number of small construction patterns emerge quickly around the original main city pattern [9,10], which manifest as "diffusion" and can be abstracted as "dots" in growth [5,11]. At the stable development stage, large urban blocks, newly constructed satellite cities, development zones, and small towns commonly appear, which are significantly smaller and are not spatially connected to the large urban blocks [12]. This phenomenon of disjoint and spatial separation can also be abstractly referred to as "dot" growth. In the traditional CA model, none of these "dot" growth processes can be effectively simulated [5,6].

According to the landscape expansion index [13–15], newly expanded urban patches can be classified into either edge, infill or outlying types [10,16]. Edge expansion refers to the extension of urban built-up areas along the original urban patch and infill expansion occurs in the internal voids patches, both of which can make the original urban built-up area more compact. Outlying expansion is a leapfrog development of land far away from the original urban land and can make the city more diffusional [7]. Few scholars have tried to improve traditional CA models to simulate outlying-type expansion. Clarke proposed an improved CA model called SLEUTH, which can randomly select newly generated points to generate extensional expansion [11], but the proliferation of new urban patches is not completely random. In fact, the occurrence of the new patches is the result of multiple driving forces [5]. As a result, SLEUTH may not be able to simulate a real urban proliferation scenario. Liu et al. proposed the SMDUGT model to simulate diffusion and aggregation in urban morphology evolution [6]. The basic idea of SMDUGT is to divide the expansion candidate areas of different expansion types based on regression results, and then consider neighborhood factors to construct the adjacent (edge and infill) and outlying (outlying) types. The cell conversion rules in the candidate regions for land-type expansion are integrated into the cellular automata model to obtain simulation methods for different types of regions. However, SMDUGT cannot determine where the "seed points" of diffusion growth are.

The maximum entropy model (MaxEnt) is a niche model based on the maximum entropy principle [17]. It takes research areas as all possible-probability distribution and finds the most likely distribution under the maximum entropy principle according to constraints given by environmental variables [18–21]. It finds an optimal distribution using single-period sample information to select the distribution with the largest entropy from various environmental variables and uses it to predict the probability of research object appearance [22]. Thanks to its simplicity and practicality, the model's application range has rapidly expanded from biological research to other fields. In the field of urban expansion simulation, Zhang et al., coupled the maximum entropy model with the cellular automata (CA), and proposed a new urban expansion model called Maxent-CA to analyze single-period training samples and obtain urban land growth suitability [18]. The proposed Maxent-CA model overcame the traditional CA model requirement of parameter correction from multi-phase remote sensing images and the request for partitioning when large-area simulations are needed. However, the simultaneous simulation of diffusional and aggregated growth processes could not be achieved in their study.

This paper attempted to use MaxEnt model to estimate the three-layers probability space: the probability layer of cell transformation from non-urban to urban status (PLCT), the probability layer of suitability distribution for adjacent patches (PLAP), and the probability layer of suitability distribution for outlying patches (PLOP). Of these, PLCT is used to sort the priority order of cells to be selected, and PLAP and PLOP are used to determine the type of the selected cell. Based on the three-layer probability space, the Maximum Entropy Category Selected CA model (Maxent Category Selected CA model, MaxEnt-CSCA) was proposed. The great advantage of the model is that it can simulate diffusional and aggregated growth simultaneously, which is closer to a real-world situation and greatly improves the model's predictive accuracy. The MaxEnt-CSCA model adopts a "divide and conquer" design paradigm, which distinguishes different features of the adjacent and outlying urban expansion type and integrates their information. First, the model identifies urban patch expansion types using the landscape expansion index, and then the MaxEnt model generates the three-layers probability space (PLCT, PLAP and PLOP), depending on patch geographic location and various kinds of social and natural environmental variables. Next, according to the different growth characteristics, two sets of local cell evolution rules are designed and integrated to the model. Finally, in the CA iterative simulation stage, the category significance (adjacent or outlying) of each geographic cell is determined according to the probability of the suitability layer and the corresponding local cell evolution rule is selected to determine the conversion state. The innovation of the MaxEnt-CSCA model is to improve the deficiencies of traditional CA in simulating outlying urban expansion, that is, to perform ternary instead of binary land use simulation (non-urban, urban-adjacent, and urban-outlying) which allows simultaneous simulation of diffusional and aggregated growth.

Luoyang City, located in Henan Province, China, was selected to test the effectiveness of MaxEnt-CSCA model. From 2009–2018, Luoyang city expanded significantly, and the growth of outlying urban patches was very obvious. According to the Chinese land use survey data, the percentage of outlying growth was 65.29%. Therefore, traditional CA models were unable to accurately capture the urban growth trajectory of Luoyang City. The objective of this paper is to verify the practicability and effectiveness of maxent-CSCA in modeling urban dynamic evolution processes where diffusional growth accounts for a significant proportion, and call on scholars related to urban modeling research to pay more attention to different types of urban expansion in the process of urban development. Our expected result is that Maxent-CSCA can achieve higher simulation accuracy than Maxent-CA and Base-CA without considering diffusional process, and thus is more suitable for the prediction and simulation of the future urban landscape of Luoyang city in 2035.

2. Study Area and Data

Luoyang city is located in the west of Henan Province, bordering Zhengzhou, the capital of Henan Province. Its latitude ranges from 118°08′ E to 112°59′ E and the longitude range ranges from 33°39′ N to 35°05′ N. The total population of the city is 7.1702 million, of which 4.091 million live in urban areas and the urbanization level is 59.10%. The population density is 455 people/km^2 and GDP is 503.49 billion yuan by 2019. Luoyang City has 15 county-level administrative districts in total. However, considering that urban built-up areas are mostly located in the main city, six administrative districts in the main urban area were selected as the research focus. The distribution of built-up areas in 2009 and 2018 is shown in Figure 1. Table 1 shows the increase in built-up area over the years from 2009 to 2018 in the study area. Using linear regression, the built-up area was predicted to be about 266.39 km^2 in 2035. In 2009, the built-up area in the main city accounted for 17.49% of the total administrative area. By 2018, this was 21.78%, and by 2035 it could reach 30.28%. The built-up area is expected to continue to expand, and the rate of urban expansion will gradually increase over time.

The 2009–2018 built-up area data was from the Chinese second national land survey data. Built-up area data for 2010–2018 was derived from a year-by-year land change survey database based on second national land survey data in 2009. These time series data had the same survey standard, with a coordinate system of CGCS_2000 and in vector format. Data surveys were led by national institutions to accurately grasp land use types and changes. Due to the combination of remote sensing images, aerial photography, and a large number of field surveys, this data source has higher authority and accuracy than the traditional land cover data based only on remote sensing image interpretation. The basic geographic data involved in this article, such as road networks, government centers, airports, water bodies, and commercial distribution, were directly extracted from this database (see these elements in Figure 1). The population density data was derived from the global 1 km grid distribution database by the National Oak Ridge Laboratory [12]. The 1 km resolution overcomes statistical deficiencies based on administrative district surveys and can express population distribution characteristics on a more refined scale [23,24]. The land price data was from the national land transaction database. The information contained in the database for each transferred land piece included the location, the specified use, the transfer area and the transfer price. The location Geocode interface provided by Gaode Maps was used to obtain the spatial positions of 1818 parcels sold in the study area from 2009 to 2018, and the average price was calculated according to sold area and price. The final spatially continuous average price distribution of land sales from 2009 to 2018 was generated using the spatial interpolation tool in ArcGIS.

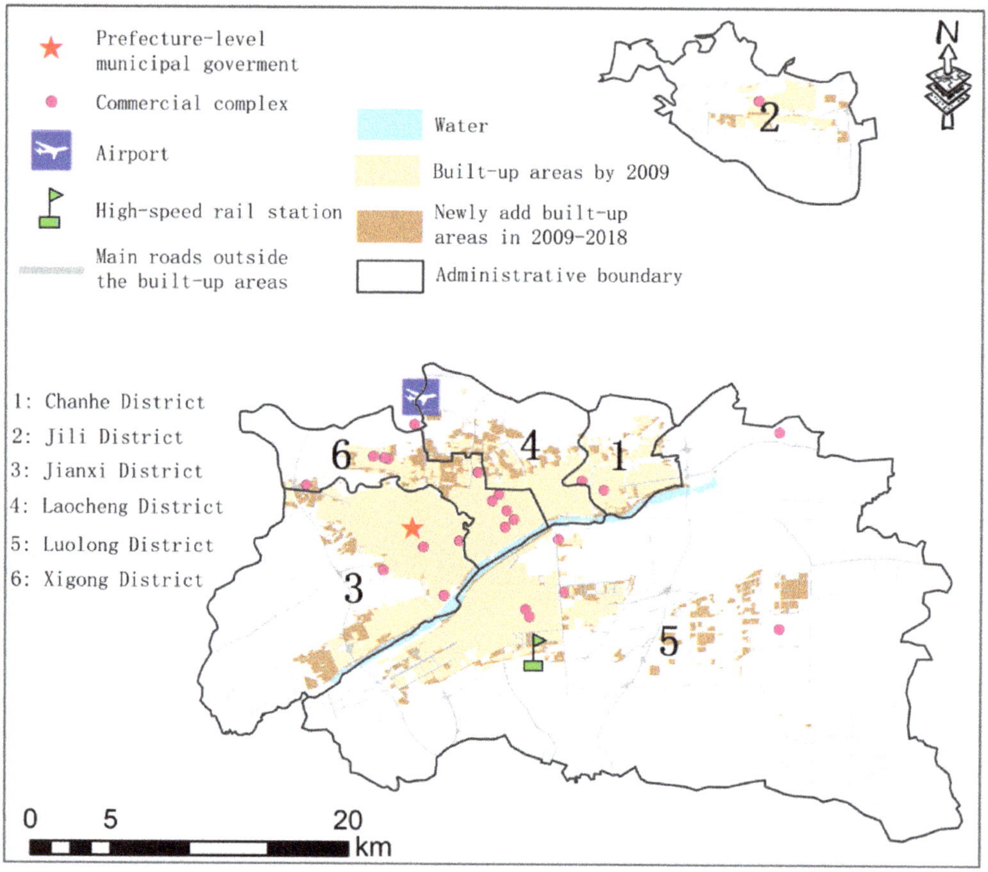

Figure 1. Distribution of built-up and basic geographical elements in the study area.

Table 1. Built-up area in the main city of Luoyang from 2009 to 2018 and projected area in 2035.

Year	2009	2010	2011	2012	2013	2014	2015	2016	2017	2018	2035 (Predicted)
Area (km^2)	153.86	161.79	166.60	173.90	178.82	180.61	181.57	189.93	190.99	191.59	266.39

3. Models and Methods

The MaxEnt-CSCA model consists of two parts: model training and spatial simulation. In the model training stage, a landscape expansion index was used to identify the expansion type of new built-up patches and several environmental variables based on spatial proximity, socio-economic development, and topography. These were selected as the driving factors that affected expansion type. Then, urban patch sample data of different types was inputted to the MaxEnt model and the three-layers probability space (PLCT, PLAP, and PLOP) was obtained by training these data to find the maximum entropy distribution. In the model training stage, all cell priorities were sorted according to their corresponding values in PLCT space. Higher values indicated higher cell selection priorities. For a selected cell, their corresponding value in PLAP and PLOP space determined whether it was an outlying type or adjacent type. The neighborhood function (NF), random factors (RF) and variability of different land types (VDT) were considered to create conversion rules for different urban patch types The details about NF and RF have been explained in detail in literature [18], and VDT will be further explained in Section 3.3. Finally, the model's parameters were corrected and used to simulate the dynamic process of future urban expansion. The MaxEnt-CSCA model framework is shown in Figure 2.

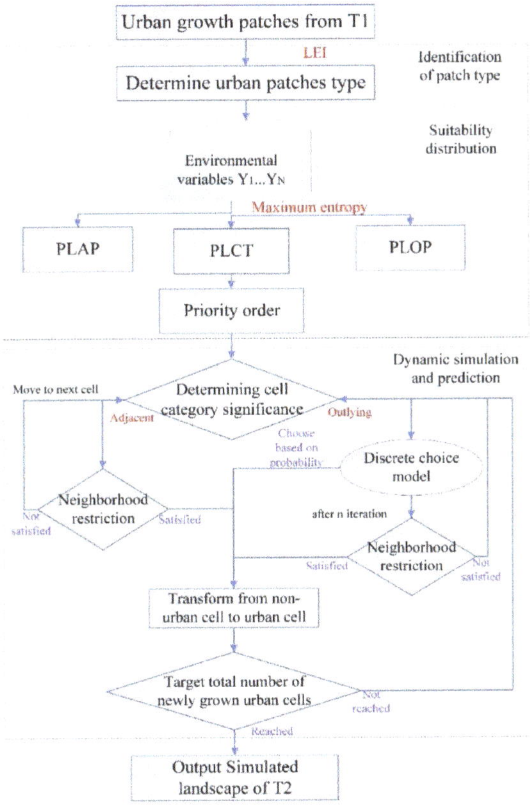

Figure 2. Algorithm flow of MaxEnt-CSCA model.

3.1. Identification of Patch Type

Urban growth type (UGT) in this article included two categories: adjacent and outlying. Adjacent UGTs refer to the expansion that transforms rural land villages within a city or cultivates unused land around the city into urban built-up area. In this type of expansion, new urban area is spatially adjacent to the original city. Outlying UGTs refer to when new urban area is spatially disconnected from the original built-up area, but the two areas are still functionally interconnected through transportation. Considering the definition of the relationship between the new and original urban patches in the literature [10], a landscape expansion index (*LEI*) was used to determine the type of each newly added urban patch and it was calculated as follows:

$$LEI = \frac{A_o}{A_o + A_v} \times 100 \tag{1}$$

In which, A_o is the intersecting area of a new urban patch with the original patch. A_v is the intersecting area of a new patch and non-urban land. According to the definition of *LEI*, when *LEI* = 0, the new urban patch is an outlying expansion. When *LEI* > 0, it is adjacent expansion.

3.2. *MaxEnt Principle and Environmental Variables Seletion*

The principle of maximum entropy (MaxEnt) is used to obtain a probability distribution with the highest entropy, which means the distribution is the most scattered or the most uniform, after satisfying the known partial information. The principle and criteria were first proposed in information theory [18], and soon were introduced and applied to various research fields. Typically, the total size of the target predicted object is unknown, so the absolute occurrence rate on a specific geographical cell is meaningless and the relative occurrence rate was introduced. In our settings, the relative occurrence rate $P(z(x_i))$ given by Maxent represents the probability that cell i is more likely to be transformed into urban (or urban-adjacent or urban-outlying) than other cells over the study area, which is calculated as follows:

$$P(z(x_i)) = \frac{exp(z(x_i)\boldsymbol{\lambda})}{\sum_i exp(z(x_i)\boldsymbol{\lambda})} \tag{2}$$

where $z(x_i)\lambda = z_1(x_i)\lambda_1 + z_2(x_i)\lambda_2 + \ldots + z_K(x_i)\lambda_K$, where x_i is the position of cell i, $z(x_i)$ is the vector of environmental variables at location x_i of cell i and λ is a vector of regression coefficients learned from the data, which indicates the relative importance of the different environmental variables. The sum of the relative occurrence rate over all cells in the study area is 1, i.e., $\sum_{i=1}^{N} P(z(x_i)) = 1$, because the denominator normalizes the relative occurrence rates.

According to the definition, the relative entropy of the target object's presence is calculated as:

$$H(x) = \sum_{i=1}^{N} P(z(x_i))log\frac{P(z(x_i))}{Q(x_i)} \tag{3}$$

where *Q(xi)* is the prior distribution, which reflects user expectations about the distribution before accounting for the known data. Here all the grid cells are a priori considered equally likely to be at the presence of the target object and *Q(xi)* was set to uniform distribution to maximize entropy in geographical space. However, in order to ensure the consistency between the predicted distribution and the known data, constraints were made so that the predicted distribution corresponded to the known known urban (or urban-adjacent or urban-outlying) cell distribution, in terms of environmental variables' moments such as mean and variance. Since the exact position $\dot{x}_1, \dot{x}_2, \ldots, \dot{x}_m$ of all urban cells present in a given year is known and MaxEnt assumes that the existence probability of the target object at a certain location (cell i) is a function of K environmental predictors, the constraints formula can be written as follows for each environmental predictor z_k:

$$\sum_{i=1}^{N} z_k(x_i)\ P(z(x_i)) = \frac{1}{M}\sum_{m=1}^{M} z_k(\dot{x}_m) \tag{4}$$

where $z(x_i)$ is the vector of environmental variables at location x_i of cell i mentioned above, $\dot{x}_m$ is the position of the known urban cells present in a given year, and $\frac{1}{M}\sum_{m=1}^{M} z_k(\dot{x}_m)$ gives the mean of the kth environmental variable of the known urban cells. The mean of the environmental value of the predicted distribution on the left side of Equation (4) is constrained to be the same value of the mean of the environmental variable at the observed locations on the right side. After the necessary constraints were generated, the maxent software minimized relative entropy within environmental space. By continuously adjusting the parameter values through a random seed generation algorithm, the optimal solution was found and output from the software. Because samples were only trained once and could introduce accidental errors, this study used the average of multiple training outcomes as the final results (PLCT, PLAP, and PLOP).

Regarding the selection of environmental variables, we referred to the previous literature and local characteristics of the study area to select spatial proximity (distance to political centers, major roads, airports, existing built-up areas, commercial centers, and high-speed rail stations), socioeconomic factors, including two variables, population density and newly added enterprise density for 2009–2018 [5,25–28]. The selection of spatial proximity has been explained in detail in previous studies and is not repeated. Although population density may not directly increase UGT, it can indirectly affect the formation of UGT through the cost of land development, which proved to be a very important factor [29]. For example, population density is related to the labor cost of demolition and resettlement. The higher the population density, the higher the investment that is compensated to the relocated population. For other density variables, such as areas with high road density with relatively complete infrastructure, additional investment required for land development will be significantly reduced, which proved beneficial to outlying growth [29]. The statistics of the above eight indicators are shown in Table 2.

Table 2. Statistics of eight environment variables.

Variable	Description	Minimum	Maximum	Standard Deviation
toBuilt	Distance to original built-up areas (km)	0	18.29	4.01
toRoad	Distance to main road (km)	0	7.73	1.60
toPCenter	Distance to political center (km)	2.12	26.28	4.92
toCCenter	Distance to commercial center (km)	2.13	34.17	6.82
toAirport	Distance to airport (km)	0	38.14	7.60
toHRailStation	Distance to high-speed rail station (km)	0	41.33	8.31
GDPDensity	GDP density (10,000 yuan/km^2)	259.48	48,295.20	2061.79
NFDensity	Newly added firm in 2009–2018 in 1 km^2	0	1059.81	35.89

3.3. Dynamic Simulation and Prediction

After the probability space and target number were calculated, the MaxEnt-CSCA model was ready to be simulated. We sorted the suitability probability of each potential cell from large to small, and selected cells with high suitability probability in each round of iteration. This iteration method is called SortCA in the literature [5], which can ensure that cells with higher suitability potential have a higher conversion probability (from non-urban to urban). For each selected cell, if it is the cell to be expanded in the study area (assuming the number of rows and columns were x, y, respectively), the significance of its expansion type will be checked. If the numerical value of the suitable probability of the adjacent category was greater than or equal to the outlying category, i.e., PLAP[x,y] $\geq$ PLOP[x,y], then the cell was judged to be the adjacent type significant, otherwise, the cell was judged to be the outlying category significant.

For a cell significant in the adjacent category, it was greatly affected by neighboring cell land use. If there were more surrounding built-up cells, it was easier for a cell to be transformed into a built-up cell. Here we defined the local cell rule. If it met the requirements that the minimum number of urban cells in the Moore neighborhood exceeded a threshold, the state of the cell was transformed from a non-urban cell to an urban cell at the probability of PLAP[x,y]. If not, the central cell moved to the next selected cell according to priority order.

For a cell significant in the outlying category, in the first n iterations (n was the set threshold), the expansion of the city was simulated based on the discrete choice model, which allowed cell transfers based on probability without meeting the minimum number of urban cells in the Moore neighborhood. According to literature [30], the probability of position i being selected was equal to the probability that position i utility was greater than or equal to the utility of any other alternative position j, namely:

$$P(i) = P(U(i) \geq U(j)) = \frac{\exp(U(i))}{\sum_{i=1}^{n} \exp(U(i))} \tag{5}$$

In the formula: U represents the utility function. In this study it represented the outlying suitability probability function, and $\sum_{i=1}^{n} \exp(U(i))$ was the sum of the candidate location utility exponential functions. The discrete selection model showed that in a location with higher probability of outlying suitability, it was more likely that a cell would be randomly selected and developed. The growth seeds of outlying category UGTs were generated according to PLOP[x,y] probabilities. After generating random seeds in the first n iterations, the condition of the minimum number of city cells in the neighborhood was added to restrict the growth of outlying-significant cells and prevent disorderly, sporadic built-up growth.

The condition to terminate the simulation was to determine whether the number of newly added urban cells reached the target urban cell quantity. If the condition was met, the simulation process ended and output a layout plan for new urban land. If the condition was not met, the next iteration began from the first cell in priority order again until the quantity of newly added construction cells was reached. The Kappa coefficient is a measure of the difference between the model's prediction result and a random guess result, and values range between 0 and 1. The higher the kappa coefficient, the higher the accuracy of the model.

4. Results

4.1. Statistical Analysis of UGT

Restricted by different natural, social and economic factors in the specific geographical location, the urban expansion size and type differed from region to region in the main urban area of Luoyang. By calculating LEI, UGTs from 2009 to 2018 in Luoyang were classified and the statistics of newly-added patches in each administrative district of the main urban area were obtained (see Figure 3). According to the UGT identification, 2373 new urban patches were found from 2009 to 2018, of which 1463 were outlying UGT patches with an area of about 24.64 km^2, accounting for 65.29% of the new urban area. Adjacent UGTs increased by 2293 patches with an area of about 13.10 km^2, accounting for 34.71% of the new urban area. Urban area growth (quantity statistics and spatial distribution) in the six districts of the main urban area of Luoyang City is shown in Table 3 and Figure 3.

Table 3. Urban area growth in the six districts of the main urban area of Luoyang City.

Type	Chanhe District (1)	Jili District (2)	Jianxi District (3)	Laocheng District (4)	Luolong District (5)	Xigong District (6)
Outlying	0.046	1.906	2.684	1.600	17.394	1.007
Adjacent	0.783	1.310	2.948	2.318	4.200	1.531

As can be seen in Table 3 and Figure 3, Luolong District (5) had the largest increase in built-up area (21.594 km^2), much more than other regions, followed by Jianxi District (3), Laocheng District (4) and Jili District (2). The newly built-up areas of these top four administrative units were all larger than 3 km^2 while Chanhe District (1) had the smallest newly built-up area, less than 1 km^2. It is worth noting that the adjacent UGT areas in four districts, namely Chanhe, Jianxi, Laocheng and Xigong, were larger than the outlying UGT areas. This was because these four districts were geographically adjacent and had relatively small jurisdictions, in which the proportion of built-up land was already high. Due to space constraints, it was difficult for the cities in these four regions to develop built-up areas in a diffusion manner. Luolong District (5), as the largest administrative district in the main urban area, had the space to build new satellite cities and industrial parks away from the existing built-up areas. In fact, Luolong District's outlying UGTs accounted for more than 80.55% in 2009–2018 and the characteristics of diffusion growth in this district were very obvious.

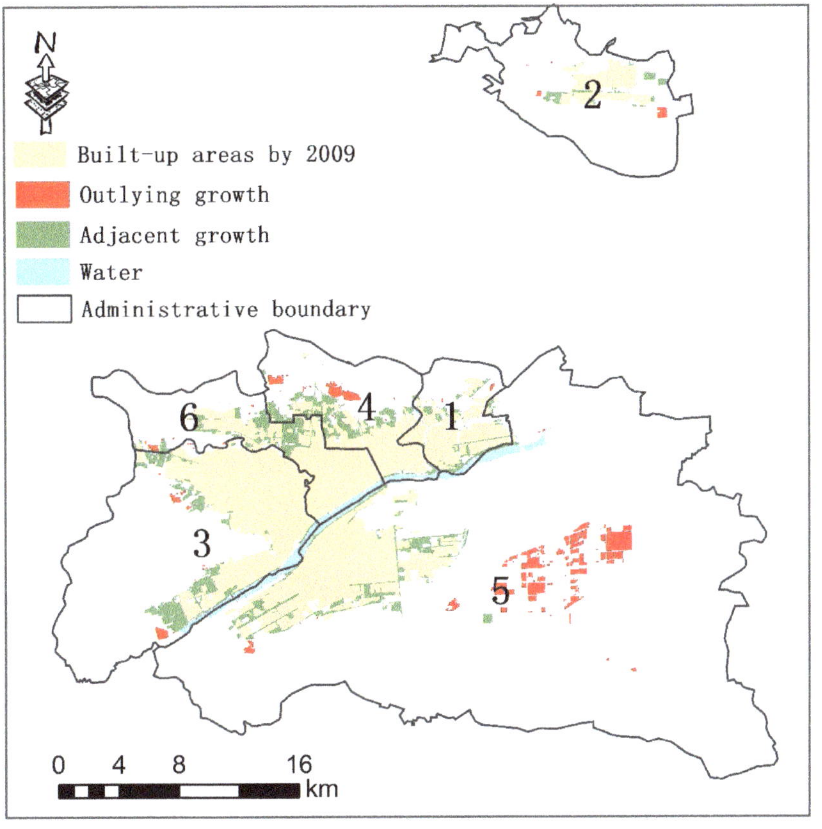

Figure 3. Spatial distribution of UGTs for new urban patches in the six districts of the main urban area of Luoyang City.

4.2. Suitability of UGT and Variable Contributions

Figure 4 shows the distributions of PLAP and PLOP based on the MaxEnt model. As can be seen in the PLOP figure, high values were distributed in areas away from the original built-up areas. Statistically, 77.37% cells with PLOP were greater than 0.70 and were located away from the original built-up area exceeding 1 km. In the PLAP figure, the high values were distributed in areas significantly closer to the original built-up than that of PLOP, which corresponded to the characteristics of adjacent or outlying urban growth.

In addition, MaxEnt also output the contributions of each variable to the result. Table 4 shows the relative contributions of environmental variables to PLOP and PLAP outputs. Percent contributions represented the normalized cumulative value of the variable gain in each iteration. Permutation importance (permutation importance) re-evaluated each environmental variable in sequence after random arrangement and displayed the AUC declines after training in the table and normalized them by percentage.

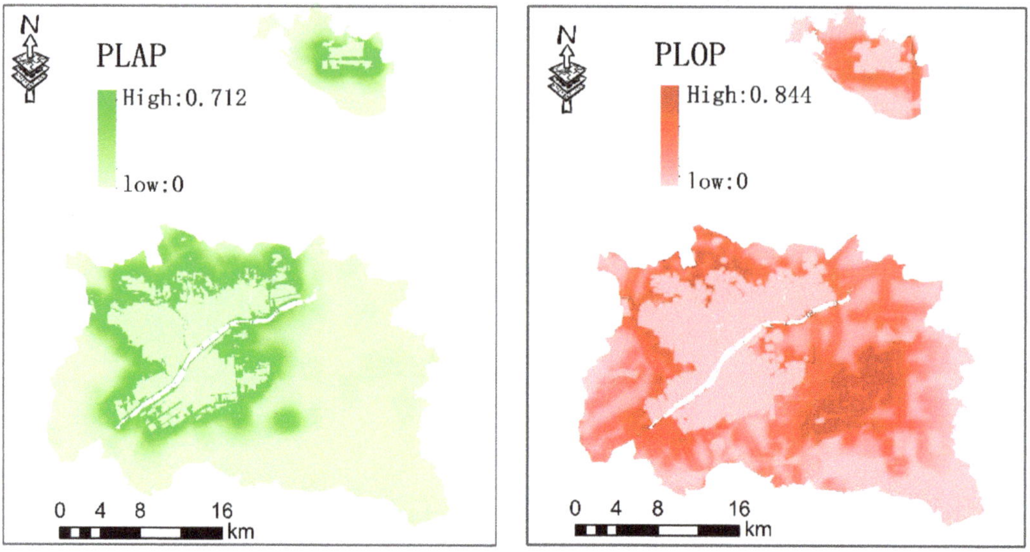

Figure 4. Spatial distributions of PLAP and PLOP.

Table 4. Variable Contributions for PLOP and PLAP.

Vibrable	PLOP		PLAP	
	Percent Contribution	Permutation Importance	Percent Contribution	Permutation Importance
toBuilt	38.4	28.6	53.6	67.6
toRoad	27.4	17.8	4.1	3.5
toAirport	14.6	6.4	0.6	3.1
NFDensity	3.9	16.7	34.7	13.7
GDPDensity	7.9	10.8	4.2	3.4
toHRailStation	4.1	12	0.5	4.9
toCCenter	3.2	1.1	2.1	0.6
toPCenter	0.5	6.5	0.2	3.2

As the table shows, the variable contributions to PLOP and PLAP is different. The variables toBuilt, toRoad and toAirport had the most significant impact on the distribution of PLOP, while variables toBuilt, NFDensity and GDPDensity had the most significant impact on PLAP. It is worth noting that the variable toAirport had a significant contribution to the PLOP distribution but no obvious effect in the PLAP distribution. This suggests that in the future, driven by airport construction, more outlying growth may appear around the airport.

4.3. Simulation Results and Accuracy Evaluation

MaxEnt-CSCA generated more realistic local evolution rules for different types of urban growth patches to determine the temporal and spatial dynamics of urban expansion. GDAL and NumPy function libraries in the Python function library were used to read the starting year's image into a two-dimensional array, arrange the priority of the cells, judge the significance of the cell category and start the cell iteration process. After iteration, the urban growth patterns in the main urban area of Luoyang from 2009 to 2018 were simulated (see Figure 5). The overall accuracy was 87.65%, the kappa coefficient was 0.78, and Figure 5 shows, the simulation distribution was similar to the actual urban pattern. Both the statistics and the visual interpretation confirmed that the model achieved good results. Not only was there clear adjacent expansion in the relatively compact area, but outlying-type growth also appeared in areas away from the original urban patches, which was a "patchy-like" expansion rather than a complete "diffusion" state, that matched the actual situation. In about 10 years from 2009 to 2018, the actual areas expanded by adjacent and outlying cities were 24.64 and 13.10 km^2, respectively, and the simulation results from MaxEnt-CSCA were 27.22 and 10.53 km^2, respectively. This verified that the MaxEnt-CSCA model can accurately predict expansion trends of urban built-up in Luoyang.

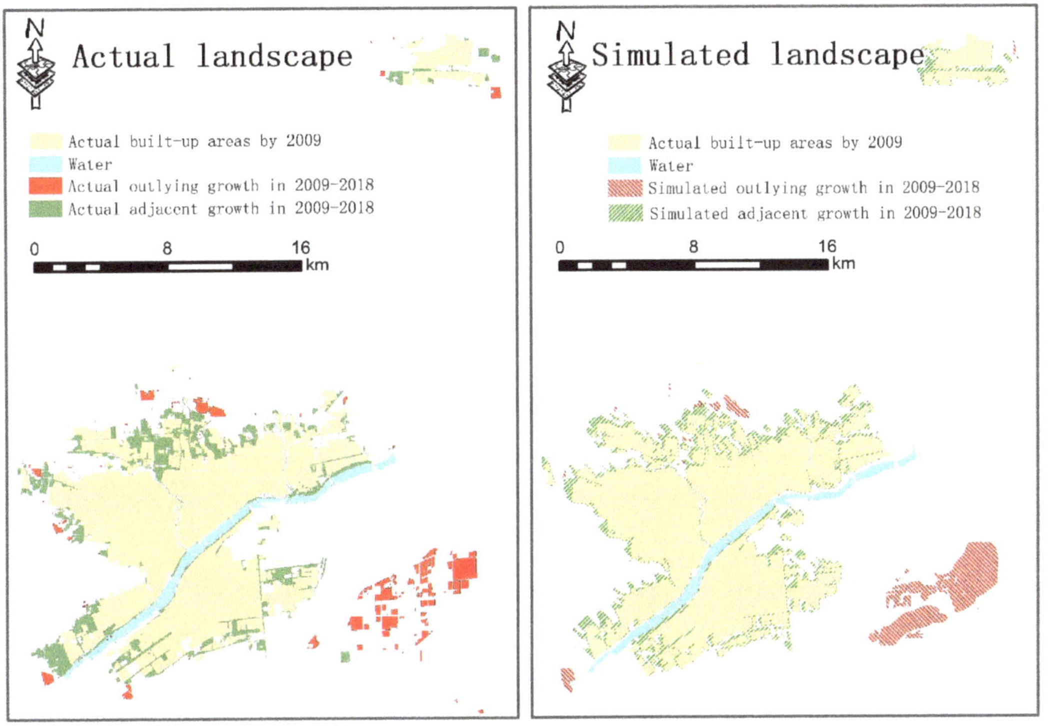

Figure 5. Actual and simulated urban landscapes of Luoyang's main urban areas by 2018.

5. Discussion

5.1. Comparison with Other CA

In simulating the historical evolution of Luoyang's urban expansion, the spatial output and several indicators of the MaxEnt-CSCA model achieved good results, but this did not indicate the advancement of the model. Only by comparing it with other models can the superiority of MaxEnt-CSCA model be confirmed. Therefore, this article continued to simulate the urban expansion of Luoyang City from 2009 to 2018 using two other CA models: MaxEnt-CA model, which considers the cellular conversion potential but does not consider the spatial zoning, and Base-CA model, which takes into account neither cellular conversion potential nor the spatial zoning. Maxent-CA still used the maximum entropy principle to obtain the probability of non-urban cells becoming urban cells and the selected environmental variables were the same to avoid subjectivity. By arranging the conversion probability from high to low and traversing each cell from non-urban to urban if the neighborhood condition was satisfied; Base-CA, on the other hand, only considering the neighborhood limit (there are enough construction land cells in the 3 $\times$ 3 neighborhood). The simulation results of the three models (Base-CA, Maxent-CA, Maxent-CSCA) were compared with the actual built-up areas in 2018. The result was shown in Figure 6. Noted that Jili District (2), which is far away from the main urban area, is not drawn in Figure 6 for visual aesthetics. In addition to spatial accuracy, Table 5 also compared the statistical accuracy, i.e., kappa coefficient and areas corrected simulated (the area of overlap between simulated landscape and actual landscape) of the three models.

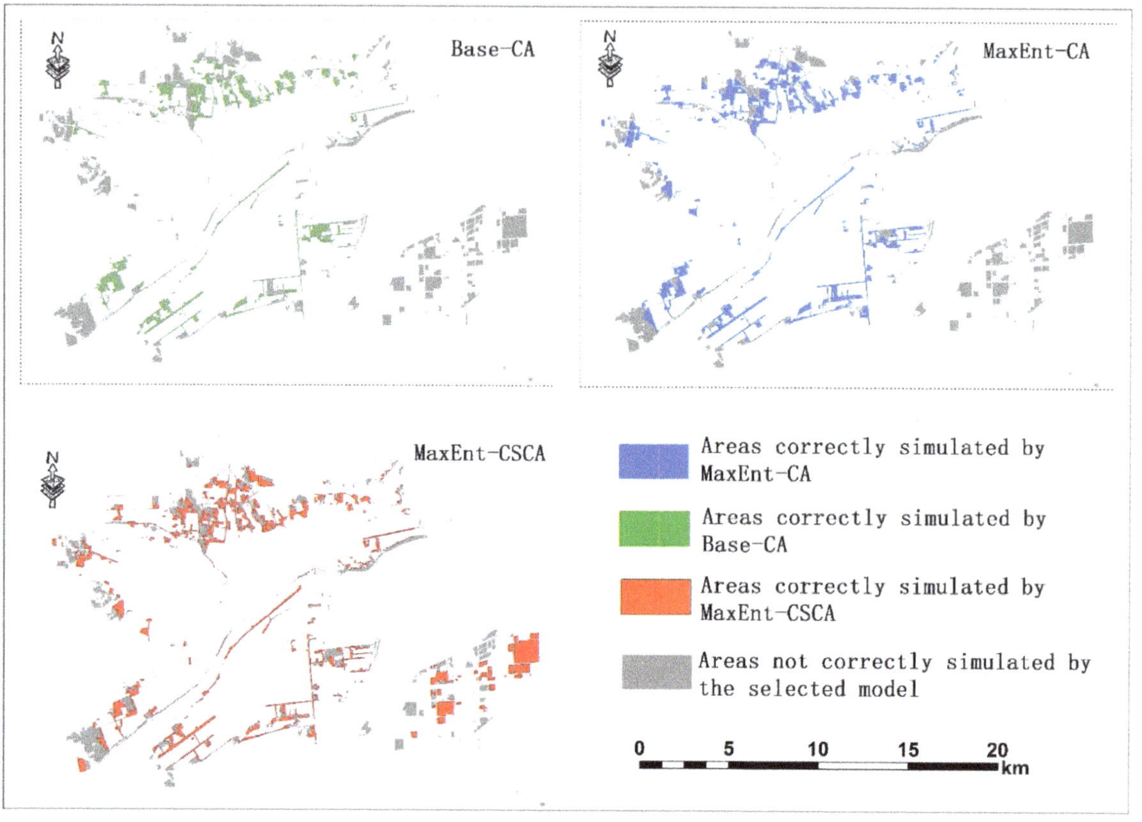

Figure 6. Comparison between the simulated landscapes by MaxEnt-CA and MaxEnt-CSCA.

Table 5. Comparison of kappa and areas corrected simulated among three models.

Indicator	Model Name		
	Base-CA	MaxEnt-CA	MaxEnt-CSCA
Kappa	0.71	0.74	0.78
Areas corrected simulated (km^2)	8.27	9.47	14.61

The Maxent-CA results obtained a kappa coefficient of 0.74, which was slightly less than MaxEnt-CSCA, and the corresponding value for Base-CA is only 0.71, much lower than that for Maxent-CSCA. Landscape simulated by Base-CA and MaxEnt-CA represented a similar compact clustered phenomenon, where outlying growth was quite different from the actual distribution (for example, the majority of Luolong District's outlying growth). However, because Base-CA only considers the limitation of neighborhood, it is more inclined to look for the internal gap of original built-up areas or the sags of the urban exterior form for growth (i.e., infilling-type growth) than Maxent-CA, and it is easier to form a regular form. MaxEnt-CA, on the other hand, had some advantages in urban aggregation growth simulation, this aggregation growth includes not only infilling growth but also edge growth. However, neither Base-CA nor MaxEnt-CA is insufficient for diffusion simulation. In contrast, as shown in Figure 6, the clustering phenomenon in the output of MaxEnt-CSCA was not as pronounced as in the other two CA model, and simulation of outlying urban growth performed better. In the adjacent UGT simulation, correctly predicted areas (e.g., spatially coincident) by base-CA, MaxEnt-CA and MaxEnt-CSCA were 54.18%, 63.05% and 47.40% respectively, while in the outlying UGT simulation, they were 2.65%, 4.92% and 34.13% respectively. This showed that the Base-CA and Maxent-CA model, which considered neighborhood limitations, performed better than MaxEnt-CSCA when simulating adjacent UGTs, but predictive ability was inferior to MaxEnt-CSCA when simulating outlying UGTs. Combining the two categories together, MaxEnt-CSCA had a higher overall kappa coefficient than Base-CA and MaxEnt-CA. From this result, it can be seen that in the actual urban expansion simulations, the city's expansion mode and compactness played a significant role when choosing different models. In this case study, the main urban area of Luoyang had a clear diffusion pattern during the study period, so it was more appropriate to choose MaxEnt-CSCA.

5.2. Potential Applications of MaxEnt-CSCA

The MaxEnt-CSCA proposed in this paper can effectively simulate urban expansion diffusion and aggregation process simultaneously, while traditional CA models mostly only show considerable simulation capabilities for aggregation growth. As a matter of fact, diffusion growth is a common phenomenon in modern urban development and it is especially pronounced in the following four situations: (1) In the early stage of urban expansion, there are a large number of diffusion-type growth situations. The most typical theory is the urban growth phase theory, which states that urban development is a process from diffusion to aggregation. The spread of small spots eventually forms several larger interconnected spots. For example, literature [5] found that the growth of Huangpi District in Wuhan City, China from 1993 to 2023 matched this situation; (2) Natural conditions restrict the continuous growth of the city. For example, cities with complex mountainous terrain, such as the "mountain city" of Chongqing, Panzhihua, Zhangjiajie, and Lhasa in China, Rio de Janeiro in Brazil, and Zurich in Switzerland. Mountainous landscapes have blocked natural continuous growth and resulted in a relatively dispersed urban pattern; For example, the urban expansion of Lhasa during 1995–2015 was mainly represented as outlying-growth [7]. The other is cities with many lakes and rivers, which also blocks the continuous expansion of urban built-up areas. Taking Wuhan (who known as "city of lakes") in China as example, diffusion urban growth accounted for a large proportion from 1995–2005 [7]; (3) Diffusion-spreading growth cities are common in Europe and America. Raleigh in the United States had low-density sprawl along suburban roads [31], which is what we usually call the "reverse urbanization" process; (4) Cities with developed county economy. For example, in Suzhou and Wuxi in the southern Jiangsu Province of China, there are many outlying-expansion of industrial parks around these cities [12]. The author believes that diffuse growth cannot be ignored in these type of urban growth simulations, and MaxEnt-CSCA is expected to be well applied in these areas.

5.3. Inspiration for Urban Planning

The MaxEnt-CSCA model proposed in this paper has been verified with good performance for predicting urban expansion patterns. Based on the urban construction patterns in Luoyang in 2018, the MaxEnt-CSCA model was adopted to predict the distribution of urban built-up areas in 2035, which is the last year of the Chinese Plan, called Long-Range Objectives Through the Year 2035. The spatial distribution is shown in Figure 7. According to the current development trend, an outlying UGT area of 8.69 km^2 and an adjacent UGT area of 66.11 km^2 are expected to occur between 2018 and 2035. The diffusion growth will decrease significantly and the urban pattern will become more compact. Luolong District (5) will have a large area of adjacent urban expansion in the future, and the newly added urban patches will fill the gap of the internal voids left by the outlying-type urban growth from 2009 to 2018. Luoyang's urban form is expected to gradually mature in the next 15 years. With the development of the city, the urban expansion of Chanhe District (1), Jili District (2), Jianxi District (3), Laocheng District (4) and Xigong District (6) will mainly occur along the edges of existing urban patches, making the existing patches continue to grow larger. Outlying-type urban expansion will almost entirely occur in Laocheng District (4) and Luolong District (5), while the urban growth process will be different. The former outlying UGT will grow outside of the existing built-up and the latter will expand on the inner gap, which will first diffuse, and then merge together. It can be predicted that in the future, land resources available for urban development in Chanhe District (1), Jili District (2), Jianxi District (3), Laocheng District (4) and Xigong District (6) will decrease, which make the urban form very compact. Measures need to be taken to increase the degree of built-up intensive use to meet the needs of future social and economic development, and attention should be paid to the protection of agricultural land around these districts. Luolong District (5) is still rich in available land resources. As the city continues to expand in the future, Luolong District (5) will form a relatively independent built-up area and serve as Luoyang's second center, which will share in urban function and the pressure in center Luoyang that are excessively concentrated.

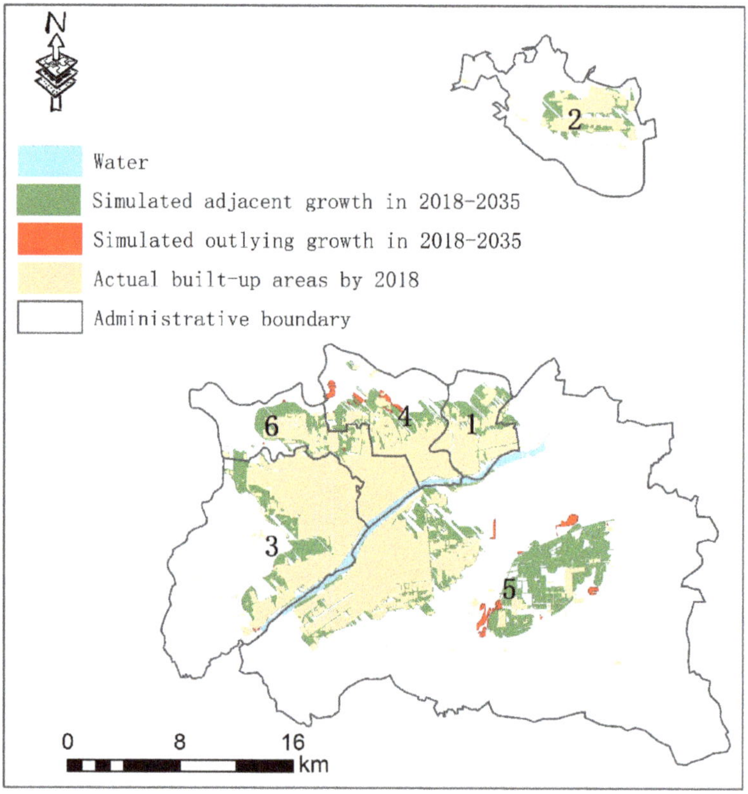

Figure 7. Predicted urban landscape in 2035 by MaxEnt-CSCA.

6. Conclusions

This paper proposed a category selection model, i.e., MaxEnt-CSCA model to simulate and predict urban expansion. This model made many improvements from the traditional cellular automata model, and expanded its hierarchical structure as well as the framework. It can simulate different urban growth categories using different local cell rules. The model of urban expansion and outlying-type urban expansion overcomes the shortcomings of the traditional CA model, which is unable to simulate diffuse outlying growth and adjacent urban sprawl simultaneously. The proposed model was verified using data from the main urban area of Luoyang between 2009 and 2018, where diffused urban expansion accounted for a large proportion. As we expected, the proposed MaxEnt-CSCA model effectively tracked the outlying built-up growth in Luoyang, which the traditional CA model was incapable of. Moreover, the MaxEnt-CSCA model was as good as the traditional CA model in terms of simulating adjacent growth. By predicting the spatial distribution of diffusion growth, it is helpful to predict the future urban growth center. Moreover, as the diffusional patch will continue to expand in the future, it will become an important carrier to undertake the spread of urban functions in the main urban area and provide choice for urban residents' real estate investment as well as the work opportunity. For future urban development, the gap between diffusional patches and original urban patches is the main place for future urban expansion and development, which is crucial for shaping and optimizing future urban form. Since urban form is associated with air quality [32], urban commuting [33], and spatial vitality [34] and so on, accurate simulation of the pattern and process of diffusional expansion patches can help decision makers to better formulate and optimize future urban form to achieve sustainable development. By further using MaxEnt-CSCA to predict and simulate the built-up landscape of Luoyang in 2035, it was found that the main urban area will continue to experience rapid urbanization, and future urban expansion will mainly occur in Luolong District and the counties surrounding it. In the near future, the urban pattern of Luolong District will grow from a relative diffusion state to a coalescing stage. Measures should be taken in these districts to make land use more efficient and intensive, and agricultural land surround the main urban areas should be protected. This paper verified the effectiveness of MaxEnt-CSCA through the case study of Luoyang, but as mentioned before, when performing spatial simulations of urban expansion for cities, the urban compactness as well as the city's development stage must be considered. Note that the proposed MaxEnt-CSCA model may not be superior in a city that is in the development stage of compact growth. Also, both the MaxEnt-CSCA and other traditional CA models have some fixed defects. For example, during simulation over a long time period, the environmental variables are assumed to be constant so that social and economic changes, such as newly built public facilities and an influx of floating population, will have no impact on the transformation of cells and thus reduce the simulated accuracy. Future research will focus on how to characterize the spatiotemporal changes of these socio-economic factors and couple their effects to MaxEnt-CSCA transformation rules.

Author Contributions: Conceptualization, R.W.; methodology, L.Z. and Y.L.; software, Y.L.; validation, W.H. and D.W.; formal analysis, L.Z.; resources, D.W.; data curation, W.H.; writing—original draft preparation, R.W. and W.H.; writing—review and editing, R.W. and L.Z.; supervision, Y.L. All authors have read and agreed to the published version of the manuscript.

Funding: This research was funded by National Natural Science Foundation of China (41801205).

Institutional Review Board Statement: Not applicable.

Informed Consent Statement: Not applicable.

Data Availability Statement: The data presented in this study are available on request from the corresponding author.

Conflicts of Interest: The authors declare no conflict of interest.

References

1. He, Q.; Tan, S.; Yin, C.; Zhou, M. Collaborative optimization of rural residential land consolidation and urban construction land expansion: A case study of Huangpi in Wuhan, China. *Comput. Environ. Urban Syst.* **2019**, *74*, 218–228. [CrossRef]
2. Li, H.; Liu, Y.; He, Q.; Peng, X.; Yin, C. Simulating urban cooperative expansion in a single-core metropolitan region based on improved ca model integrated information flow: Case study of Wuhan urban agglomeration in China. *J. Urban. Plan. Dev.* **2018**, *144*, 05018002. [CrossRef]
3. van Vliet, J.; White, R.; Dragicevic, S. Modeling urban growth using a variable grid cellular automaton. *Comput. Environ. Urban Syst.* **2009**, *33*, 35–43. [CrossRef]
4. Liu, X.; Ma, L.; Li, X.; Ai, B.; Li, S.; He, Z. Simulating urban growth by integrating landscape expansion index (LEI) and cellular automata. *Int. J. Geogr. Inf. Sci.* **2013**, *28*, 148–163. [CrossRef]

5. Liu, Y.; He, Q.; Tan, R.; Liu, Y.; Yin, C. Modeling different urban growth patterns based on the evolution of urban form: A case study from Huangpi, Central China. *Appl. Geogr.* **2016**, *66*, 109–118. [CrossRef]
6. He, Q.; Liu, Y.; Zeng, C.; Chaohui, Y.; Tan, R. Simultaneously simulate vertical and horizontal expansions of a future urban landscape: A case study in Wuhan, Central China. *Int. J. Geogr. Inf. Sci.* **2017**, *31*, 1907–1928. [CrossRef]
7. He, Q.; Song, Y.; Liu, Y.; Yin, C. Diffusion or coalescence? Urban growth pattern and change in 363 Chinese cities from 1995 to 2015. *Sustain. Cities Soc.* **2017**, *35*, 729–739. [CrossRef]
8. Jiao, L.; Liu, J.; Xu, G.; Dong, T.; Gu, Y.; Zhang, B.; Liu, Y.; Liu, X. Proximity Expansion Index: An improved approach to characterize evolution process of urban expansion. *Comput. Environ. Urban Syst.* **2018**, *70*, 102–112. [CrossRef]
9. Dietzel, C.; Herold, M.; Hemphill, J.J.; Clarke, K.C. Spatio-temporal dynamics in California's Central Valley: Empirical links to urban theory. *Int. J. Geogr. Inf. Sci.* **2005**, *19*, 175–195. [CrossRef]
10. Liu, X.; Li, X.; Chen, Y.; Tan, Z.; Li, S.; Ai, B. A new landscape index for quantifying urban expansion using multi-temporal remotely sensed data. *Landsc. Ecol.* **2010**, *25*, 671–682. [CrossRef]
11. Clarke, K.C.; Hoppen, S.; Gaydos, L. A self-modifying cellular automaton model of historical urbanization in the San Francisco Bay area. *Environ. Plan. B Plan. Des.* **1997**, *24*, 247–261. [CrossRef]
12. He, Q.; Zhou, J.; Tan, S.; Song, Y.; Zhang, L.; Mou, Y.; Wu, J. What is the developmental level of outlying expansion patches? A study of 275 Chinese cities using geographical big data. *Cities* **2019**, *105*, 102395. [CrossRef]
13. Zheng, L.; Chen, K.; Tan, S.; Yin, C.; He, Q.; Wu, J. Landscape expansion-place quality nexus revisited: How fundamental and transitory growth impact in China? *Land Use Policy* **2021**, *103*, 105294. [CrossRef]
14. Jiao, L.; Mao, L.; Liu, Y. Multi-order Landscape Expansion Index: Characterizing urban expansion dynamics. *Landsc. Urban Plan.* **2015**, *137*, 30–39. [CrossRef]
15. Xia, C.; Zhang, A.; Yeh, A.G.-O. Shape-weighted landscape evolution index: An improved approach for simultaneously analyzing urban land expansion and redevelopment. *J. Clean. Prod.* **2019**, *244*, 118836. [CrossRef]
16. He, Q.; Zeng, C.; Xie, P.; Tan, S.; Wu, J. Comparison of urban growth patterns and changes between three urban agglomerations in China and three metropolises in the USA from 1995 to 2015. *Sustain. Cities Soc.* **2019**, *50*, 101649. [CrossRef]
17. Jaynes, E.T. Information Theory and Statistical Mechanics. *Phys. Rev.* **1957**, *106*, 620–630. [CrossRef]
18. Zhang, Y.; Liu, X.; Chen, G.; Hu, G. Simulation of urban expansion based on cellular automata and maximum entropy model. *Sci. China Earth Sci.* **2020**, *63*, 701–712. [CrossRef]
19. Holder, A.M.; Markarian, A.; Doyle, J.M.; Olson, J.R. Predicting geographic distributions of fishes in remote stream networks using maximum entropy modeling and landscape characterizations. *Ecol. Model.* **2020**, *433*, 109231. [CrossRef]
20. Yan, H.; Feng, L.; Zhao, Y.; Zhu, C.; Qu, Y.; Wang, H. Predicting the potential distribution of an invasive species, Erigeron canadensis L., in China with a maximum entropy model. *Glob. Ecol. Conserv.* **2019**, *21*, e00822. [CrossRef]
21. Zhang, K.; Zhang, Y.; Zhou, C.; Meng, J.; Sun, J.; Zhou, T.; Tao, J. Impact of climate factors on future distributions of Paeonia ostii across China estimated by MaxEnt. *Ecol. Inform.* **2019**, *50*, 62–67. [CrossRef]
22. Liu, Z.S.; Gao, H.; Teng, L.W.; Su, Y.; Wang, X.Q.; Kong, F.Y. Habitat suitability assessment of blue sheep in Helan Mountain based on MAXENT modeling. *Acta Ecol. Sin.* **2013**, *33*, 7243–7249. [CrossRef]
23. Liu, X.; Derudder, B.; Wang, M. Polycentric urban development in China: A multi-scale analysis. *Environ. Plan. B Urban Anal. City Sci.* **2017**, *45*, 953–972. [CrossRef]
24. Li, Y. Towards concentration and decentralization: The evolution of urban spatial structure of Chinese cities, 2001–2016. *Comput. Environ. Urban Syst.* **2019**, *80*, 101425. [CrossRef]
25. Tan, R.; Liu, Y.; Zhou, K.; Jiao, L.; Tang, W. A game-theory based agent-cellular model for use in urban growth simulation: A case study of the rapidly urbanizing Wuhan area of central China. *Comput. Environ. Urban Syst.* **2015**, *49*, 15–29. [CrossRef]
26. Firozjaei, M.K.; Sedighi, A.; Argany, M.; Jelokhani-Niaraki, M.; Arsanjani, J.J. A geographical direction-based approach for capturing the local variation of urban expansion in the application of CA-Markov model. *Cities* **2019**, *93*, 120–135. [CrossRef]
27. Li, X.; Zhou, W.; Ouyang, Z. Forty years of urban expansion in Beijing: What is the relative importance of physical, socioeconomic, and neighborhood factors? *Appl. Geogr.* **2013**, *38*, 1–10. [CrossRef]
28. Tan, R.; Liu, Y.; Liu, Y.; He, Q.; Ming, L.; Tang, S. Urban growth and its determinants across the Wuhan urban agglomeration, central China. *Habitat Int.* **2014**, *44*, 268–281. [CrossRef]
29. Wu, J.; Wu, Y.; Yu, W.; Lin, J.; He, Q. Residential landscapes in suburban China from the perspective of growth coalitions: Evidence from Beijing. *J. Clean. Prod.* **2019**, *223*, 620–630. [CrossRef]
30. Mcfadden, D. Modelling the choice of residential location. *Transp. Res. Rec.* **1978**, *673*, 72–77.
31. Dong, T.; Jiao, L.; Xu, G.; Yang, L.; Liu, J. Towards sustainability? Analyzing changing urban form patterns in the United States, Europe, and China. *Sci. Total Environ.* **2019**, *671*, 632–643. [CrossRef] [PubMed]
32. Mou, Y.; Song, Y.; Xu, Q.; He, Q.; Hu, A. Influence of Urban-Growth Pattern on Air Quality in China: A Study of 338 Cities. *Int. J. Environ. Res. Public Health* **2018**, *15*, 1805. [CrossRef] [PubMed]
33. Rao, Y.; Yang, J.; Dai, D.; Wu, K.; He, Q. Urban growth pattern and commuting efficiency: Empirical evidence from 100 Chinese cities. *J. Clean. Prod.* **2021**, *302*, 126994. [CrossRef]
34. He, Q.; He, W.; Song, Y.; Wu, J.; Yin, C.; Mou, Y. The impact of urban growth patterns on urban vitality in newly built-up areas based on an association rules analysis using geographical 'big data'. *Land Use Policy* **2018**, *78*, 726–738. [CrossRef]

sustainability

MDPI

Article

Urban Growth Boundaries Delineation under Multi-Objective Constraints from the Perspective of Humanism and Low-Carbon Concept

Yan Yu [1], Chenhe Zhang [1], Weilin Ma [2,*], Yaxin Xu [1] and Xinxin Gao [1]

[1] School of Resources and Environmental Engineering, Wuhan University of Technology, Wuhan 430070, China
[2] Patent Examination Cooperation Guangdong Center of the Patent Office, CNIPA, Guangzhou 510000, China
* Correspondence: 253567@whut.edu.cn; Tel.: +86-15871438261

Abstract: Urban growth boundaries (UGBs) play an important role in controlling urban sprawl and protecting natural ecosystems. Traditional methods mainly focus on the heterogeneity of regional resources and environment rather than residents' behavioral activities. However, residents' behavioral activities are one of the most important factors influencing urban spatial form. Fortunately, the emergence of big data, especially phone signaling data, provides alternative data sources to understand the dynamic resident behavior activity space, which is significant for people-oriented urban development. Therefore, we propose a novel framework for UGB delineation based on multi-source big data and multi-objective constraints, which emphasizes humanism and the low-carbon concept in urban expansion simulation. The multi-objective constraints are constructed from the evaluation of resident activity space expansion potential, the evaluation of urban construction suitability, the evaluation of ecological conservation importance, and the human survival materials limitation. We apply the framework to Ningbo, and the results show that the framework under multi-objective constraints from a people-oriented and low-carbon perspective is more reliable and comprehensive than that without constraints. The findings also show that the UGB delineation based on multi-source big data has higher accuracy and better performance. The conceptual and methodological advances of this study are also applicable to other cities to help UGBs delineation.

Keywords: urban growth boundaries (UGBs); humanism; low-carbon concept; multi-source big data; resident activity space

Citation: Yu, Y.; Zhang, C.; Ma, W.; Xu, Y.; Gao, X. Urban Growth Boundaries Delineation under Multi-Objective Constraints from the Perspective of Humanism and Low-Carbon Concept. *Sustainability* **2022**, *14*, 16100. https://doi.org/10.3390/su142316100

Academic Editor: Anna Visvizi

Received: 14 October 2022
Accepted: 29 November 2022
Published: 1 December 2022

1. Introduction

The amount of land that has been utilized to satisfy urban growth has increased dramatically during rapid urbanization, resulting in urban sprawl, degradation of ecosystem functions, and over-exploitation of natural resources, which restricts sustainable urban development [1,2]. Meanwhile, carbon emissions caused by urban sprawl are contributing to climate changes. Thus, how to scientifically promote people-oriented and low-carbon development of city and reasonably coordinate the balanced relationship between development and protection have become urgent problems in urban planning and management [3–5]. To resolve these issues, UGB provides an effective method for sustainable urban management and is regarded as a useful planning tool [3,6,7].

The UGB is the regional boundary that can carry out urban construction and allow urban construction land expansion under different planning objectives. It is a tool that can be used to guide and limit urban growth, controlling urban sprawl with rigid boundaries [8]. UGB was originally proposed in 1976 by the city of Salem to resolve the conflicts with neighboring counties Polk and Marion over the management of urban space [9,10]. Subsequently, UGBs were implemented in some cities, such as Portland and Melbourne, effectively controlling the urban sprawl caused by a rapidly rising population and a rapidly growing number of cars [11,12]. By the 1990s, more than 100 cities and regions in the United

States had employed UGBs to manage urban development [6]. In China, with the advancement of reform and opening up and rapid economic development, the urbanization process has been accelerating, and the urbanization rate of the population has increased from 17.9% in 1978 to 63.8% in 2020, exceeding the world average. The rapid urban development has also brought about many urban problems [13]. In order to solve these problems, researchers introduced UGBs into China in the early 21st century [14], and UGBs became an important policy tool to control urban sprawl [15,16].

Methods of UGB delineation are mainly divided into two types: static spatial analysis methods and dynamic spatial growth simulations [17]. The former involves Frey's qualitative delineation method and Portland's qualitative delineation method [18]. Frey's qualitative delineation method delineates UGBs by identifying regional development issues and predicting future construction land expansion based on population, infrastructure, and development costs [19]. Portland's qualitative delineation method is to determine the urban development pattern and then refines UGB by combining factors that affect urban development. These factors include land use, service center's location, environmentally sensitive areas, and unsuitable land for development, etc. However, the classic static spatial analysis methods failed to simulate the spatio-temporal change of the urban boundary. Thus, dynamic models for urban growth simulation attracted considerable attention [20], such as cellular automaton (CA) [21] and an agent-based model [22]. In particular, CA has been extensively employed in urban growth simulation due to its advantages in dynamic simulation of urban development. Traditional CA focuses on the interaction of land units and does not consider the influence of environmental and economic factors on land. Therefore, quite a few models have been proposed to consider the above-mentioned factors affecting urban development, such as the SLEUTH model, the CLUE-S model, the CA–Markov model, and the constrained CA model [23,24]. Among them, the constrained CA model has certain advantages and application prospects in urban simulations. For example, Long et al. (2009) simulated Beijing's UGB using the constraint CA [25]. Ma et al. (2014) utilized this model to pre-evaluate land-use planning schemes in Guangzhou and analyzed the conflict areas between simulation results and planning results [26]. The constraints of these existing models have considered factors that influence urban expansion, such as natural conditions, spatial location from ecological perspective, etc. However, limited by the absence of data that can characterize human activities at fine scales, few studies have included humanistic indicators such as human behavioral activities in the constrained CA [27,28]. However, humanism is the criterion for the appropriateness of urban development and is necessary and significant in urban growth boundary's delineation [29]. Thus, indicators like residents' activities, living environment, and so on should be integrated into the constrained CA.

Most of the existing studies are based on traditional data, such as satellite images, data on transportation, and natural features. These data are ineffective in describing the characteristics of human behavior activities, and the description of the refinement of urban simulation is yet insufficient. The reason is that it is difficult to obtain fine-scale data to study the impact of residential activities on urban sprawl. The emergence and development of big data technology provides good conditions to solve the above problems. Compared with traditional data, big data has advantages in describing the human–land relationship and discovering spatial problems, which provides an effective method for people-oriented urban planning [30]. In recent years, it is worth noting that big data have attracted considerable attention in urban planning. It has been widely employed to study the urban system, urban spatial structure zoning, etc. For instance, using 2.5 million communication data, obtained from Belgian mobile operators, Kring analyzed the links between 571 cities in Belgium to build a social network between cities and to describe the city hierarchy [31]. Gong identified the maintenance activity space, commuting activity space, and recreational activity space at multiple geographic scales by using mobile phone data and analyzed the relationship between the built environment and activity space [32]. However, in the field of UGB delineation, big data is not yet widely used. The above-mentioned urban measures, which use big data, have also been rarely introduced into the UGB delineation. However, big data, especially mobile phone signaling data, not only provides a new perspective for the study of human activities and urban space but also provides a new method for UGB delineation [33]. Thus, it is an excellent data source for the future urban expansion model [34]. The UGB delineation based on big data is more humanized and scientific.

At the same time, low-carbon urban development is a topic that must be considered to balance climate change and human development. The consideration of low-carbon concepts in urban development can better guide cities to reduce carbon emissions and promote carbon peaking and carbon neutrality goals. Lei Chen employed the DID method for panel data from 2000 to 2019 in China to shed light on the effects on carbon emissions. Results show that the UGB can reduce carbon emissions considerably, and the carbon emissions of the pilot cities decreased by 23.91%. [35] Thus, incorporating low-carbon concepts in the UGB delineation can promote carbon emission reduction in urban development and ultimately reduce carbon emissions from urban human activities to better contribute to solving climate change issues.

Therefore, we attempt to establish a framework for the delineation of UGBs that couples humanism with the concept of low-carbon development. In this framework, multi-source big data, such as phone signaling data, POI, and night-light data, are used to analyze the spatial expansion potential of residents' activities as a way to reflect the role of people in urban development. Then, combined with other multi-objective constraints like urban construction suitability, ecological conservation importance, and human survival materials, we optimize the constrained CA model to make it more comprehensive and humanistic and use the improved constrained CA model to simulate urban growth. The proposed framework aims to develops more humanized urban boundary and promote low-carbon development.

2. Study Area and Data

2.1. Study Area

Ningbo, a rapidly urbanizing city, is located in the southeastern coast of China. It is one of the economically central cities in Yangtze River Delta. Ningbo consists of six districts (Haishu, Jiangbei, Beilun, Zhenhai, Yinzhou, and Fenghua), two counties (Ninghai and Xiangshan), and two county-level cities (Cixi and Yuyao) (Figure 1). During the period from 1978 to 2020, Ningbo experienced rapid development in terms of population and urban built-up area: the population in 2020 was 9.4 million, which was approximately 2.05 times that in 1978 (4.58 million); the urban built-up area in 2019 was 526 km^2, which was over 28 times that in 1978 (18.3 km^2) [36]. Ningbo's rapid urbanization has brought great pressure to the ecological environment and to human settlements. In addition, human needs and carbon reduction will receive more attention in urban growth management over the next five years (2020–2025) according to the Ningbo development strategy. Thus, UGB delineation from the perspective of humanism and the low-carbon concept should be considered in Ningbo's future growth.

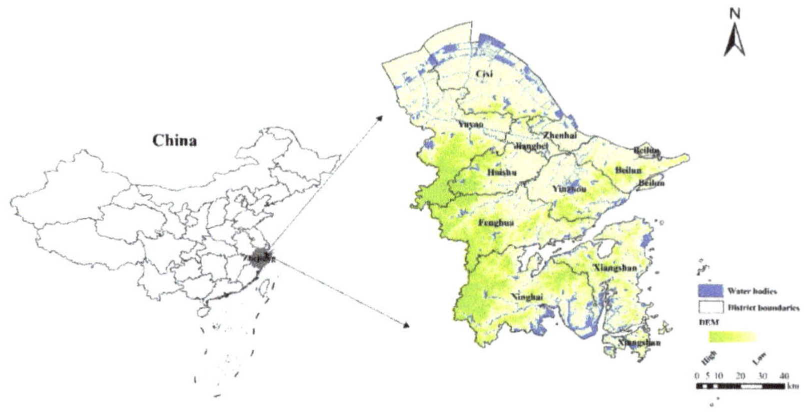

Figure 1. Location of the study area.

2.2. Data

There are considerable data needed for UGB delineation in this research. They are classified into administrative boundaries, land-use data, and constraint factor data (Table 1). Ningbo's administrative boundary is used as a reference to clip all geospatial data from Ningbo. The land-use data served as the foundation in the constraint CA model. The constraint factor data is divided into four groups: residential activity space constraint, human settlement suitability constraint, human settlement ecological constraint, and human settlement security constraint. The coordinate system of spatial data is set as a CGCS2000 national geodetic coordinate system, and all data are resampled to the same resolution of 30 m × 30 m.

Table 1. Data types and sources.

Data Types			Data Sources
Administrative boundaries			https://www.webmap.cn/ (accessed on 5 July 2020)
Land-use data			http://www.globallandcover.com/ (accessed on 5 July 2020)
Constraint factor data	Residential activity space constraint factor	Mobile signaling data	http://www.smartsteps.com/ (accessed on 2 August 2020)
		Point of interest (POI)	https://lbs.amap.com/api (accessed on 5 July 2020)
		Night light data	https://www.noaa.gov/ (accessed on 8 July 2020)
	Human settlement suitability constraint factor	Digital elevation model (DEM)	http://www.gscloud.cn/ (accessed on 1 July 2020)
		Total water consumption control index	Ningbo Water Conservancy Bureau
		Meteorological data (including temperature, precipitation, wind speed, air relative humidity)	http://data.cma.cn/ (accessed on 11 May 2020)
		River and lake data Road data Residential data	https://www.webmap.cn/ (accessed on 11 May 2020)
	Human settlement ecological constraint factor	Ecosystem types	http://www.ecosystem.csdb.cn/ (accessed on 11 May 2020)
		Species distribution data	https://www.gbif.org/ (accessed on 14 May 2020)
		Normalized difference vegetation index (NDVI)	https://www.resdc.cn/ (accessed on 14 May 2020)
		Soil attribute data	http://www.fao.org/soils-portal/soil-survey/soil-maps-and-databases/harmonized-world-soil-database-v12/en/ (accessed on 14 May 2020)
		Evaporation (ET)	http://www.geodoi.ac.cn/WebCn/Default.aspx (accessed on 20 May 2020)
	Human settlements security constraint factor	Permanent basic farmland data	Ningbo natural resources and Planning Bureau

3. Methodology

3.1. Research Framework

In this study, Figure 2 demonstrates the proposed framework of UGB delineation. We adopt the constraint CA model to conduct UGB delineation by coupling the multi-objective constraints from humanism and the low-carbon concept. The humanism is reflected by the residential activity space constraint and the human settlements suitability constraint. The low-carbon concept is embodied in the human settlements ecological constraint and the human survival conditions constraint. The above multi-objective constraints are derived from three evaluations and a limitation (see Sections 3.2–3.5 for details). Then, they are integrated into the constrained CA model to simulate urban expansion. Finally, the urban growth boundary is delimited using the expansion and erosion algorithm, because this algorithm has advantages in boundary smoothing and extraction [37].

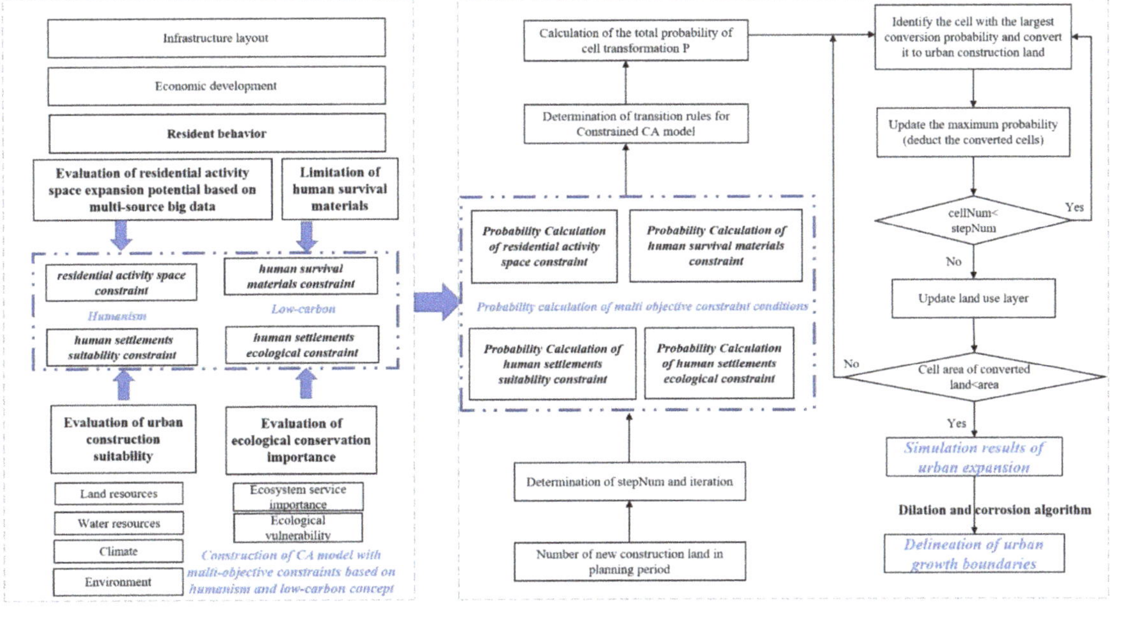

Figure 2. The framework of UGBs delineation.

3.2. Evaluation of Residential Activity Space Expansion Potential Based on Multi-Source Big Data

Evaluation of residential activity space expansion potential is used to reflect the willingness of residents' behaviors. The stronger this willingness is, the higher the probability that the area will develop into a city. Thus, resident behavior, such as commuting, leisure consumption, etc., play an important role in driving urban development and forming an urban boundary [38]. Previous studies focus on factors in the economic, social and natural environment. However, there are few studies incorporating human behavior into analyzing the urban form. Thus, it is necessary to consider the influence of human activities on urban expansion. In this study, the evaluation was conducted from the aspects of resident behavior, economic development, and infrastructure layout by using telephone signal data, POI, and nighttime-lighting data, respectively (Table 2) [39].

Table 2. Resident activity spatial potential evaluation system based on multi-source big data.

Category		Indicator Name	Data
Resident behavior	Commuting	Average commuting distance to residents' workplaces	Phone signaling data
	Leisure consumption	Average travel distance for leisure consumption	
	Jobs—housing space	Job—resident ratio	
	Population distribution	Actual service population density	
	People flow	The total number of people flow	
Economic development		The total nighttime lighting index	Night time Nighttime-lighting data
Infrastructure layout	Transportation	Distance to bus stop Distance to station Distance to station	POI
	Medical	Distance to hospitals, clinics	
	Education	Distance to schools, colleges, and universities	
	Commercial Servicesservices	Distance to the commercial center, and office buildings	
	Recreation facilities	Distance to recreational facilities	

In terms of resident behavior, studying urban expansion through residents 'daily activities (e.g., commuting, leisure consumption) and evaluating residents' activity space is helpful to discover new urban spatial characteristics. Using phone signaling data, we evaluate the impact of residents' behavior characteristics on urban sprawl from five aspects: commuting, leisure consumption, the jobs–housing space, population distribution, and people flow. Specifically, (1) the average commuting distance to residents' workplaces is calculated to characterize the attractiveness of jobs. The farther away it is, the greater is the potential of residential activity space expansion, because residents tend to choose employment in economically developed neighboring regions. (2) the average travel distance for leisure consumption is calculated to reflect the convenience of the residents' leisure. The closer the distance, the greater the potential of residential activity space expansion because of residents' leisure consumption proximity. (3) The job–resident ratio is calculated to judge the balance between work and housing. It also reflects the happiness of residents' commuting. The closer the job–resident ratio is to 1, the greater the potential of residential activity space expansion. (4) The actual service population density is calculated to reflect the level of urban public service management. The bigger the actual service population density, the greater the potential of residential activity space expansion. (5) The total amount of people flow is evaluated to reflect the scale of the city's hierarchy. The greater the total amount of people flow, the greater the potential of residential activity space expansion.

As far as economic development is concerned, the total nighttime-lighting index extracted from nighttime-light data reflects the impact of socio-economic factors on urban growth. The greater the total nighttime-lighting index, the greater the potential of residential activity space expansion [40].

In terms of infrastructure layout, using POI data, the distance to service facilities characterizes the impact of infrastructure layout on urban expansion. The more developed the infrastructure, the lower the development cost of urban space and the easier it is to convert non-urban land into urban land [41]. That is, the closer the distance, the greater the potential of residential activity space expansion.

After obtaining these indicators based on multi-source big data, the weight of each evaluation index is assigned due to different impacts of the index on urban expansion. The potential of residential activity space expansion is calculated using Equation 1, which is used to construct the residential activity space constraint.

$$S_{ij} = \sum_{k=1}^{n} \beta_k * D_k \quad (1)$$

In Equation (1), S_{ij} is the residential activity space expansion potential; β_k is the weight of impact factors in urban expansion; D_k is the impact factor on urban expansion; n is the different dimension.

3.3. Evaluation of Urban Construction Suitability

The evaluation of urban construction suitability characterizes human settlement livability. Creating a good human settlement is essential for people-oriented sustainability development. In our study, the urban construction suitability indicates how natural environment elements influence human settlement and urban construction [42,43]. This evaluation mainly considers factors such as the livability of land resources, water resources, environment, and climate. After evaluating each factor, the results of the suitability of urban construction are synthesized, which is used to construct the constraints of human settlements suitability.

3.4. Evaluation of Ecological Conservation Importance

The evaluation of ecological conservation importance aims to protect important ecosystems, such as forests and wetlands. It is vital for enhancing the capacity of ecological carbon sinks, which should be considered in the UGB delineation. In this study, the ecological protection importance is evaluated from two aspects: ecosystem service importance and ecosystem vulnerability (Table 3) [44]. The ecosystem service importance is assessed regarding water resources retention capacity conservation, soil and water conservation, and biodiversity conservation. The ecosystem vulnerability is estimated from soil erosion, land desertification, and rock desertification [45].

Table 3. Indicator system for evaluating the importance of ecological protection.

Evaluation Content	
Ecosystem service importance	Water conservation
	Soil and water conservation
	Biodiversity conservation
Ecological vulnerability	Soil erosion
	Land desertification
	Rock desertification

3.5. Limitation of Human Survival Materials

The limitation of human survival materials is mainly used to protect the most basic resources for human survival and low-carbon development. Due to the expansion of urban construction land, limited land resources will inevitably lead to farmland occupation, water pollution, and other problems. However, water bodies and farmlands, which are important for food security, are the most basic elements of human survival and are also helpful in reducing carbon emission. At the same time, if farmland or water bodies are converted into construction land, the carbon emissions from human activities such as building will increase tremendously compared to their current state. Thus, their conservation makes sense not only for human survival but also for the sustainable development of low-carbon cities. However, such protection can act as a disincentive to urban expansion and must be considered in the constraints. At the same time, the government, as the supply side, considers its own low-carbon development requirements, and it restricts the conversion of basic agricultural land and water bodies into construction land, especially in the housing market. In order to reflect this requirement, we set farmland and water bodies as limitations from the low-carbon perspective and the supply perspective to ensure basic human survival needs and the sustainable development of low-carbon cities.

3.6. Constrained CA Model with Multi-Objective Constraints

According to the above three evaluations and a limitation, we establish four multi-objective constraints from humanism and the low-carbon concept. They are utilized to optimize the transition rules in the constrained CA model. Furthermore, the improved model's delineation result helps to promote the carbon reduction capacity and quality of life for residents.

The multi-objective constraints for our proposed constrained CA model are expressed as follows:

In terms of residential activity space constraint, it reflects human behavior characteristics affecting urban expansion on the state transition of cells in the constrained CA model. According to the evaluation of residential activity space expansion potential, the probability of the residential activity space constraint is calculated using Equation (2).

$$P_{\{i,j\}loc} = 1/\left\{1 + Exp\left(-S_{\{i,j\}z}\right)\right\} \tag{2}$$

In Equation (2), $P_{\{i,j\}loc}$ is the probability of residential activity space constraint. $S_{\{i,j\}z}$ is the residential activity space expansion potential.

With regards to the human settlements suitability constraint, it is established to reflect the suitability of urban natural environment elements for human-concentrated living. The probability of the human settlements suitability constraint is determined by the result of urban construction suitability evaluation using Equation (3):

$$P_{\{i,j\}urb} = P_{\{i,j\}U} \tag{3}$$

In Equation (3), $P_{\{i,j\}urb}$ is the probability of the human settlements suitability constraint, and $P_{\{i,j\}U}$ is obtained from the normalized result of urban construction suitability evaluation.

Regarding the human settlements ecological constraint, it is important to prioritize the protection of important ecological areas for carbon storage. In the constrained CA model, the probability of the human settlements ecological constraint is computed based on the result of ecological conservation importance evaluation using Equation (4):

$$P_{\{i,j\}eco} = 1 - P_{\{i,j\}A} \tag{4}$$

In Equation (4), $P_{\{i,j\}eco}$ is the probability of the human settlements ecological constraint, and $P_{\{i,j\}A}$ is obtained from the normalized result of ecological conservation importance evaluation.

In terms of the human survival materials constraint, it is important to protect farmland and water bodies from being occupied by urbanization. In the constrained CA model, its probability is computed using Equation (5):

$$P_{\{i,j\}ins} = con\left(c_{ij} = suitable\right) \tag{5}$$

In Equation (5), *con* is a function that determines whether the cell is located within the areas of limitation of human survival materials. If the cell is located within the areas of limitation of human survival materials, $P_{\{i,j\}ins}$ is assigned a value of 0, which means it cannot be converted to an urban area, and if it is located outside the areas of limitation of human survival materials, $P_{\{i,j\}ins}$ is assigned a value of 1, which means it is allowed to be developed as urban land.

For the constrained CA model, the change of cell state is not only related to the above constraints but is also affected by the surrounding-cell state, that is, the neighborhood effect [46], because it helps to enhance the compactness of urban forms [47]. It is calculated by using Equation (6):

$$P_{\{i,j\}nei} = \frac{\sum_{n\times n} con(c_{ij} = urban)}{n \times n - 1} \tag{6}$$

In Equation (6), n is the neighborhood size, c_{ij} is the cell state, and *con* is the function that counts the amount of urban land in neighborhoods with a particular size.

Based on the above constraints and neighborhood effect, we propose a comprehensive transition rule. Equation (7) is used to determine the final transition probability of the cell state.

$$P^t_{\{i,j\}} = P_{\{i,j\}loc} * P_{\{i,j\}urb} * P_{\{i,j\}eco} * P_{\{i,j\}ins} * P_{\{i,j\}nei} \tag{7}$$

where $P^t_{\{i,j\}}$ denotes the final transition probability of the cell state, $P_{\{i,j\}loc}$ is the probability of the residential activity space constraint, $P_{\{i,j\}urb}$ is the probability of the human settlements suitability constraint, $P_{\{i,j\}eco}$ is the probability of the human settlements ecological constraint, $P_{\{i,j\}ins}$ is the probability of the human settlements security constraint, and $P_{\{i,j\}nei}$ is the neighborhood effect.

Finally, the constrained CA model is optimized by the comprehensive transition rules and utilized to simulate urban growth.

3.7. UGB Delineation Based on the Dilation and Erosion Algorithm

According to the simulation results in Section 3.6, there are some fragmented patches of the future urban boundary, which is not conducive to the implementation of management policies. UGB delineation requires eliminating these patches and obtaining a continuous urban-land polygon. The common methods in boundary processing include the artificial drawing method, moving window method, ant colony algorithm, dilation and erosion algorithm, etc. Among them, the dilation and erosion algorithm has advantages in edge extraction and image processing, which is beneficial to delineating UGB [48]. Therefore, we utilize it to extract urban boundaries. It involves two basic operations: dilation and erosion. The dilation operation convolves the image X with an arbitrarily shaped kernel (B), which has a defined anchor point, usually a square or a circle. As kernel B is scanned over the image, the maximal pixel value overlapped by B is computed to replace the image pixel in the anchor point position. In contrast to the dilation operation, the erosion operation focuses on the minimal pixel value overlapped by B. In fact, based on them, the opening and closing operations are used for boundary smoothing and interior filling [37]. The opening operation is an erosion operation followed by a dilation operation (Equation (8)), while the closing operation is a dilation operation followed by an erosion operation (Equation (9)). After dealing with the simulation results, the urban boundary is smoother and more suitable for urban managers' decision making.

$$X \circ B = (X \ominus B) \oplus B \tag{8}$$

$$X \cdot B = (X \oplus B) \ominus B \tag{9}$$

4. Implementations and Results

4.1. Results of the Residential Activity Space Expansion Potential Evaluation

Using the collected multi-source big data, the expansion potential of residents' activity space in Ningbo is evaluated from three dimensions: resident behavior, economic development, and infrastructure layout, according to the method in Section 2.2.

In the dimension of resident behavior, this paper uses mobile signaling data to analyze Ningbo residents' behavior such as commuting, consumption, and travel. The data were from Ningbo and were provided by a large telecom operator in China. They covered the records of about 5.3 million subscribers for one month (December, 2019) for this operator. Each record contained information including user location status (daily/monthly residence location, travel information records), user attribute information (gender, age, home, etc.), and so on. By setting different rules, these data are filtered and used to analyze different characteristics of residents' activities. To be specific, the characteristics of residents' behavior in Ningbo are estimated from five aspects: commuting, leisure consumption, the jobs–housing space, population distribution, and people flow. (1) In terms of commuting, the average commuting distance to residents' workplaces is calculated to characterize the attractiveness of jobs. We set 0:00–8:00 and 20:00–24:00 every day as rest periods and set 8:00–20:00 on weekdays as working periods according to the characteristics of residents' travel behavior in Ningbo. Then, all base stations visited by users in December, 2019 are classified according to the user records of different periods. The location of the base station where the user stays for the longest time during the rest period and stays for more than 16 days is regarded as the user's residence. The location where the user stays the longest and stays for more than 11 days during the working period is considered the user's working location. According to the above location, the average commuting distance of the residents' workplace is calculated by district and county. (Figure 3a). (2) In terms of leisure consumption, the average travel distance for leisure consumption is calculated to reflect the convenience of the residents' shopping and leisure, and 15:00–19:00 on weekends and holidays is set as the recreation period because of the large population size during this period. The location of the base station where the user stays the longest during the recreation period more than three times is regarded as their recreation place. Then, the average travel distance is calculated for leisure consumption by county and district (Figure 3b). (3) In terms of the jobs–housing space, the job–resident ratio is calculated to judge the balance between work and housing. Based on the identification of the place of residence and employment, the amount of the resident population and the employed population are counted, respectively, by county and district. Then the job–resident ratio is obtained by dividing the amount of employment by the number of residents. (Figure 3c). (4) In terms of population distribution, the actual service population density reflects the level of urban public service management. The actual service population is identified by the population residing in a certain place for more than three days. It is divided by the administration district area to obtain the actual service population density (Figure 3d). (5) In terms of people flow, the total amount of people flow reflects the scale of the city's hierarchy. The residence place is regarded as the origin of residents' travel. The districts and counties with the farthest travel distance is regarded as travel destinations. Then, the OD (origin to destination of residents' travel) volume is counted to calculate the total amount of people flow in each district and county (see Figure 3e).

In the dimension of economic development, the total night-lighting index (TNL) is extracted from the night-lighting data (Figure 4), which is used to measure the economic development characteristics of Ningbo. To be specific, the night-lighting data from the Ningbo area are corrected and noise-reduced, firstly based on the invariant target area method and then the sum of the digital number (DN) of the night-lighting data is calculated, which is just the total night-lighting index (TNL). After obtaining the TNL, the distribution is compared with the economic statistics values from the Ningbo Bureau of Statistics, and it was found that the volume was basically the same as the level of economic development of the region. The higher the TNL value is, the higher the economic development degree is, then the characteristics of Ningbo's economic development can be reflected from the spatial distribution of the TNL value.

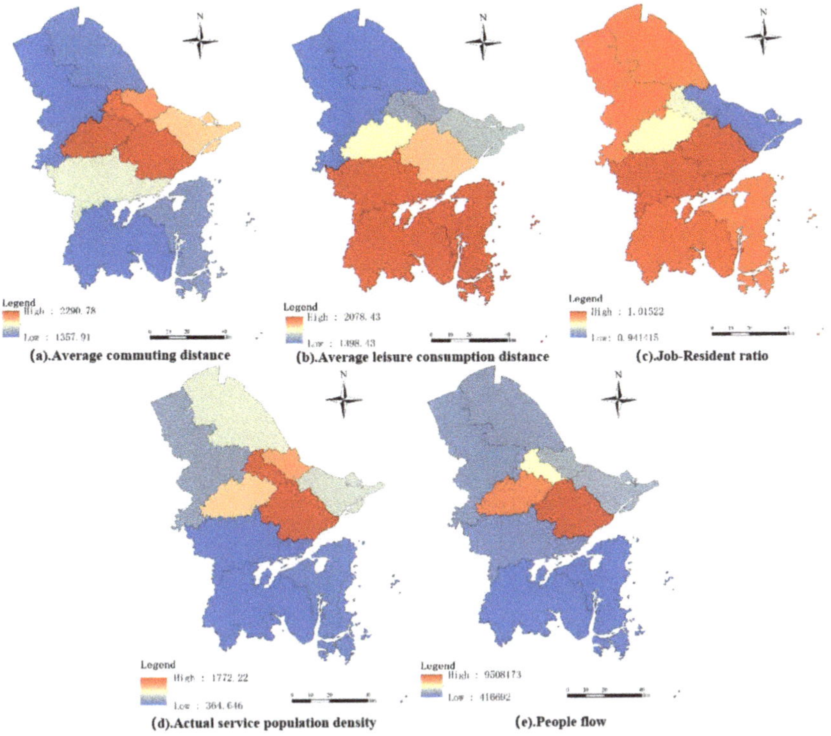

Figure 3. The results of resident behavior analysis. (**a**) shows the result of commuting distance calculations; (**b**) shows the result of leisure consumption calculations; (**c**) shows the result of the job–resident ratio calculations; (**d**) shows the result of the actual service population density calculations; (**e**) shows the result of people flow calculations.

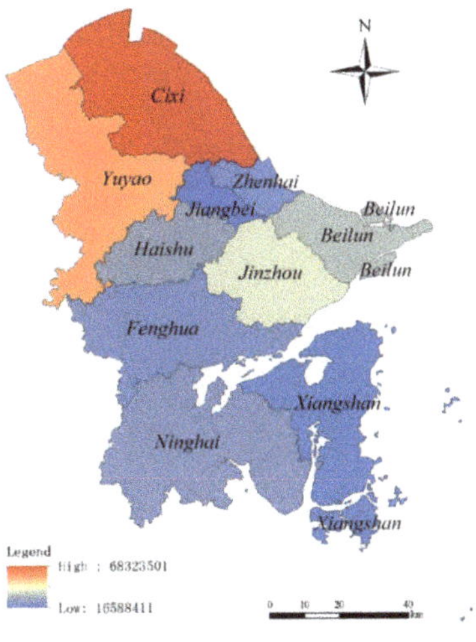

Figure 4. The total night-lighting index of Ningbo.

In the dimension of infrastructure layout, the Euclidean distance from pixels in each raster to the nearest POI point is considered to reflect the infrastructure perfection required for residents' daily activities. Seven types of infrastructure, including bus stations, subway stations, railway stations, medical facilities, schools, commercial facilities, and recreation areas, are collected, and the distance between each raster's pixels and the above seven types of infrastructure POI is calculated to represent the level of infrastructure layout improvement (Figure 5a–g).

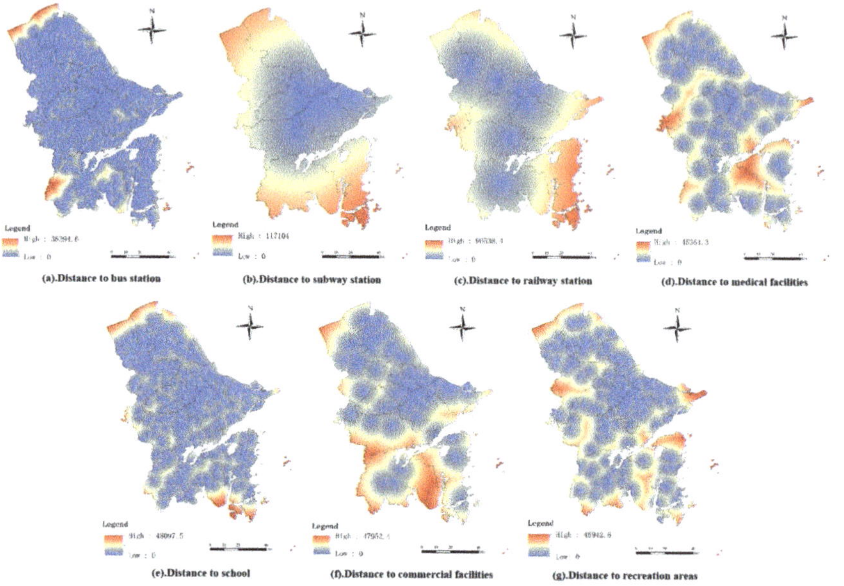

Figure 5. Distance to different POI. (**a**) shows the distance to bus stations; (**b**) shows the distance to subway stations; (**c**) shows the distance to railway stations; (**d**) shows the distance to medical facilities; (**e**) shows the distance to schools; (**f**) shows the distance to commercial facilities; (**g**) shows the distance to recreation areas.

Following completing the calculation of indicators under the three dimensions, their weights need to be assigned. In this study, the analytic hierarchy process (AHP) is utilized to calculate the weights of each indicator. After checking for consistency, the weights that reflect the importance of different indicators are obtained (Table 4). Then the sum of weights and three-dimensional indicators is used to establish the residential activity space constraint. The evaluation results are obtained as shown in Figure 6. It shows that the potential is greatest in the north and smallest in the south of Ningbo. By comparing with the current urban planning scheme of Ningbo government, it is found that the development pattern of Ningbo in our study is generally consistent with previous research results. Specifically, the regions with high potential values in the evaluation results are mainly distributed in Yuyao and Cixi, because the residents' activities in these regions are relatively rich and the degree of economic development is better. In addition, the areas with the smallest potential are mainly located in Xiangshan, Ninghai, and Fenghua because their urban growth is constrained by ecological land in the south.

Table 4. The weight of each indicator.

Category		**Indicator Name**	**Weight**
Resident behavior	Commuting	Average commuting distance to residents' workplaces	0.0667
	Leisure consumption	Average travel distance for leisure consumption	0.0667
	Jobs–housing space	Job–resident ratio	0.0667
	Population distribution	Actual service population density	0.2
	People flow	The total number of people flow	0.2
Economic development		The total nighttime-lighting index	0.2
Infrastructure layout	Transportation	Distance to bus stop	0.0286
		Distance to station	0.0286
		Distance to station	0.0286
	Medical	Distance to hospitals, clinics	0.0286
	Education	Distance to schools, colleges, and universities	0.0286
	Commercial services	Distance to commercial center and office buildings	0.0286
	Recreation facilities	Distance to recreational facilities	0.0286

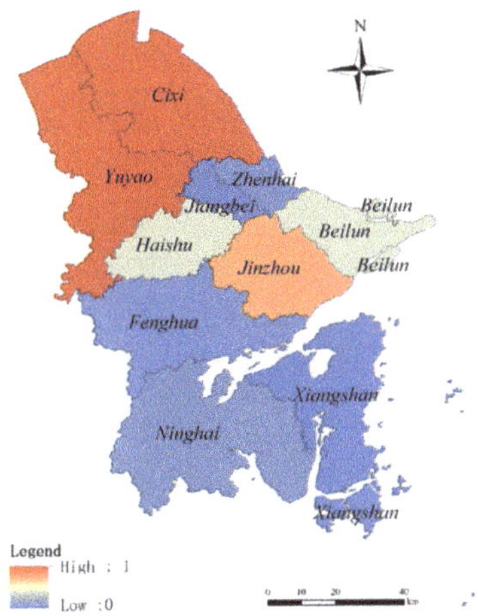

Figure 6. The results of the evaluation of the potential for the expansion of residential activity space.

4.2. Results of the Urban Construction Suitability Evaluation

Good natural environmental conditions are important for human settlement livability and urban development. The concept of humanism also emphasizes the suitability of the living environment in urban development. Thus, we evaluate urban construction suitability according to a standard technical guideline [45]. This guideline defines an evaluation method including four sub-evaluations: land resources evaluation, water resources evaluation, environment evaluation, and climate evaluation. The above four sub-evaluations are evaluated respectively from topographic conditions, water supply conditions, climatic comfort, and atmospheric environmental capacity [45]. The calculation of each factor was performed according to the guidelines. As space is limited, the details of the calculations are not described here. Then the results of sub-evaluation (Figure 7a–d) are integrated to obtain the urban construction suitability in Ningbo (Figure 7e), which are used to establish the human settlements suitability constraint in the constrained CA model. As can be seen from Figure 7e, the urban construction suitability is high in the center and north of Ningbo, as well as the coastal areas, which have flat terrain, good water-supply conditions, comfortable and pleasant climate, and the good quality of the atmospheric environment. The area southwest of Ningbo is not suitable for urban construction because of its high topography and ecological importance

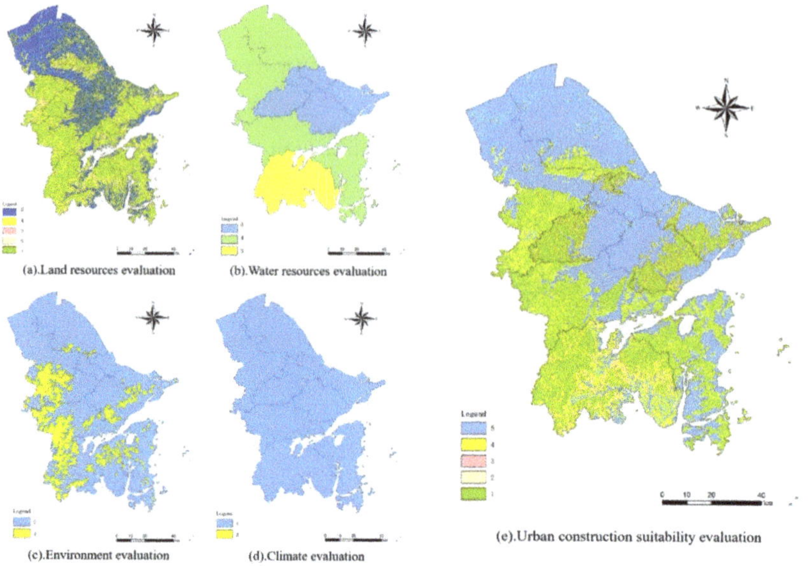

Figure 7. The results of urban construction suitability evaluation. (**a**) shows the result of land resources evaluation; (**b**) shows the result of water resources evaluation; (**c**) shows the result of environment evaluation; (**d**) shows the result of climate evaluation; (**e**) shows the result of urban construction suitability evaluation.

4.3. Results of the Ecological Conservation Importance Evaluation

A stable ecosystem and healthy ecological environment, which can expand the space of urban carbon sink, is an important foundation and contributes to green low-carbon development. Therefore, ecologically important areas need to be protected in urban development. To identify priority areas for ecological protection in urban expansion, we evaluate ecosystem service importance and ecological vulnerability in Ningbo according to the standard technical guideline mentioned in Section 4.2. This guideline also defines the methods of the above two evaluations.

The evaluation of ecosystem service importance is based on the results of three sub-evaluations, namely, the evaluation of water-resources retention capacity, the evaluation of soil and water conservation, and the evaluation of biodiversity conservation (Figure 8a–c). Figure 8d shows that ecosystem service importance in the southwest is higher than that in the northeast of Ningbo.

The evaluation of ecological vulnerability is integrated by the results of soil erosion, land desertification, and rock desertification evaluation (Figure 9a–c). After synthesizing the results of the above evaluation, Figure 9d shows that the ecological vulnerability of Ningbo is generally fragile.

Thus, the result of Ningbo ecological conservation importance evaluation (Figure 10) is derived from the combination of the above two evaluations, and the ecological importance level of Ningbo was divided into three grades: very important, important, and generally important. Figure 10 shows that the very important areas are located in the southwest. Most of the important areas were found in Fenghua, Ninghai, and Xiangshan. These areas are important for water connotation and soil conservation, necessitating the limitation of urban construction. The generally important areas are mainly located in Cixi, Yuyao, and the main urban area. There are less ecological constraints on human habitation, and the area can be prioritized for urban construction. The result is consistent with the layout of ecological land in urban planning (Ningbo Urban Master Plan 2006–2020), which proves the ecological conservation importance evaluation result is valid.

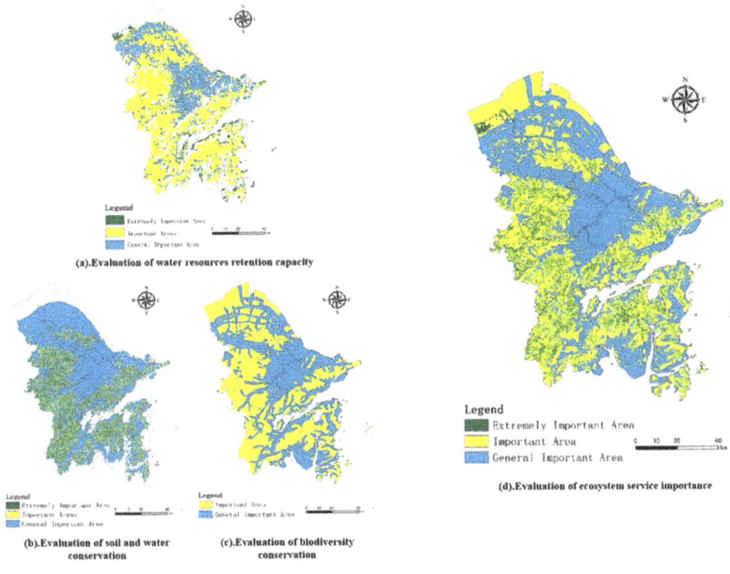

Figure 8. The results of the evaluation of ecosystem service importance. (**a**) shows the result of water-resources retention capacity evaluation; (**b**) shows the result of soil and water conservation evaluation; (**c**) shows the result of biodiversity conservation evaluation; (**d**) shows the result of ecosystem service importance evaluation.

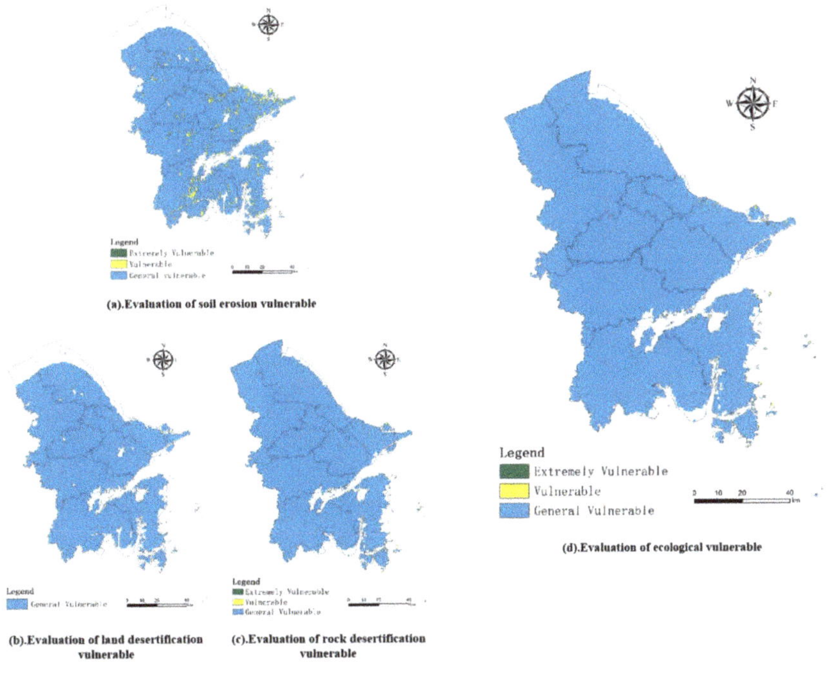

Figure 9. The results of ecological vulnerability evaluation. (**a**) shows the result of soil erosion vulnerability evaluation; (**b**) shows the result of land desertification vulnerability evaluation; (**c**) shows the result of rock desertification vulnerability evaluation; (**d**) shows the result of ecological vulnerability evaluation.

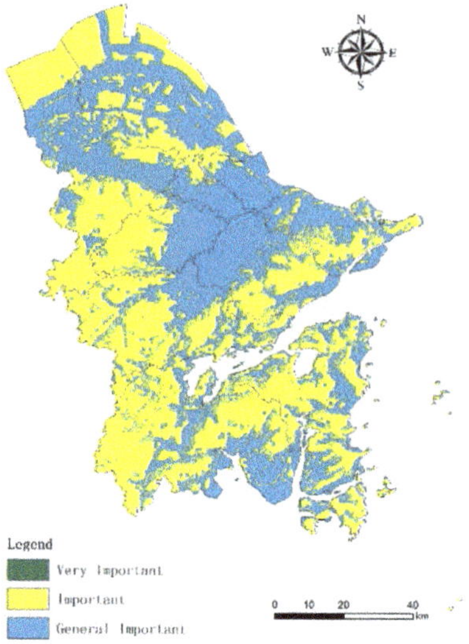

Figure 10. The results of ecological conservation importance evaluation.

4.4. Result of the Limitation of Human Survival Materials

With the high-density development and the growth in urban construction land, urban problems such as farmland loss and water pollution have become increasingly serious. The conversion of these land uses to construction land also increases the city's carbon emissions. In order to improve environment quality and promote low-carbon development, high-quality farmland and water bodies are protected from urban sprawl. In our study, these farmlands and water bodies are chosen to be free from invasions by urban build-up land; these sites are a human settlements security constraint (Figure 11).

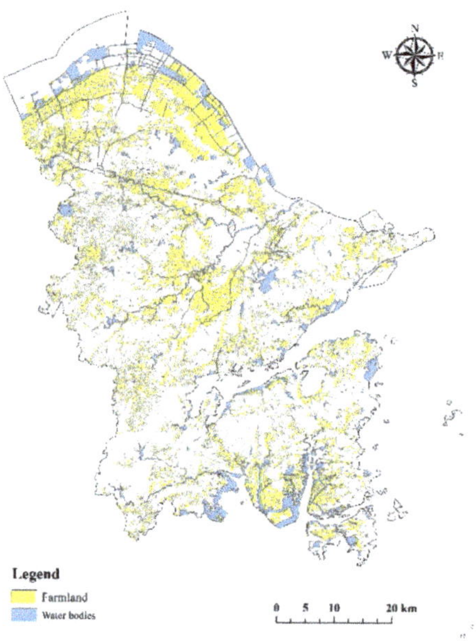

Figure 11. The areas of limitation of human survival materials.

4.5. Simulation Results

Based on above results in Sections 4.1–4.4 three constraints and a limitation are obtained, respectively. Using these constraints and limitation, the multi-objective constrained CA model from the perspective of humanism and the low-carbon concept is developed to simulate the urban expansion in Ningbo. As for the model validation, we use the kappa coefficient to assess the relationship between the simulation results compared to the actual situation, and a larger coefficient value indicates that the simulation results are more reliable. For details, we choose the land-use map of Ningbo in 2010 and 2015 to simulate urban boundaries in 2015 and 2020, respectively (Figures 12 and 13), and the kappa value is 0.84 and 0.85 for 2015 and 2020, respectively. Generally, a kappa coefficient between 0.8 and 1 indicates a high degree of consistency between the simulated and real values. Thus, the results of our study show that the model can be used to predict urban growth to the year 2025.

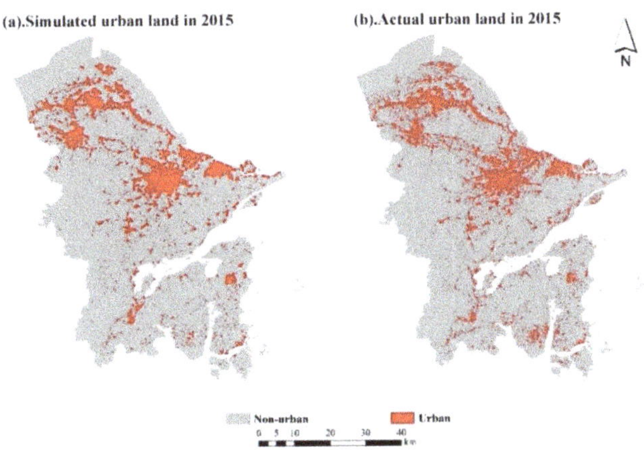

Figure 12. Simulated urban land and actual urban land in 2015. (**a**) shows simulated urban land; (**b**) shows actual urban land (data source from http://data.ess.tsinghua.edu.cn/ accessed on 15 January 2021).

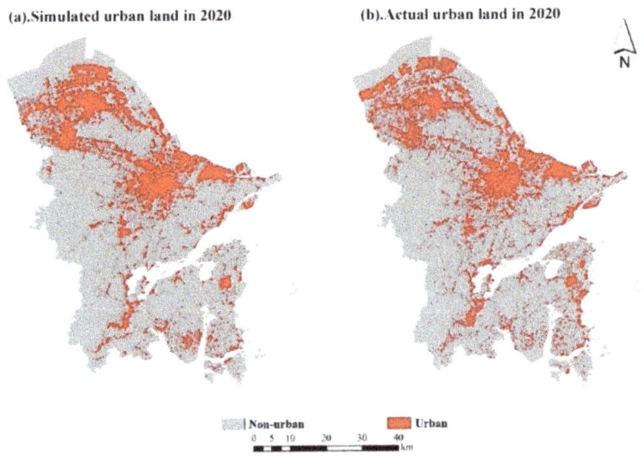

Figure 13. Simulated urban land and actual urban land in 2020. (**a**) shows simulated urban land; (**b**) shows actual urban land (data source from http://www.globallandcover.com/ accessed on 15 January 2021).

4.6. UGB Delineation

The urban boundary in 2025 obtained from our simulations using the model in Section 4.4 has some fragmented patches that are not conducive to the consolidated management of land. In order to ensure that the UGB is as smooth and continuous as possible, morphological function in the FLUS–UGB module is used to determine the UGB. A 7∗7 window is selected as the structural element for dilation and erosion. Then, the raster format UGB generated by FLUS is converted to vector format using GIS software, and small patches of UGB less than 2 km^2 are removed to obtain the final UGB of Ningbo in 2025 (Figure 14).

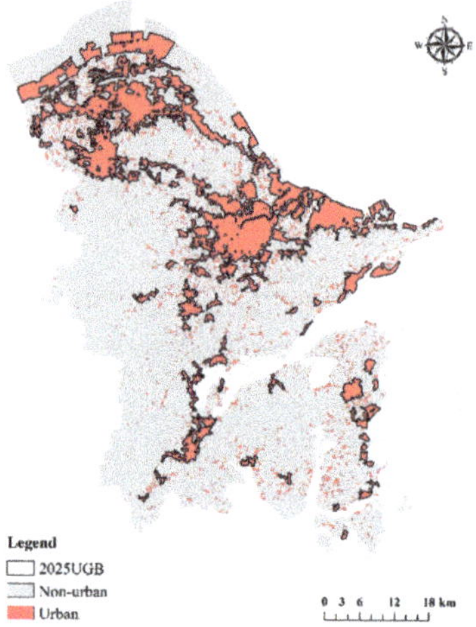

Figure 14. Urban growth boundary of Ningbo in 2025.

5. Discussion

5.1. The Necessity of the UGB Delineation Framework from a People-Oriented and Low-Carbon Perspective

UGB delineation is a complex decision-making issue and should consider multiple objectives and constraints. Through reviewing the existing studies, we find that most of the existing studies on UGB delineation are limited to the objective perspectives like economic development and natural conditions, and few of them consider human perspective constraints. However, people are the center of urban development, and the people-oriented perspective is an indispensable core concept of urban development. The delineation of UGB should definitely consider multiple constraints such as humanism and low-carbon development. Further, in order to verify the necessity of these multi-objective constraints, we compare urban boundary simulation in 2020 with and without multi-objective constraints. Figure 15 shows that urban growth without multi-objective constraints leads to the encroachment on water bodies (see Figure 15a), such as the Yuyao River, Fenghua River, Yong River, etc., as well as agricultural land (see Figure 15b) and forested land (see Figure 15c). In contrast, urban expansion based on multi-objective constraints avoids water bodies and effectively protects ecological and agricultural land, which is helpful for carbon emission reduction. Figure 16 shows that there is a leap-type cluster in the north due to the increase in residents' activities, which is consistent with the actual construction of Qianwan New District. The comparison results show that the multi-objective constraint from a people-oriented and low-carbon perspective is necessary and that the proposed framework is more comprehensive.

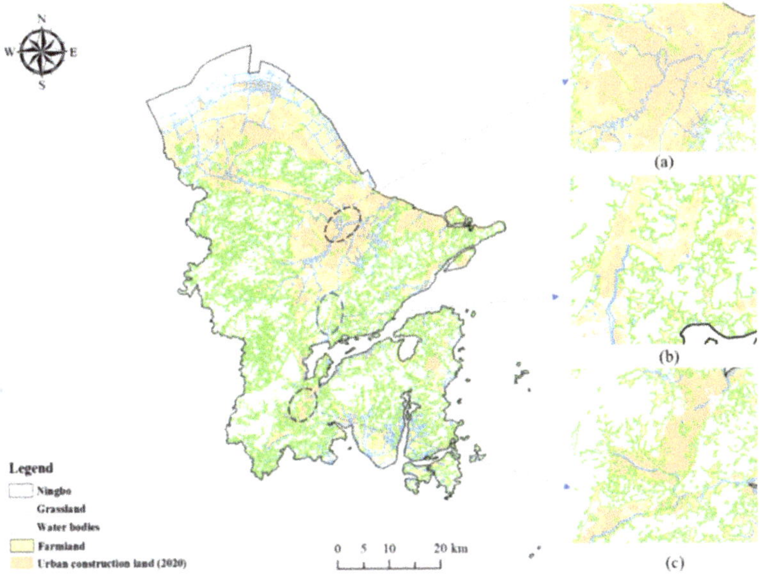

Figure 15. Simulated urban lands without constraints in 2020. (**a**) shows that multi-objective constraints leads to encroachment on water bodies; (**b**) shows that multi-objective constraints leads to encroachment on agricultural land; (**c**) shows that multi-objective constraints leads to encroachment on forested land.

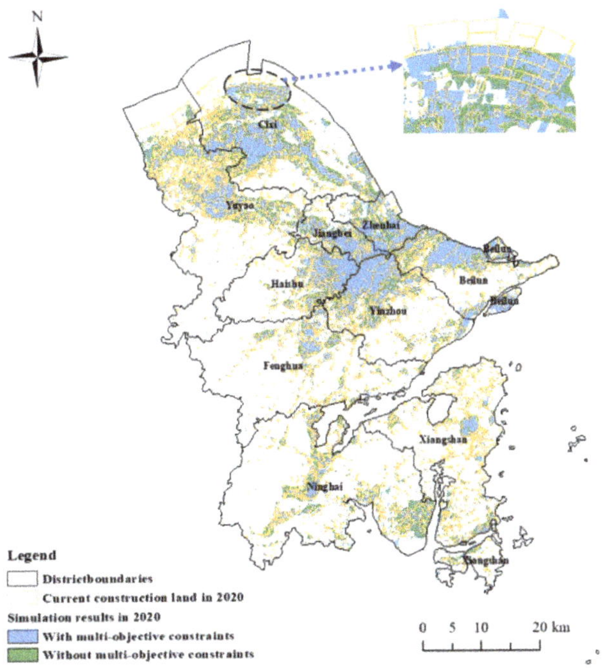

Figure 16. Simulated urban lands with multi-objective constraints in 2020.

5.2. Comparison in UGBs Delineation with and without Big Data

Most previous studies on urban sprawl simulations have utilized traditional data, which are limited by the problem of an insufficiently fine scale and can only reflect the human activities at large scale but cannot finely characterize human activities at small scale. This makes it difficult for the simulation results to reflect the real situation. The emergence of big data, such as cell phone signaling, has provided a solution to this problem. Therefore, we innovatively use multi-source big data from three perspectives to demonstrate human activities and introduce it into the CA model. Among them, mobile phone signaling big data is used to measure the characteristics of residents' behavior activities, night-light remote-sensing big data is used to measure the level of urban economic development, and POI big data is used to measure the layout of urban infrastructure. These big data, especially mobile signaling data, provide information on the impact of human behavior on urban expansion. In order to verify the necessity and feasibility of residential activity space constraints based on multi-source big data, this study compared the delineation results of UGBs obtained with and without the use of multi-source big data for residential activity space constraints under the same other two constraints and one limitation. The result shows that the delineated UGB based on multi-source big data is roughly consistent with the current situation of urban land in Ningbo in 2020. The Kappa value of urban extension simulation results considering residential activity space constraint is 0.855, while, without the residential activity space constraint, it is 0.831. The results show that the framework based on multi-source big data has higher accuracy and better performance in urban expansion simulation. The findings also show that big data provide an important data basis for refined urban land simulation. In addition, the international community has promoted the application of multi-source big data in territorial spatial planning [49,50], which also shows that multi-source big data has a high potential for application in spatial planning in the future.

6. Conclusions

The existing UGB delineation methods are limited by the lack of data granularity, and most of them can only simulate urban development based on the perspectives of economic development and natural resources, lacking consideration for human activities. To this end, we propose a new framework for the delineation of UGBs. This framework integrates the concepts of humanism and low-carbon development and sets constraints, including the potential of human activities, natural resource background, ecological environment protection, and subsistence conditions, to comprehensively consider the influence of human activities in low-carbon urban development. In terms of model construction, this framework utilizes a constrained CA model supported by multiple sources of big data. The CA model takes the patch change as the base unit of urban development simulation and uses the results of three evaluations and one limitation as constraints to simulate urban development. In particular, the support of big data makes it possible to calculate the expansion potential of human activities, and high fine-scale multi-source big data, such as mobile phone signaling data, also provide the possibility of a solution to the better evaluation of human activities. That makes the simulation results more accurate and reliable. The application of our proposed UGB delineation framework in a rapidly growing city in China demonstrates the applicability and reliability of this framework. Considering multi-objective constraints and big data support, the framework is also applicable in other cities.

However, our study still has some limitations. The range of mobile phone signaling data used in the case study only includes December 2019, which may not fully reflect the characteristic patterns of residents' behaviors. In future studies, the data will be analyzed with a longer time series to obtain a more objective and comprehensive characteristic pattern. Meanwhile, this paper only discusses low-carbon development in terms of the evaluation of ecological conservation importance and the limitation of human survival materials constraints, and in the future, we will add quantitative measurements to make our framework more detailed.

Author Contributions: Conceptualization, Y.Y. and W.M.; methodology, Y.Y.; software, C.Z. and W.M.; validation, C.Z., Y.X. and X.G.; formal analysis, C.Z. and W.M.; investigation, C.Z.; data curation, Y.X. and X.G.; writing—original draft preparation, W.M.; writing—review and editing, Y.Y. and C.Z.; visualization, Y.X. and X.G. All authors have read and agreed to the published version of the manuscript.

Funding: This research was funded by National Natural Science Foundation of China (42171260, 52079101).

Conflicts of Interest: The authors declare no conflict of interest.

References

1. Wang, W.; Jiao, L.; Zhang, W.; Jia, Q.; Su, F.; Xu, G.; Ma, S. Delineating urban growth boundaries under multi-objective and constraints. *Sustain. Cities Soc.* **2020**, *61*, 102279. [CrossRef]
2. Hoekstra, A.Y.; Wiedmann, T.O. Humanity's unsustainable environmental footprint. *Science* **2014**, *344*, 1114–1117. [CrossRef] [PubMed]
3. Long, Y.; Han, H.; Lai, S.-K.; Mao, Q. Urban growth boundaries of the Beijing metropolitan area: Comparison of simulation and artwork. *Cities* **2013**, *31*, 337–348. [CrossRef]
4. Wu, X.; Liu, X.P.; Liang, X.; Chen, G. Multi-scenarios simulation of urban growth boundaries in Pearl River Delta based on FLUS-UGB. *J. Geo-Inf. Sci.* **2018**, *20*, 532–542. [CrossRef]
5. Yu, Y.; Han, Q.; Tang, W.; Yuan, Y.; Tong, Y. Exploration of the Industrial Spatial Linkages in Urban Agglomerations: A Case of Urban Agglomeration in the Middle Reaches of the Yangtze River, China. *Sustainability* **2018**, *10*, 1469. [CrossRef]
6. Tayyebi, A.; Pijanowski, B.C.; Tayyebi, A.H. An urban growth boundary model using neural networks, GIS and radial parameterization: An application to Tehran, Iran. *Landscape Urban Plan.* **2011**, *100*, 35–44. [CrossRef]
7. Ball, M.; Cigdem, M.; Taylor, E.; Wood, G. Urban growth boundaries and their impact on land prices. *Environ. Plan.* **2014**, *46*, 3010–3026. [CrossRef]
8. He, X.; Mai, X.; Shen, G. Delineation of Urban Growth Boundaries with SD and CLUE-s Models under Multi-Scenarios in Chengdu Metropolitan Area. *Sustainability* **2019**, *11*, 5919. [CrossRef]
9. Gerrit, K.; Arthur, N. *The Regulated Landscape: Lessons on State Land Use Planning from Oregon*; Lincoln Institute of Land Policy: Cambrige, MA, USA, 1992.
10. He, Q.; Tan, R.; Gao, Y.; Zhang, M.; Xie, P.; Liu, Y. Modeling urban growth boundary based on the evaluation of the extension potential: A case study of Wuhan city in China. *Habitat Int.* **2018**, *72*, 57–65. [CrossRef]
11. Jun, M. The effects of portland's urban growth boundary on urban development patterns and commuting. *Urban Stud.* **2004**, *41*, 1333–1348. [CrossRef]
12. Ma, S.; Zhao, Y.; Tan, X. Exploring Smart Growth Boundaries of Urban Agglomeration with Land Use Spatial Optimization: A Case Study of Changsha-Zhuzhou-Xiangtan City Group, China. *Chin. Geogr. Sci.* **2020**, *30*, 665–676. [CrossRef]
13. Wang, J.; Fang, C. Growth of Urban Construction Land: Progress and Prospect. *Prog. Geogr.* **2011**, *30*, 1440–1448.
14. Zhang, J. Urban Growth Management in the United States. *Urban Plan. Overseas* **2002**, *2*, 37–40. (In Chinese)
15. Ma, S.; Li, X.; Cai, Y. Delimiting the urban growth boundaries with a modified ant colony optimization model. *Comput. Environ. Urban Syst.* **2017**, *62*, 146–155. [CrossRef]
16. China's State Council. Several Opinions on Establishing Territorial Spatial Planning System and Supervising for Implementation. (In Chinese). Available online: http://www.gov.cn/zhengce/2019-05/23/content_5394187.htm (accessed on 23 May 2019).
17. Li, Y.H. Method of determining urban growth boundary from the view of ecology: A case study of Hangzhou. *City Plan. Rev.* **2011**, *35*, 83–90. (In Chinese)
18. Wang, Y.; Gu, C.; Li, X. Research progress of urban growth boundary at home and abroad. *Urban Plan. Int.* **2014**, *29*, 1–11.
19. Tan, R.; Liu, Y.; Liu, Y.; He, Q. A literature review of urban growth boundary: Theory, modeling, and effectiveness evaluation. *Prog. Geogr.* **2020**, *39*, 327–338. [CrossRef]
20. Feng-Ming, X.I.; Yuan-Man, H.U.; Hong-Shi, H.E.; Tie-Mao, S.H.I.; Ren-Cang, B.U.; Xiao-Qing, W.U.; Jing-Hai, Z.H.U. Urban planning based on SLEUTH model in Shenyang-Fushun metropolitan area. *J. Grad. Sch. Chin. Acad. Sci.* **2009**, *26*, 765–774. (In Chinese)
21. Long, Y.; Shen, Z.; Mao, Q. Retrieving spatial policy parameters from alter-native plans using constrained cellular automata and regionalized sensitivity analysis. *Environ. Plan. B Plan. Des.* **2012**, *39*, 586–604. [CrossRef]
22. Alghais, N.; Pullar, D. Modelling future impacts of urban development in Kuwait with the use of ABM and GIS. *Trans. GIS* **2018**, *22*, 20–42. [CrossRef]
23. Ying, L.; Jin, X.; Yang, X.; Zhou, Y. Reconstruction of historical arable land use patterns using constrained cellular automata: A case study of Jiangsu, China. *Appl. Geogr.* **2014**, *52*, 67–77.
24. Wu, H.; Li, Z.; Clarke, K.; Shi, W.; Fang, L.; Lin, A.; Zhou, J. Examining the sensitivity of spatial scale in cellular automata Markov chain simulation of land use change. *Int. J. Geogr. Inf. Sci.* **2019**, 33, 1040–1061. [CrossRef]
25. Long, Y.; Han, H.; Mao, Q. Establishing urban growth boundaries using constrained CA. *Acta Geogr. Sin.* **2009**, *64*, 999–1008. (In Chinese with English Abstract)

26. Ma, S.; Ai, B.; Nian, P. Pre-assessment and Warning of Land Use Planning with Constrained Cellular Automata. *Geogr. Geo-Inf. Sci.* **2014**, *30*, 51–55+2. (In Chinese)
27. Sun, Z.; Deal, B.; Pallathucheril, V.G. The land-use evolution and impact assessment model: A comprehensive urban planning support system. *Urisa J.* **2009**, *21*, 57–68.
28. Liu, X.; Li, X.; Ai, B.; Tao, H.; Wu, S.; Liu, T. Multi- agent systems for simulating and planning landuse development. *Acta Geogr. Sin.* **2006**, *61*, 1101–1112.
29. Qin, X.; Zhen, F.; Li, Y. Discussion on the application framework of big data in territorial spatial planning. *J. Nat. Resour.* **2019**, *34*, 2134–2149. (In Chinese) [CrossRef]
30. Kong, Y.; Zhen, F.; Zhang, S. Evaluation on High-quality Utilization of Territorial Space Based on Multi-source Data. *China Land Sci.* **2020**, *34*, 115–124. (In Chinese)
31. Krings, G.; Calabrese, F.; Ratti, C.; Blondel, V.D. Urban Gravity: A Model for Intercity Telecommunication Flows. *J. Stat. Mech. Theory Exp.* **2009**, *2009*, L07003. [CrossRef]
32. Gong, L.; Jin, M.; Liu, Q.; Gong, Y.; Liu, Y. Identifying urban residents activity space at multiple geographic scales using mobile phone data. *Int. J. Geo-Inf.* **2020**, *9*, 241. [CrossRef]
33. Zhang, Q.; Hu, Y.; Liu, J. Identification of Urban Clusters in China Based on Assessment of Transportation Accessibility and Socio-Economic Indicators. *Acta Geogr. Sin.* **2011**, *66*, 761–770.
34. He, Q.; He, W.; Song, Y.; Wu, J.; Yin, C.; Mou, Y. The impact of urban growth patterns on urban vitality in newly built-up areas based on an association rules analysis using geographical 'big data'. *Land Use Policy* **2018**, *78*, 726–738. [CrossRef]
35. Chen, L.; Qin, J.; Xu, L. Urban growth boundary: A revolution for low-carbon development. *Environ. Sci. Pollut. Res. Int.* 2022. [CrossRef]
36. National Bureau of Statistics (NBS). Available online: http://www.stats.gov.cn/ (accessed on 3 December 2020).
37. Liang, X.; Liu, X.; Li, X.; Chen, Y.; Tian, H.; Yao, Y. Delineating multi-scenario urban growth boundaries with a CA-based FLUS model and morphological method. *Landsc. Urban Plan.* **2018**, *177*, 47–63. [CrossRef]
38. Tu, W.; Cao, J.; Gao, Q.; Cao, R.; Fang, Z.; Yue, Y.; Li, Q. Sensing Urban Dynamics by Fusing Multi-sourced Spatiotemporal Big Data. *Geomat. Inf. Sci. Wuhan Univ.* **2020**, *45*, 1875–1883. (In Chinese)
39. Mei, M.; Chen, Z. Study on the Delineation Method of Urban Growth Boundary under the Coordination of Residential Activity Space and Ecological Constraint: A Case Study of Changsha, China. *Resour. Environ. Yangtze Basin.* **2018**, *27*, 2472–2480.
40. Xu, K.; Chen, F. The Truth of China Economic Growth: Evidence from Global Night-time Light Data. *Econ. Res. J.* **2015**, *50*, 17–29+57.
41. Zhou, C.; Zhang, R.; Jin, W. Urban Growth Boundary Delineation Oriented by Comprehensive Service level of Infrastructure: A case study of Guangzhou. *Geogr. Geo-Inf. Sci.* **2017**, *33*, 42–49. (In Chinese)
42. Wang, H.; Huang, J.; Zhou, H.; Deng, C.; Fang, C. Analysis of sustainable utilization of water resources based on the improved water resources ecological footprint model: A case study of Hubei Province, China. *J. Environ. Manag.* **2020**, *262*, 110331. [CrossRef]
43. Hao, Q.; Shan, J.; Deng, L. Evaluation on Natural Suitability of Human Settlement in the Context of Territorial Space Planning. *China Land Sci.* **2020**, *34*, 86–93.
44. Ministry Of Natural Resources of the People's Republic of China. Technical Guide for the Evaluation of Resource and Environment Carrying Capacity and Territory Spatial Development Suitability (In Chinese). 2019. Available online: http://gui-hua.com/post/33.html (accessed on 19 January 2020).
45. *Guidelines for Evaluation of Resource and Environment Carrying Capacity and Suitability of Territorial Spatial Development (for Trial Implementation)*; Ministry of Natural Resources of the People's Republic of China: Beijing, China, 2019.
46. Long, Y.; Mao, Q.; Shen, Z.; Du, L.; Gao, Z. Comprehensive Constrained CA urban model: Instructional Constraints and Urban Growth Simulation. *Urban Plan. Forum* **2008**, *6*, 83–91.
47. Xia, C.; Wang, H.; Zhang, A.; Deng, Y. Multi-scenario simulation and policyanalysis of urban space under the effects of coupling and control ling. *Hum. Geogr.* **2017**, *32*, 68–76.
48. Dai, Q.Y.; Yu, Y.L. The advances of mathematical morphology in imageprocessing. *Control Theory Appl.* **2001**, *18*, 478–482.
49. Zeng, L.; Yu, W.; Cui, Y.; Liu, J. Innovative Method of Urban Planning Based on Intelligent City. In Proceedings of the 2021 International Conference on Intelligent Transportation, Big Data & Smart City (ICITBS), Xi'an, China, 27–28 March 2021.
50. Luo, J. Exploring the Rational Management and Control Model of Land Development Intensity in the Era of Big Data. In Proceedings of the 2nd International Symposium on Economics, Management, and Sustainable Development (EMSD 2021), Hangzhou, China, 27–28 November 2021; Available online: https://bcpublication.org/index.php/BM/article/view/227 (accessed on 10 October 2022).

sustainability

MDPI

Article

Exploring Differentiated Conservation Priorities of Urban Green Space Based on Tradeoffs of Ecological Functions

Huiying Li [1], Dianfeng Liu [1,2,*] and Jianhua He [1,2]

[1] School of Resources and Environmental Sciences, Wuhan University, Wuhan 430079, China; lihuiying@whu.edu.cn (H.L.); hjianh@whu.edu.cn (J.H.)

[2] Key Laboratory of Digital Mapping and Application of Land Information Engineering, Ministry of Natural Resources, Wuhan 430079, China

* Correspondence: liudianfeng@whu.edu.cn; Tel.: +86-134-8707-4270; Fax: 0086-027-6877-8893

Abstract: Urban green space (UGS) can simultaneously provide social and ecological benefits for humans. Although numerous studies have evaluated the multifunctional benefits of urban green space, few of them have determined the differentiated conservation priorities of UGS towards the tradeoff relationship of multiple UGS functions. Here, we proposed an integrated framework to explore the targeted conservation strategies of UGS patches. Specifically, the circuit theory model and gravity floating catchment area method were adopted to evaluate ecological connectivity and spatial accessibility of UGS under multiple scenarios in terms of different species dispersal distances and resident travelling modes, and Pareto ranking analysis was utilized to identify conservation priorities of UGS. Wuhan City in central China was taken as a case study. The results show that Wuhan exhibits low synergic relationship of ecological connectivity and spatial accessibility of UGS, and only approximately 7.51% of UGS patches on average rank high. Based on the frequency of UGS Pareto ranks under different scenarios, the differentiated conservation strategy was developed, which identified 10 key green areas that need to be protected and 11 green areas that need to be restored. This work is expected to provide an applicable framework to identify key UGS patches and assist in urban planning and layout optimization of multifunctional UGS in Wuhan, China.

Keywords: urban green space; pareto ranking; ecological connectivity; spatial accessibility

Citation: Li, H.; Liu, D.; He, J. Exploring Differentiated Conservation Priorities of Urban Green Space Based on Tradeoffs of Ecological Functions. *Sustainability* **2022**, *14*, 1845. https://doi.org/10.3390/su14031845

Academic Editor: Richard Hauer

Received: 17 January 2022
Accepted: 3 February 2022
Published: 5 February 2022

1. Introduction

Urban green spaces (UGS) are a key component of urban ecosystems that can simultaneously provide ecological and social benefits [1,2]. With the acceleration of the urbanization process, however, urban green spaces are increasingly encroached upon, and the urban ecological quality has shown a significant downward trend [3,4]. At the same time, people's vision of pursuing a better life is increasing, and urban green spaces are needed to provide better services to meet the leisure and cultural needs of residents. In this context, how to make comprehensive use of the scarce UGS resources to realize their maximum social and ecological benefits has become a challenging issue and also a key agenda in promoting harmonious human–nature relationships.

As the scarce natural resource within cities, urban green space has significant versatility [5], e.g., a range of regulation, supporting, and cultural services such as water cycle supporting, climate regulation, air purification, and recreation [6,7]. According to existing studies, improving versatility of urban green spaces can effectively coordinate the conflict between urban development and ecological protection and then enhance resilience and sustainability of urban ecosystems [8–10]. Various functions of urban green space are closely related to different land use stakeholders, and the differences in the interests of different stakeholders will cause tradeoffs of multiple UGS functions. In traditional urban planning, the decision processes of different government departments in terms of planning and management of different UGS functions are usually separated, and the interrelationship between the multi-functions of urban green spaces is rarely considered. Thus, how to improve the multi-functionality of urban green spaces is still one of great challenges of urban planning. To face this challenge, many scholars have explored the trade-off/synergy between urban green spaces [11] and hope to coordinate multiple UGS functions to determine its conservation priorities to maximize the potential ecosystem services of urban green spaces [6,12]. Existing research mostly adopted a multi-criteria evaluation method to conduct a comprehensive assessment of the versatility of urban green space or utilized spatial overlay analysis to reveal conflicts between multiple functions of green space [9,10]. However, the weighted aggregation of multiple criteria is subject to weighting uncertainty, while ignoring the tradeoff relationship between multi-functionalities of green space. It is urgent to address the multi-functionality evaluation issue from the perspective of UGS function tradeoffs.

Among various social and ecological benefits of urban green space, biodiversity conservation and recreation provision are two essential functions of urban green space, which have attracted increasing attention in spatial allocation of urban green space [13,14]. Complex tradeoffs and synergies exist between two UGS functions. For example, biodiversity conservation of UGS may significantly improve the recreation experience and health well-being of urban residents [15] while resident recreation behavior will impose a certain disturbance on the protection of urban green habitats [16]. In this context, accurately identifying the tradeoffs and synergies of two functions of urban green spaces can provide a scientific basis for formulating multi-functional protection strategies for urban green spaces. In order to effectively measure the functional benefits of urban green space, a large number of studies have selected ecological connectivity and spatial accessibility to characterize the level of biodiversity conservation and the level of recreation service of UGS. Specifically, scholars chose the ecological connectivity index to represent the impact of landscape on species movement among landscape patches and the influence of landscape fragmentation on biodiversity [3,17,18]. The measurement methods of ecological connectivity mainly include empirical studies on the diffusion behaviors of species, the least cumulative resistance model based on graph theory, and the connectivity model based on circuit theory [19,20], among which the circuit theory model was able to identify multiple alternative dispersal pathways for species to migrate between landscape patches [21,22]. On the other hand, spatial accessibility is an effective indicator that reflects the ability of urban green space to provide recreation service to residents [23]. Spatial accessibility analysis of urban green space will help to identify the differences in recreation service of green space, and optimizes the spatial layout of green space to meet the needs of different types of social groups, thereby improving the overall social benefit of green space [24,25]. Current spatial accessibility analysis approaches mainly comprise spatial buffering analysis, minimum proximity distance analysis, and the floating catchment area method, of which the floating catchment area method was proven to be more practical due to its integration with the urban road network and the gravity model of spatial interaction [26]. Although the existing studies provide solid support for quantitative analysis of ecological and social benefits of urban green space, the differentiated conservation priorities related to the socio-ecological interactions of UGS still require further exploration, especially the impact of different travelling modes and species dispersal distances on the tradeoff relationship of two UGS functions were rarely taken into account.

The purpose of this work is to coordinate the ecological connectivity and the spatial accessibility of UGS from the functional tradeoff aspect, and to take Wuhan City in central China as a case study to identify the conservation priorities of UGS. To achieve the win-win goal of maximizing the level of biodiversity maintenance and the level of recreation services of urban green space, the Pareto ranking method is adopted to analyze the tradeoff relationship between ecological connectivity and spatial accessibility of urban green spaces in the main urban area of Wuhan under the scenarios of different travel modes and species dispersal distances. The key conservation and restoration UGS patches will be identified to provide practical basis for urban planning and the optimal adjustment of green space of Wuhan City.

2. Study Area and Data Sources

2.1. Study Area

The main urban area of Wuhan City is selected as the study area, with the total area of 678 km^2 (Figure 1). Although the green area in the entire city reached 209 km^2, with a coverage rate of 39.55%, it only occupies 4% in the study area due to high land use intensity. The green space ratio of the study area is significantly lower than its required standard as regulated by the urban planning policy in China (i.e., 30%, which will evidently contribute to the sustainable development of the urban ecosystem). The mixed forest composed of evergreen broad-leaved forest and deciduous broad-leaved forest is a typical vegetation type in the city, and the main tree species include camphor, masson pine, and fir. As the inventory of urban green space reported, there are more than 400 bird species in the study area and the brown-headed crow finches, white-headed bulbuls, and gray magpies dominate.

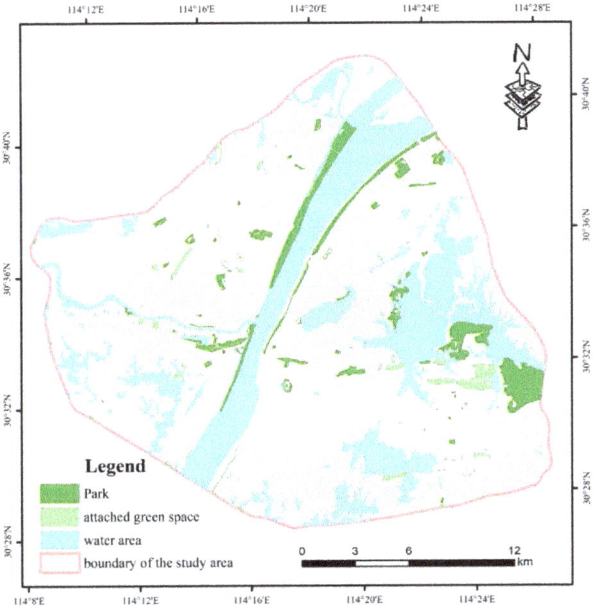

Figure 1. Location of the study area and distribution of urban green spaces.

However, the green space has become increasingly fragmented under the influence of urban expansion, which seriously affects the equitable access of urban residents to green space resources [14,27]. The decline of urban green space will also exert a negative impact on ecological connectivity of urban ecosystems and then cause the loss of urban biodiversity [3]. It is of great importance to identify key UGS patches that play an ecological and social role in urban planning.

2.2. Data and Processing

We collected the urban green space data based on the 2018 POI data from Baidu Map and the Open Street Map (http://download.geofabrik.de/ (accessed on 5 May 2021)), and a total of 150 UGS patches were extracted, which were mainly divided into three types: park green space, road auxiliary green space, and other auxiliary green space attached to urban construction land such as residential areas, commercial service facilities, and utility land. To identify the tradeoff between biodiversity conservation and resident welfare services, we utilized the following criteria to identify green spaces: (1) From the perspective of ecological connectivity, we determined 3 ha as the area threshold for species habitat due to the minimum conditions of typical urban species for habitat selection [17,28]; (2) From the perspective of daily travel accessibility of urban residents, green spaces for non-daily leisure use (e.g., UGS patches along streets) were excluded from candidate green spaces. Then, 71 urban green spaces with an area of more than 3 ha and accessible to residents for their daily leisure were selected as research subjects, which include 45 park green spaces and 26 auxiliary green spaces such as universities and squares, with a total area of 2735.29 ha.

Moreover, land use data, population distribution data, and road networks were also included in this study. The land use map in 2018 was obtained from the Resources and Environmental Science Data Center of the Chinese Academy of Sciences (http://www.resdc.cn (accessed on 8 March 2021)); the population data at the community level were obtained from the sixth census of Wuhan City; the road networks were downloaded from the Open Street Map (http://download.geofabrik.de/ (accessed on 6 May 2021)), and roads suitable for walking and cycling were extracted for network analysis.

3. Methodology

3.1. Analysis of UGS Ecological Connectivity Based on Circuit Theory

Urban green spaces are key habitat patches for urban species (birds, arthropods, etc.), which provide habitats and food sources for species survival [17]. Exploring the ecological connectivity between urban green spaces can provide scientific support for urban biodiversity protection. Circuit theory, a fundamental concept in physics that was first applied to landscape ecology by McRae, can adopt the random travel of electrons in circuits to simulate the migration or dispersal of species, genes, or populations in complex landscapes [29]. An ecological connectivity model based on circuit theory treats species populations or gene streams as electrons, habitat patches as nodes, and conductivities as landscape resistance surfaces, and characterizes the probability of dispersal of species in landscape as the current density generated by electron travel [21].

In this article, the circuit theory model considers UGS patches as the source nodes of species and uses the average current density of the patches to quantitatively evaluate the level of ecological connectivity of green patches. Considering that the dispersal capabilities of species are mainly affected by their own size, different area thresholds of the source nodes, i.e., 1 ha, 3 ha, and 5 ha, were selected to simulate the dispersal behaviors of different species in the study area. In the construction of landscape conductive surfaces, we used habitat quality to represent the ecological resistance to species dispersal.

The habitat quality was calculated using the InVEST model and the corresponding parameters were set as shown in Tables 1 and 2 [30], the specific calculation follows Equations (1)–(3). The Circuitscape 4.0 software was used to simulate species dispersal between green patches, and the cumulative and average current density through each UGS node was then calculated iteratively.

$$D_{xj} = \sum_{r=1}^{R}\sum_{y=1}^{Y_r} r_y \left(\frac{w_r}{\sum_{r=1}^{R} w_r}\right) i_{rxy}\beta_x S_{jr} \tag{1}$$

$$i_{rxy} = 1 - \left(\frac{d_{xy}}{d_{rmax}}\right) \tag{2}$$

$$Q_{xj} = H_j\left[1 - \left(\frac{D_{xj}^2}{D_{xj}^2 + k^2}\right)\right] \tag{3}$$

where, D_{xj} is the habitat degradation degree of grid x in land type j, and Q_{xj} is the habitat quality; r is the threat source of the habitat, and y is the grid in the threat source r; r_y, w_r, and i_{rxy} are the human disturbance intensity, weight, and impact on the habitat of the grid where the threat source is located, respectively; β_x and S_{jr} are the anti-disturbance ability of the habitat and its relative sensitivity to various threat sources; d_{xy} is the distance between the habitat grid x and the threat source grid grid y, and d_{rmax} is the maximum influence range of the threat source r; H_j is the habitat suitability of land use type j; k is the half-saturation constant, $k = 0.5$; z is the default parameter, $z = 2.5$ in this study.

Table 1. Threats of species and their weights.

Threat Source	Influencing Distance/km	Weights	Spatial Decay Type
Urban and rural construction land	2.0	1.0	Exponential
Transportation land	1.0	0.9	Linear
Cultivated land	0.5	0.5	Linear
Barren land	1.5	0.6	Exponential

Table 2. Habitat suitability and their relative sensitivity to different threat sources.

Land Use Type	Habitat Suitability	Urban and Rural Construction Land	Transportation Land	Cultivated Land	Barren Land
Woodland	1.00	0.90	0.65	0.85	0.65
Grassland	0.60	0.65	0.35	0.50	0.50
Water areas	0.90	0.85	0.50	0.75	0.50
Cultivated land	0.50	0.50	0.25	0.35	0.30
Urban and rural construction land	0.00	\	\	\	\
Transportation land	0.00	\	\	\	\
Barren land	0.00	\	\	\	\

3.2. Spatial Accessibility Analysis of Green Space Based on Gravity-Based Floating Catchment Area Model

The potential population size within the service range of a certain urban green space represents its capability to satisfy the leisure demand of urban residents, and the UGS patches that have greater potential serving population will be easier to be accessed by people [31]. Compared with the traditional spatial accessibility analysis approach, the gravity based floating catchment area model incorporate the distance decay effect into the spatial accessibility of UGS patches as well as to indicate the spatially heterogeneity of human demand to UGS ecosystem services [26]. The calculation is as follows:

$$R_j = \sum\nolimits_{k \in \{d_{kj} \le d_0\}} G\left(d_{kj}, d_0\right) P_k \quad (4)$$

$$G\left(d_{kj}, d_0\right) = \begin{cases} \frac{e^{-(\frac{1}{2}) \times (\frac{d_{kj}}{d_0})^2} - e^{-(\frac{1}{2})}}{1 - e^{-(\frac{1}{2})}} & , d_{kj} \le d_0 \\ 0 & , d_{kj} > d_0 \end{cases} \quad (5)$$

where, R_j is the spatial accessibility level of urban green space j; d_{kj} is the minimum travel distance from community k to green space j; d_0 is the longest distance a resident in a certain community travels to a green space; P_k is the weighted population size of community k within the service range of a UGS patch ($d_{kj} \le d_0$); $G(d_{kj}, d_0)$ is a Gaussian equation that takes into account distance decay of UGS accessibility.

Specifically, this study assumed that urban residents go to urban green spaces by walking or cycling, with travelling speeds of 5 km/h and 15 km/h, respectively. Based on the community living circles and travelling limit time, the thresholds of travelling time were set to 15 min and 30 min [14,27].

3.3. Identification of UGS Protection Priorities Based on Pareto Ranking Analysis

The ecological connectivity and spatial accessibility patches are key indicators of urban green space to provide biodiversity conservation and cultural services, which usually exhibit tradeoff relationships with each other. Pareto ranking analysis was proposed to address multi-objective optimization issues. Compared with traditional linear weighted models, Pareto ranking approach can avoid the process of determining parameters such as weights and threshold values and can reach the optima of individual solutions that balance the conflicts of multiple objectives [32,33]. Therefore, we used Pareto ranking method to identify UGS patches that can maximize the comprehensive benefits of ecological connectivity and spatial accessibility in the context of scarcity of UGS resources.

According to the definition of Pareto ranking, the ecological connectivity level and spatial accessibility level of urban green space were taken as two maximum development targets of green space patches, and sequentially identify Pareto solution sets with different Pareto ranks. The ranking process is as follows:

$$f_i(x) \geq f_i(x') \ \forall i = 1, \ldots, k \ and \ f_i(x) > f_i(x') \quad for \ some \ i \tag{6}$$

where, k is the number of maximum development targets ($k = 2$ in this study), x is a certain UGS patch, x' indicates any patch of green space except x. By definition, a solution with a lower Pareto rank will dominate a solution with a higher Pareto rank. In this work, UGS patches in the lower rank indicate that they can exert the comprehensive benefits of ecological connectivity and spatial accessibility, and conversely, the comprehensive benefits of green patches at high Pareto ranks are low.

4. Results and Analysis

4.1. Ecological Connectivity of Urban Green Spaces

Three ecological networks were constructed based on different area thresholds of UGS habitat patches, i.e., 1 ha, 3 ha, and 5 ha. Figure 2 presents current density in the three networks. As the area threshold of UGS habitat patches increases, the patch amount gradually decreases and will drop by 50% when the area threshold is 5 ha. Overall current density at regional scale will reach up to 189.14 when the minimum area threshold is fixed (1 ha), while it will decrease to 79.59 when UGS patches with an area greater than 5 ha are considered as ecological sources. The changes in current density through UGS patches demonstrate the significance of small-sized urban green space for species dispersal in urban center areas with high development intensity. As we can see in Figure 2, the distribution pattern of current density present similar under three ecological connectivity scenarios with different area thresholds of UGS patches. There are many remarkable species dispersal corridors in the outer areas of the study area such as Tingtao Scenic Area of the East Lake, the Moshan Mountain Scenic Area, the Ma'anshan Forest Park, especially the adjacent relationship of these areas will further facilitate species survival.

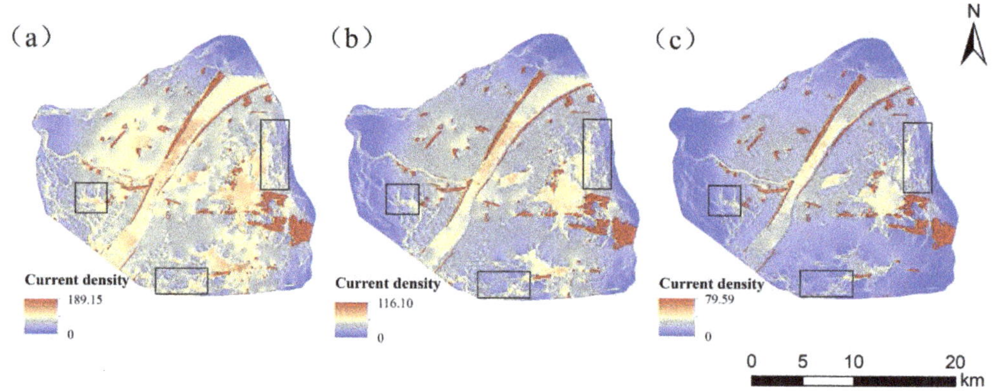

Figure 2. Current density in the urban center of Wuhan under different area thresholds of UGS patches (area threshold of nodes in ecological networks, (**a**) 1 ha; (**b**) 3 ha; (**c**) 5 ha. Black rectangles indicate the remarkable species dispersal corridors in landscapes.

The average current density of park and attached green patches in three ecological networks was statistically analyzed (Table 3). In all three networks, the proportion of attached green space below the average of current density is greater than that of parks, indicating the superior connectivity level of parks over attached green patches. With the increase of area thresholds, the ecological connectivity level of individual patches in urban green space gradually decreases. Specifically, when the area threshold equals 1 ha, the ecological connectivity level of UGS patches is the best, and the average current density of patches reaches up to 162.52. When the threshold increases to 5 ha, the maximum value of the average current density decreases by more than 55%.

Table 3. Ecological connectivity level of urban green patches in Wuhan under different habitat area thresholds.

	Habitat Area Threshold	Current Density through UGS Patches				
		Maximum	**Minimum**	**Mean**	**Standard Deviation**	**Proportion below the Average**
Parks	1 ha	162.52	65.09	123.24	18.05	37.78%
	3 ha	103.32	69.61	88.97	5.75	42.22%
	5 ha	71.55	1.50	52.18	24.84	20.00%
Attached green space	1 ha	155.73	64.29	125.14	19.86	50.00%
	3 ha	101.36	70.35	89.94	6.33	61.54%
	5 ha	68.82	1.09	45.67	28.92	30.77%

The average current density of patches under different ecological networks were divided into three levels using the natural breaking analysis. In comparison, the green spaces affiliated to Simeitang Park, Baodao Park, and Zhongnan University of Finance and Economics have always been at a relatively low level of ecological connectivity, while parks such as Hankou beach park, Hanyang beach park, and Tingtao Scenic Area of Donghu Lake, as well as affiliated green spaces such as Wuhan University, Huazhong Agricultural University, and Huazhong University of Science and Technology, have always been at a high level of ecological connectivity, among which Guishan Scenic Area, Yellow Crane Tower Park, and Luojia Mountain always ranked in the top three in terms of ecological connectivity. From the perspective of spatial distribution, the green space patches with a relatively high ecological connectivity level are mainly concentrated in the eastern part of the study area, e.g., Donghu Lake Scenic Area (Figure 3).

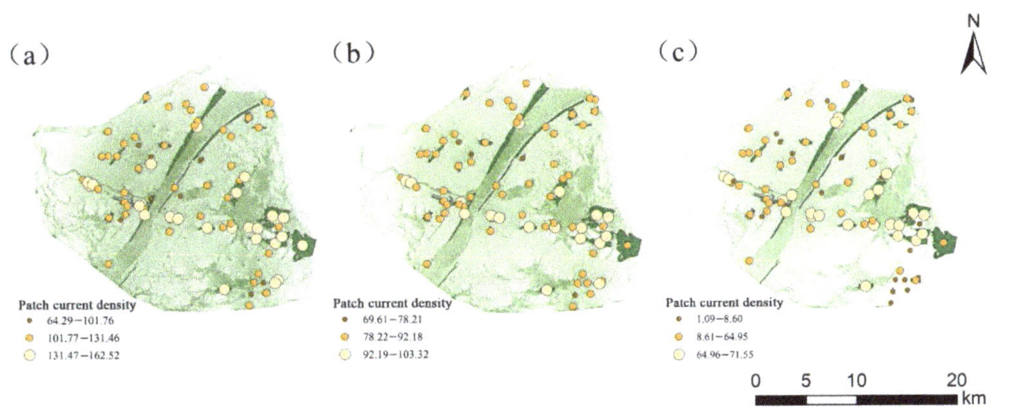

Figure 3. Spatial distribution of ecological connectivity of UGSs in the urban center of Wuhan (area threshold of nodes in ecological networks, (**a**) 1 ha; (**b**) 3 ha; (**c**) 5 ha).

4.2. Spatial Accessibility of Urban Green Space

The spatial accessibility level of urban green space under four travelling scenarios is presented in Figure 4. In comparison, the accessible areas of urban green space will cover more than 80% of the study area within 30 min by bicycle, while within 15 min by walking, the accessible rate will be less than 20%. Comparing different types of green spaces, there is no doubt that the spatial accessibility of parks is always better than that of attached green spaces under different travelling scenarios, and the total size of the potential population served by parks is always twice as large as that of attached green spaces (Figure 4). However, from the perspective of the service capacity per unit area of green patches, the attached green spaces are likely to meet more leisure and recreation needs of residents, and with the increase of the service thresholds, this superiority of the attached green spaces will become more evident. It can be observed that the population size covered by per unit area of parks is only 50% of that of the attached green space under the scenario of 30 min by cycling.

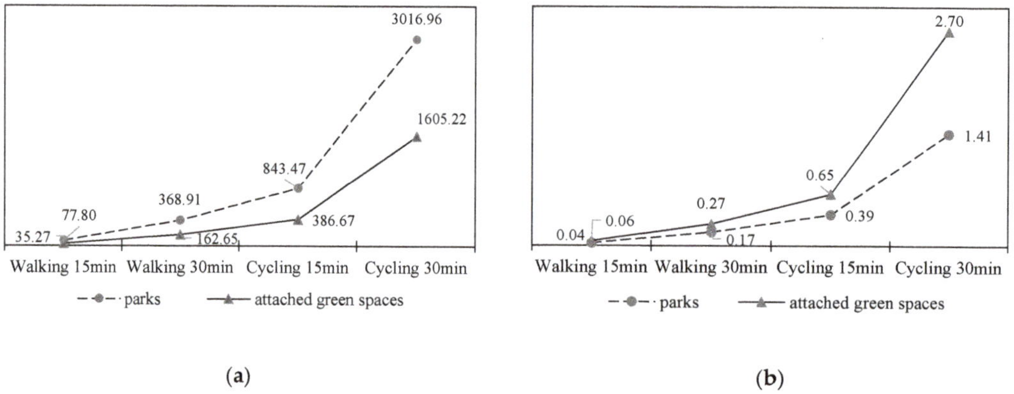

Figure 4. Spatial accessibility level of UGSs under different travelling scenarios (**a**) Potentially serving population (10,000 people); (**b**) Population served per unit area of green space (10,000 people/m^2)).

For spatial distribution, under different travel modes, the spatial accessibility level of urban green spaces in the central and western parts of the study area is relatively high, while that of eastern green spaces exhibits relatively low (Figure 5). Among them, Zhongshan Park, Baodao Park and Lingjiaohu Park, and Wuhan Youth Palace always have high spatial accessibility under different travelling scenarios, while the river beach park in the center and the green space in the east periphery are relatively poorly accessible.

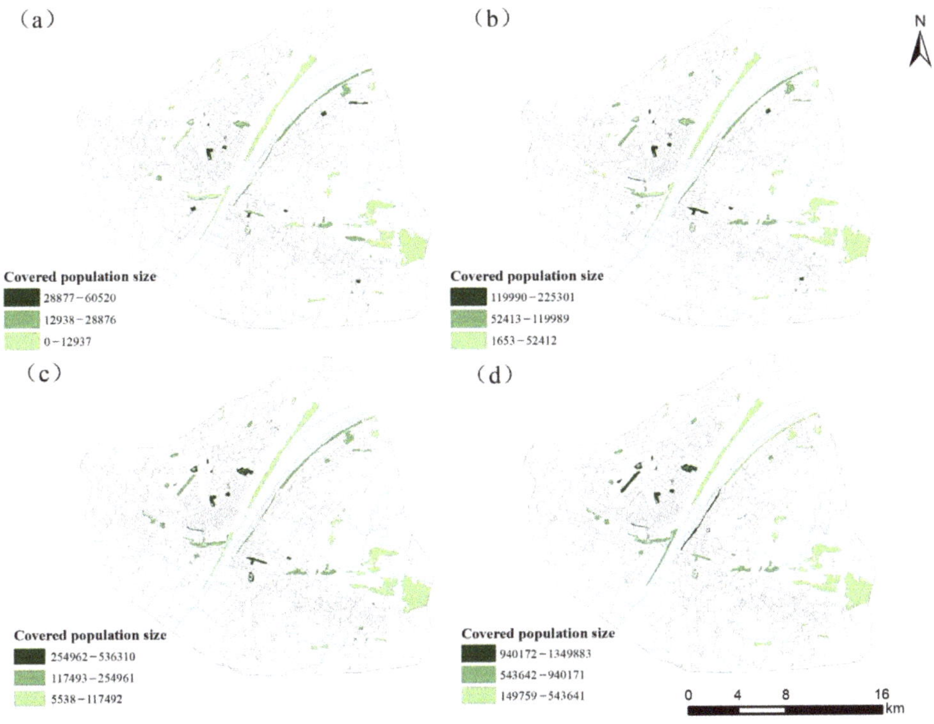

Figure 5. Spatial distribution of spatial accessibility of UGSs in the urban center of Wuhan ((**a**) Accessible by walking 15 min; (**b**) Accessible by walking 30 min; (**c**) Accessible by cycling 15 min; (**d**) Accessible by cycling 30 min).

4.3. Key Conservation Patches

Based on the ecological connectivity of urban green patches under three area thresholds and the spatial accessibility of green space under four kinds of travelling modes, Pareto ranking analysis was used to compare the comprehensive benefit level of urban green patches in 12 potential combinations of different travelling modes and connectivity types.

Figure 6 shows the Pareto ranks of comprehensive benefits of UGS patches. When the area threshold is less than 3 ha, a significant convex Pareto boundary can be observed, which indicates the tradeoff between ecological and social functions of urban green space, for example, to facilitate species dispersal and resident recreation. When the area threshold is increased to 5 ha, the relative levels of ecological connectivity of the two types of green patches are remarkably different, resulting in the separate distribution of the scattered spots in Figure 6.

According to the Pareto ranking theory, urban green spaces at the first two ranks have higher comprehensive benefits of ecological connectivity and spatial accessibility, and vice versa. Overall, the comprehensive benefits of urban green space in the study area are relatively low, and only approximately 7.51% of green patches on average rank the first. The frequency of each UGS patch at different Pareto ranks in 12 combination scenarios was calculated; a total of 27 patches appears at Pareto level 1–2 at least once, and 28 patches rank at the last three levels at least once (Figure 7a). In space, the comprehensive benefit level of green spaces exhibits a pattern of "high in the center and the south, and low in the periphery and the north" (Figure 7a). Urban green spaces with high comprehensive benefits are mostly distributed in the central area, among which Guishan Scenic Area, Yellow Crane Tower Park, Shouyi Park, and Wuhan Youth Palace always rank the first. As for urban green space in the periphery of the study area, e.g., Zhongnan University of Finance and Economics, Chufeng Park, and Dijiao Park, their comprehensive benefits are relatively low, and the frequency of ranking at the last three levels exceeds 11 times. Compared with different types of green space, the comprehensive benefits of parks are higher than that of attached UGS, and nearly 38% of parks appears in the first two levels and less than 35% in the last three levels, but for attached green space, the proportion of low ranked patches exceeds 46%.

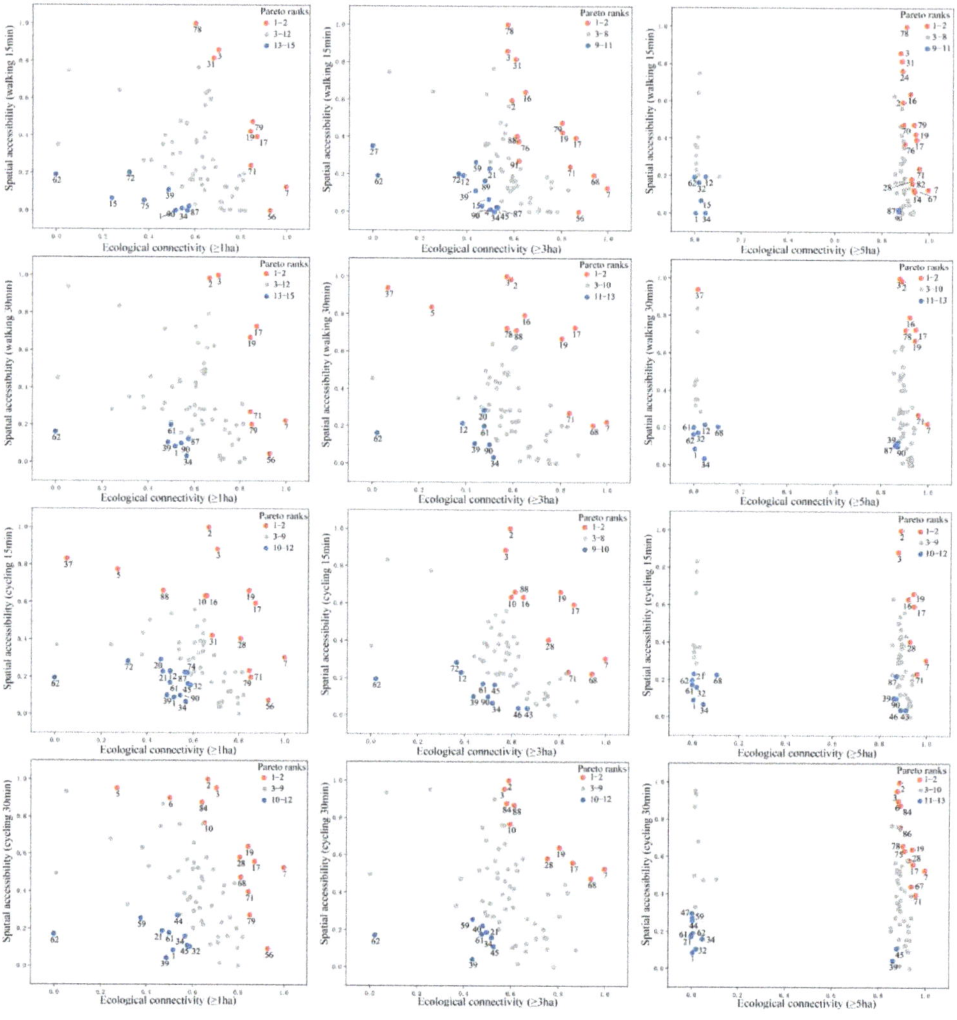

Figure 6. Pareto ranks of UGSs in the urban center of Wuhan.

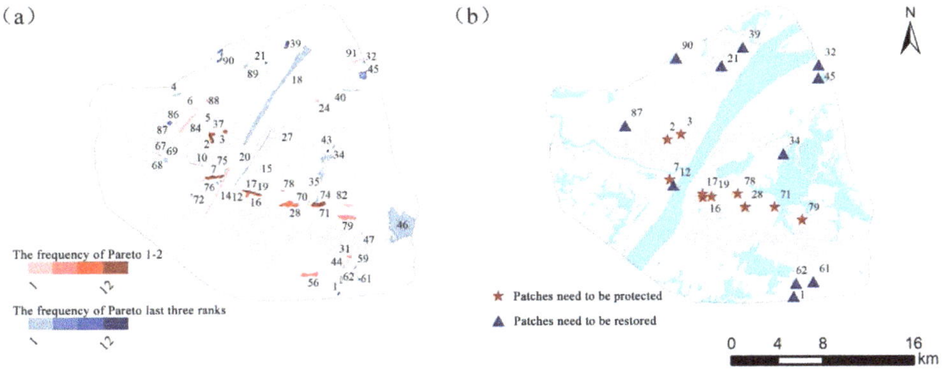

Figure 7. The comprehensive benefit of UGSs in the urban center of Wuhan ((**a**) Frequency of patches in different Pareto ranks; (**b**) Distribution of key green patches).

Urban green spaces with a frequency of more than six times at the first two levels are regarded as key protected patches due to their high comprehensive benefits of ecological connectivity and spatial accessibility, while UGS patches with a frequency of more than 50% in the last three levels are considered as restoration areas. In total, 10 green patches, including seven parks and three attached green spaces, need to be protected, and seven parks and four attached green spaces were identified as restoration patches (Table 4). The key protective patches are mainly concentrated in the middle of the study area, while the key restoration ones are mostly distributed in the periphery (Figure 7b).

Table 4. Key UGS patches in the main urban area of Wuhan.

	Park			Attached Green Space		
	No.	Name	Frequency at the First Two Levels	No.	Name	Frequency at the First Two Levels
Green space to be protected with high comprehensive benefits	7	Guishan Scenic Area	12	3	Wuhan Youth Palace	12
	17	Yellow crane tower	12	71	Luojia Mountain	11
	19	Shouyi Park	12	79	Nanwangshan Mountain	6
	2	Zhongshan Park	11			
	16	Shouyi Cultural Park	7			
	28	Hongshan Park	7			
	78	Hongshan Square	6			
Green space to be restored with low comprehensive benefits	34	Chufeng Garden	12	62	University affiliated green space	12
	39	Dijiao Park	11	61	Optics Valley	9
	1	Jinxiulongcheng Park	8	90	Attached green space in residential area	9
	21	Baibuting Garden	7	87	Attached green space in residential area	8
	12	Lotus Lake Park	6			
	32	Tianxingzhou Bridge Park	6			
	45	Qingshan Park	6			

5. Discussion

5.1. *Strengths of Trade-Off Analysis of Urban Green Spaces*

UGS multifunctionality has attracted increasing attention from urban planners. The coordination of different UGS functions may maximize their potential benefits in both social and ecological aspects [6,8,10]. However, the existing multifunctional assessment of UGSs based on weighted aggregation of UGS merits may largely ignore the coordinated relationships of multiple UGS functions [8,10]. This phenomenon is mainly caused by the separated governance from different departments of the local government. For example, ecology and environment management may highlight ecological functions of UGSs, while housing and urban development will emphasize their capabilities of providing cultural tourism and leisure services. As one type of scarce ecological components in urban systems, UGS provides both social and ecological benefits for human beings. It is a new attempt to rearrange the UGS pattern from a multifunctional perspective in the context of human–nature harmonious development. Therefore, we propose a Pareto ranking method to identify tradeoff relationship of UGS patches in terms of ecological and social functions. Different from previous studies, this work will help to develop a differentiated conservation strategy for UGS management. For example, UGS patches with high comprehensive socio-ecological benefit will be protected as a priority, while those with low comprehensive benefit will be restored as soon as possible.

5.2. Implications for Urban Landscape Planning

The case study will provide solid support for urban landscape planning in Wuhan City. As it can be observed, within an intensive development city, a large river beach and forest parks can better facilitate species dispersal (e.g., brown-headed crow finches, white-headed bulbuls, and gray magpies in the study area) and maintain urban biodiversity. However, the ecological significance of small sized urban green spaces cannot be ignored, for example, the attached green spaces within residential areas may exert a steppingstone effect in the process of species dispersal.

Moreover, the spatial accessibility of urban green space in the main urban area is severely limited, and the construction of a "15-min community life circle" in Wuhan still has plenty of room for improvement. Although the spatial accessibility of the attached green spaces shows lower than that of parks, their service capability per unit area is likely to be higher than that of parks. Therefore, on the one hand, it is necessary to pay attention to the design of pedestrian roads; on the other hand, the service capacity of the attached green space should be promoted in combination with the community living circle planning and the optimal utilization of the attached green space such as residential areas and universities in the living circle.

Overall, 21 key UGS patches were identified in the main urban area of Wuhan (Table 4). In the process of urban development, 10 key patches such as Guishan Scenic Area, Yellow Crane Tower, and Shouyi Park require better protection from the occupation of urban expansion. Meanwhile, 11 key green patches are considered as the hotspots of restoration in further UGS planning and construction, which can be improved in terms of habitat quality, landscape pattern, spatial accessibility, and surrounding transportation conditions.

6. Conclusions

For the urban areas with high land intensity and large population density, it is impossible to improve the comprehensive benefits of urban green spaces through large-scale UGS rearrangement. In this context, comprehensive use of the existing UGS patches will be highlighted, and targeted policies need to be developed accordingly. In this work, the differentiated conservation priorities of UGSs are expected to be identified from the perspective of multi-functional tradeoffs. The comprehensive benefits of the ecological connectivity and spatial accessibility of urban green space were explored using Pareto ranking analysis, and the differentiated priorities of green space patch, i.e., protection and restoration, are determined based on the ranking levels. This work is expected to provide an applicable framework for the quantitative identification of key urban green space patches.

Author Contributions: Conceptualization, H.L. and D.L.; methodology, H.L.; formal analysis, H.L.; writing—original draft preparation, H.L.; writing—review and editing, D.L. and J.H. All authors have read and agreed to the published version of the manuscript.

Funding: This research was funded by National Natural Scientific Foundation of China, grant number 41771429.

Institutional Review Board Statement: Not applicable.

Informed Consent Statement: Not applicable.

Conflicts of Interest: The authors declare no conflict of interest.

References

1. Aronson, M.F.J.; Lepczyk, C.A.; Evans, K.L.; Goddard, M.A.; Lerman, S.B.; MacIvor, J.S.; Nilon, C.H.; Vargo, T. Biodiversity in the city: Key challenges for urban green space management. *Front. Ecol. Environ.* **2017**, *15*, 189–196. [CrossRef]
2. Douglas, O.; Lennon, M.; Scott, M. Green space benefits for health and well-being: A life-course approach for urban planning, design and management. *Cities* **2017**, *66*, 53–62. [CrossRef]
3. Kong, X.S.; Fu, M.X.; Zhao, X.; Wang, J.; Jiang, P. Ecological effects of land-use change on two sides of the Hu Huanyong Line in China. *Land Use Policy* **2022**, *113*, 105895. [CrossRef]
4. Haaland, C.; van den Bosch, C.K. Challenges and strategies for urban green-space planning in cities undergoing densification: A review. *Urban For. Urban Green.* **2015**, *14*, 760–771. [CrossRef]
5. Madureira, H.; Andresen, T. Planning for multifunctional urban green infrastructures: Promises and challenges. *Urban Des. Int.* **2014**, *19*, 38–49. [CrossRef]
6. Belmeziti, A.; Cherqui, F.; Kaufmann, B. Improving the multi-functionality of urban green spaces: Relations between components of green spaces and urban services. *Sustain. Cities Soc.* **2018**, *43*, 1–10. [CrossRef]
7. Kabisch, N.; Qureshi, S.; Haase, D. Human-environment interactions in urban green spaces—A systematic review of contemporary issues and prospects for future research. *Environ. Impact Assess. Rev.* **2015**, *50*, 25–34. [CrossRef]
8. Sun, Y.; Liu, D.; Wang, P. Urban simulation incorporating coordination relationships of multiple ecosystem services. *Sustain. Cities Soc.* **2022**, *76*, 103432. [CrossRef]
9. Charoenkit, S.; Piyathamrongchai, K. A review of urban green spaces multifunctionality assessment: A way forward for a standardized assessment and comparability. *Ecol. Indic.* **2019**, *107*, 105592. [CrossRef]
10. Meerow, S.; Newell, J.P. Spatial planning for multifunctional green infrastructure: Growing resilience in Detroit. *Landsc. Urban Plan.* **2017**, *159*, 62–75. [CrossRef]
11. Derkzen, M.L.; van Teeffelen, A.J.A.; Verburg, P.H. REVIEW Quantifying urban ecosystem services based on high-resolution data of urban green space: An assessment for Rotterdam, the Netherlands. *J. Appl. Ecol.* **2015**, *52*, 1020–1032. [CrossRef]
12. Dai, E.; Wang, X.; Zhu, J.; Xi, W. Quantifying ecosystem service trade-offs for plantation forest management to benefit provisioning and regulating services. *Ecol. Evol.* **2017**, *7*, 7807–7821. [CrossRef]
13. Guo, S.Y.; Saito, K.; Yin, W.D.; Su, C. Landscape Connectivity as a Tool in Green Space Evaluation and Optimization of the Haidan District, Beijing. *Sustainability* **2018**, *10*, 1979. [CrossRef]
14. Xing, L.J.; Liu, Y.F.; Liu, X.J. Measuring spatial disparity in accessibility with a multi-mode method based on park green spaces classification in Wuhan, China. *Appl. Geogr.* **2018**, *94*, 251–261. [CrossRef]
15. Ekkel, E.D.; de Vries, S. Nearby green space and human health: Evaluating accessibility metrics. *Landsc. Urban Plan.* **2017**, *157*, 214–220. [CrossRef]
16. Sarah, P.; Zhevelev, H.M. Effect of visitors' pressure on soil and vegetation in several different micro-environments in urban parks in Tel Aviv. *Landsc. Urban Plan.* **2007**, *83*, 284–293. [CrossRef]
17. Zhao, H.; Liu, D.; Li, F.; Liu, X.; Niu, J.; He, J.; Liu, Y. Incorporating spatio-temporal connectivity for prioritized conservation of individual habitat patches in a dynamic landscape. *Ecol. Indic.* **2021**, *124*, 107414. [CrossRef]
18. Tian, Y.H.; Liu, Y.Q.; Jim, C.Y.; Song, H.Z. Assessing Structural Connectivity of Urban Green Spaces in Metropolitan Hong Kong. *Sustainability* **2017**, *9*, 1653. [CrossRef]
19. Ersoy, E.; Jorgensen, A.; Warren, P.H. Identifying multispecies connectivity corridors and the spatial pattern of the landscape. *Urban For. Urban Green.* **2019**, *40*, 308–322. [CrossRef]
20. Gao, Y.; Ma, L.; Liu, J.; Zhuang, Z.; Huang, Q.; Li, M. Constructing Ecological Networks Based on Habitat Quality Assessment: A Case Study of Changzhou, China. *Sci. Rep.* **2017**, *7*, 1–11. [CrossRef]
21. Liu, X.; Liu, D.; Zhao, H.; He, J.; Liu, Y. Exploring the spatio-temporal impacts of farmland reforestation on ecological connectivity using circuit theory: A case study in the agro-pastoral ecotone of North China. *J. Geogr. Sci.* **2020**, *30*, 1419–1435. [CrossRef]
22. Merrick, M.J.; Koprowski, J.L. Circuit theory to estimate natal dispersal routes and functional landscape connectivity for an endangered small mammal. *Landsc. Ecol.* **2017**, *32*, 1163–1179. [CrossRef]
23. Li, Z.; Wei, H.; Wu, Y.; Su, S.; Wang, W.; Qu, C. Impact of community deprivation on urban park access over time: Understanding the relative role of contributors for urban planning. *Habitat Int.* **2019**, *92*, 102031. [CrossRef]
24. Tan, C.D.; Tang, Y.H.; Wu, X.F. Evaluation of the Equity of Urban Park Green Space Based on Population Data Spatialization: A Case Study of a Central Area of Wuhan, China. *Sensors* **2019**, *19*, 2929. [CrossRef] [PubMed]
25. Rahman, K.; Zhang, D.F. Analyzing the Level of Accessibility of Public Urban Green Spaces to Different Socially Vulnerable Groups of People. *Sustainability* **2018**, *10*, 3917. [CrossRef]
26. Li, L.J.; Du, Q.Y.; Ren, F.; Ma, X.Y. Assessing Spatial Accessibility to Hierarchical Urban Parks by Multi-Types of Travel Distance in Shenzhen, China. *Int. J. Environ. Res. Public Health* **2019**, *16*, 1038. [CrossRef]
27. Xing, L.J.; Liu, Y.F.; Liu, X.J.; Wei, X.J.; Mao, Y. Spatio-temporal disparity between demand and supply of park green space service in urban area of Wuhan from 2000 to 2014. *Habitat Int.* **2018**, *71*, 49–59. [CrossRef]

28. Bowman, J. Is dispersal distance of birds proportional to territory size? *Can. J. Zool. Rev. Can. Zool.* **2003**, *81*, 195–202. [CrossRef]
29. McRae, B.H.; Dickson, B.G.; Keitt, T.H.; Shah, V.B. Using circuit theory to model connectivity in ecology, evolution, and conservation. *Ecology* **2008**, *89*, 2712–2724. [CrossRef]
30. Wang, B.; Liu, Z. Functional connectivity analyses and construction strategies of the structure of green space network in haidian district of Beijing. *Landsc. Archit. Front.* **2019**, *7*, 34–51. [CrossRef]
31. Kwan, M. Recent advances in accessibility research: Representation, methodology and applications. *J. Geogr. Syst.* **2003**, *5*, 129–138. [CrossRef]
32. Cao, K.; Zhang, W.; Liu, S.; Huang, B.; Huang, W. Pareto law-based regional inequality analysis of PM2.5 air pollution and economic development in China. *J. Environ. Manag.* **2019**, *252*, 109635. [CrossRef] [PubMed]
33. Yang, W.; Jin, Y.; Sun, T.; Yang, Z.; Cai, Y.; Yi, Y. Trade-offs among ecosystem services in coastal wetlands under the effects of reclamation activities. *Ecol. Indic.* **2018**, *92*, 354–366. [CrossRef]

sustainability

Article

Transformation for Feature Upgrades or Higher Property Prices: Evidence from Industrial Land Regeneration in Shanghai

Fan Yang [1,*], Peihong Tao [1], Xiao Cai [2] and Jiayin Wang [1]

[1] Department of Urban Planning, College of Architecture and Urban Planning, Tongji University, Shanghai 200092, China; pehont@163.com (P.T.); 18916099869@163.com (J.W.)
[2] Department of Urban & Regional Planning, Faculty of the Built Environment, University of Malaya, Kuala Lumpur 50603, Malaysia; andrew_chua@126.com
* Correspondence: fanyangsh@tongji.edu.cn

Abstract: In recent years, an increasing amount of industrial land has been left idle in China. This gave rise to a wide range of urban issues hindering sustainable urban development. To solve this problem, industrial land transformation has been promoted. However, factors affecting industrial land transformation have not been adequately explored. To fill this gap, this study employs the bivariate K-function to analyze the spatial association between agglomeration patterns of industrial land parcels and living quarters. Moreover, a series of discrete choice models (i.e., the LOGIT, PROBIT, and IVPROBIT model) are adopted to examine empirically complicated relationships between industrial land transformation and its influencing factors. This study argues that the land price and its rising expectations are major determinants of industrial land transformation. The results revealed that transformation-oriented industrial land tended to be located next to accessible living quarters with higher prices. A higher-level industrial park typically had less possibilities for industrial land transformation. The findings also indicated that production efficiency served as a moderator variable to regulate the transformation process. Implications are formulated for policymakers to guide industrial land transformation in an appropriate manner.

Keywords: industrial land use change; land prices; K-function; econometrics; urban regeneration; Shanghai

Citation: Yang, F.; Tao, P.; Cai, X.; Wang, J. Transformation for Feature Upgrades or Higher Property Prices: Evidence from Industrial Land Regeneration in Shanghai. *Sustainability* **2022**, *14*, 5280. https://doi.org/10.3390/su14095280

Academic Editors: Wenwu Tang, Dianfeng Liu and Jianxin Yang

Received: 24 March 2022
Accepted: 25 April 2022
Published: 27 April 2022

1. Introduction

The industry, as one of the most critical urban economic foundations, is a long-term tax source for local governments [1,2]. The increase in industrial land-leasing scales can produce a greater potential to create fiscal income for local governments because China's urban construction land is state-owned [3]. However, industrial development has not kept up with what is expected by governments, so a large amount of industrial land either stays idle or ceases production. The inefficient occupation of industrial land by enterprises creates the illusion that industrial land is in short supply [4]. Thus, local governments tend to lease new industrial land rather than consolidate inefficient industrial land for further industrial production. Even worse, the corresponding planning control to prevent low-efficiency industrial land use is not in place [5,6]. As a result, the capital density [7,8], land input–output efficiency [9,10], and industrial land productivity [11] of industrial land in China are relatively low, which fails to meet governments' expectations of gaining revenue from industrial development. Local governments cannot help adopting necessary measurements to promote the transformation of low-efficiency industrial land, and thus avoid an economic loss. Financial benefits (e.g., conveyance fees) are generated from the transformation process, which further stimulates governments' industrial land transformation motivation.

In addition to interventions from governments to carry forward industrial land transformation, the intrinsic driving force is the expectation from industrial land developers that land prices will rise. Harvey [12] argued that land value-added pursuit reshaped urban space and produced various urban renewal projects. Land users often expect a high return from land redevelopment because urban housing prices and land prices are generally considered to be continually rising [13]. Huge profits in real estate investments can discourage developers from engaging in manufacturing on industrial land [14]. Some developers even tend to leave industrial land idle and await future gains from land transformation [13]. Moreover, in recent years, industrial enterprises have been gradually migrating to peripheral urban areas in China. This is partly because commercial developers can afford higher land rents compared to industrial developers. This movement constantly and repetitively takes place with urban sprawl [15]. However, this iterative pattern may familiarize industrial developers, and thus raise their expectations of industrial land redevelopment so as to speculate in the industrial land market. To solve the aforementioned problems, many previous studies empirically analyzed relationships between industrial land and local economic development at the regional level using various econometric models [2,16]. Some of them also explored industrial land or enterprises' features using city- or subdistrict-level data with case studies [1] or from the perspective of industrial efficiency [8–10]. There is still a lack of empirical research to identify the motivation (i.e., land prices rising expectations) for promoting industrial land transformation at the parcel level [15,17,18]. Meanwhile, existing studies have verified that many links in the industrial land use cycle (e.g., land conveyance, enterprise production, and operation) have contributed to local fiscal revenue. Governments and enterprises' behaviors in the process of industrial land-leasing and production have also been fully studied. However, the reason why a large amount of industrial land is not used for manufacturing but transformed into the commercial and residential use is still under-explored. There is a range of questions that need answering. For instance, is this phenomenon driven by the pursuit of land and property price appreciation? Has the contribution of this market process to local fiscal revenue prompted the support of local governments? To fill this knowledge gap, this study aims to empirically examine what affects industrial land transformation in the Minhang District of Shanghai from the perspective of expectations of rising land prices. Specifically, this study is expected to verify the following four hypotheses:

Hypothesis 1 (H1). *There are significant differences in characteristics and locations between transformation-oriented and continued-production land;*

Hypothesis 2 (H2). *The higher land price around industrial land contributes to its transformation;*

Hypothesis 3 (H3). *The administrative grade of industrial parks and land-leasing policies can influence industrial land transformation;*

Hypothesis 4 (H4). *The production efficiency of industrial enterprises has a mediate effect on industrial land transformation.*

2. Materials and Methods

2.1. Study Area

Shanghai is a world city, which has the highest degree of industrialization and fastest urbanization process in China (Figure 1a). Minhang District of Shanghai was selected as the study area (Figure 1b). It is located in the southwest of Shanghai's central area, with a total area of 372 km^2, covering nearly 6% of the Shanghai territory. As is shown in Figure 1c, in 2014, the construction land in Minhang District covered 285 km^2, nearly 76.6% of the total area of Minhang District. Noticeably, industrial land accounted for 25.8% of entire construction land in Minhang District, Shanghai in 2014.

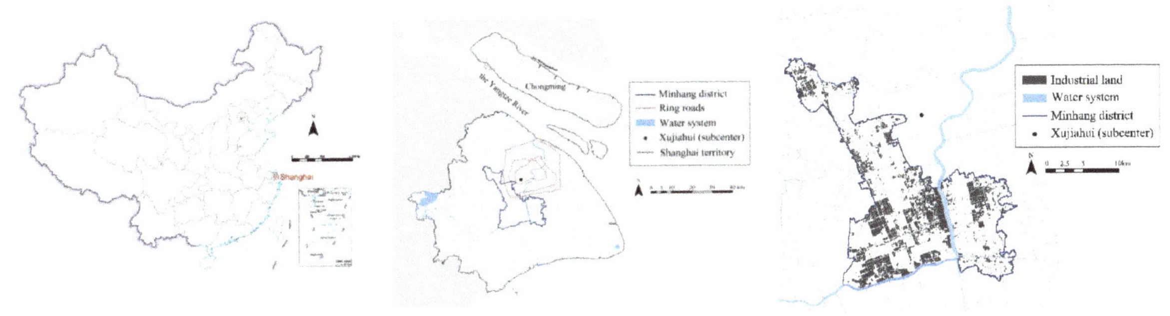

(a) (b) (c)

Figure 1. The location and distribution of industrial land in Minhang District, Shanghai: (**a**) The location of Shanghai in China; The map is based on the standard map from the standard map service website GS (2019) No. 1651 of the Ministry of Natural Resources, China (http://bzdt.ch.mnr.gov.cn/index.html, accessed on 27 April 2020), and the base map boundary is not modified. (**b**) The location of Minhang District in Shanghai. (**c**) The distribution of industrial land in Minhang District.

In past decades, continuous factory relocation occurred in the suburb of Shanghai due to the functional mix. To maintain a certain proportion of industrial space (Shanghai Municipal Commission of Economy and Informatization, 2016), the Shanghai Municipal Government guided industrial enterprises moving into designated industrial parks according to the "104,195,198 Industrial Land Policy". It can promote the redevelopment of idle industrial land in a centralized construction area (CCA), an area for urban development and centralized construction within a certain period. Minhang District is one of the traditional industrial agglomeration areas in Shanghai in different urbanization periods. It is the nearest industrial zone to downtown Shanghai. The above information shows that Minhang District of Shanghai is exceptionally suitable for this study.

2.2. Data and Pre-Processing

2.2.1. Industrial Land and Enterprises

The industrial land parcel data (n = 5610) come from the land rights data set of Minhang District in 2014. It contains the area, location, year assigned, and stakeholder name of each industrial land parcel. From February to October in 2019, this study conducted a series of detailed on-site investigations to collect the manufacturing condition data of each industrial land parcel. All industrial land in the study area was classified into four types: (1) transformed land, (2) under-construction land, (3) idle land, (4) continued-production land (Table 1).

The output data of industrial enterprises in Minhang District were derived from the 2013 Annual Survey of Industrial Companies. Each industrial land parcel's gross output value was computed by summing all output values of industrial enterprises in the identical land parcel. However, some enterprises are not involved in the above survey because their annual output is below CNY 20 million. To solve this problem, this study estimated their outputs by subtracting the known industrial output value from the total gross value of industrial production in Minhang District, Shanghai.

2.2.2. Land Prices and Distribution

This study employed an appropriate interpolation method (i.e., ordinary Kriging) in ArcGIS to estimate the land price condition at the parcel level using the housing price data, considering the strong correlation between land prices and housing prices [19,20]. Kriging is one of the interpolation methods that can be used to predict the distribution of housing prices [21]. Based on the housing price data and its known locations, Kriging provides absolute housing prices at unobserved locations [22]. The housing price data were collected from the Lianjia website (https://sh.lianjia.com/chengjiao/, accessed on 27 April 2020), including all second-hand housing transaction records published on the website in 2015 and 2019 in Shanghai. Previous studies have proved that this kind of open data is reliable to represent housing price levels [23,24]. It is worth noting that the land price estimated in this paper is a comprehensive replacement of location, facilities, road network level, greening, and other elements in the built environment. Thus, the transformation driven by the land price also reflects the complex driving mechanism of spatial elements.

Table 1. The descriptions and examples of various industrial land parcels.

Categories	Descriptions	Examples	
Transformed land	It refers to the industrial land that has been transformed from industrial usage to other usages.		
Under-construction land	It refers to industrial land where the production activity is suspended and/or some buildings are being constructed.		
Idle land	It refers to industrial land that is disused and awaiting redevelopment.		
Continued-production land	It refers to industrial land where production activities stay active.		

The following steps show the specific procedure to obtain the spatial distribution of the land price and its growth rate (Figure 2). First, the latitude and longitude coordinates of each living quarter were obtained through the Map Location website (https://maplocation.sjfkai.com/, accessed on 27 April 2020). Second, the average housing prices in 2015 and 2019 were assigned to the corresponding living quarter, respectively. Third, the spatial distribution of the average housing price in 2015 and its growth rate from 2015 to 2019 were acquired, respectively, using the ordinary Kriging interpolation method with a patch size of 25 m × 25 m (Figure 2). To strengthen the interpolation's validity in the study area, this study took the housing prices data of all living quarters in Shanghai except Chongming District into the mean value calculation. This is because the spatial segmentation of the Yangtze River has a significant influence on interpolation results.

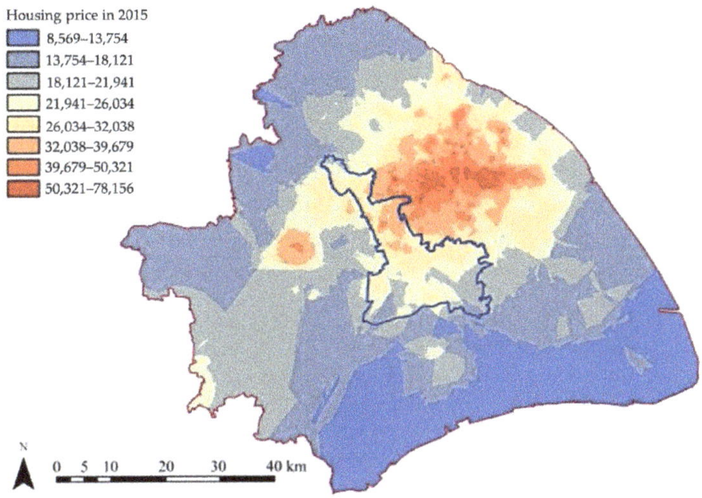

(**a**)

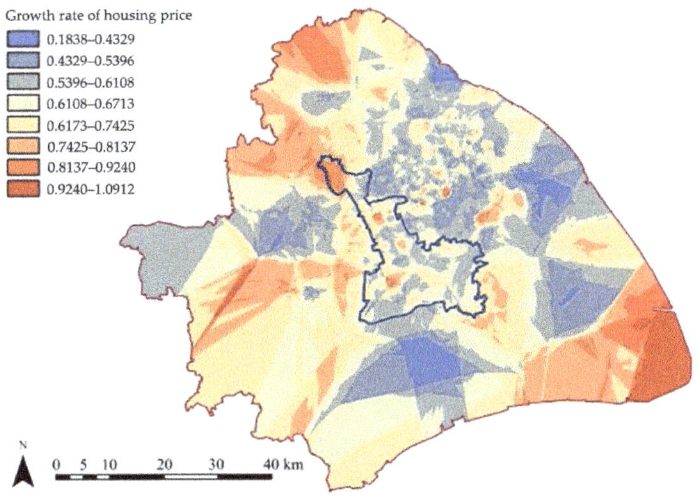

(**b**)

Figure 2. Kriging interpolation of the housing price and its growth rate: (**a**) The spatial distribution of the housing price in 2015; (**b**) The spatial distribution of the housing price growth rate from 2015 to 2019.

2.2.3. Classification of Industrial Land and Living Quarters

This study conducted a detailed classification of industrial land and living quarters to facilitate the analysis of land prices' effect on industrial land transformation (Table 2).

Table 2. Classification of industrial land and living quarters.

Titles	Categories	Descriptions
Industrial land	Transformation-oriented land (L1)	Transformed land Under-construction land Idle land
	Continued-production land (L2)	Continued-production land
Living quarters	Residence with the higher growth rate (P1)	The growth rate of average housing price that is higher than the median rate from 2015 to 2019
	Residence with the lower growth rate (P2)	The growth rate of average housing price that is lower than the median rate from 2015 to 2019

2.3. Methods

Based on the aforementioned data, this study manifested differences in location and characteristics between L1 and L2 according to a Student's t-test. Then, the aggregation level between industrial land (L1 and L2) and living quarters (P1 and P2) was examined using the bivariate K-function method. This study also employed an econometric method to explore the influence of several important factors (i.e., land prices, industrial parks' administrative grades, land-leasing policies, and industrial enterprises' production efficiency) on industrial land transformation. The framework of the methodology is shown in Figure 3.

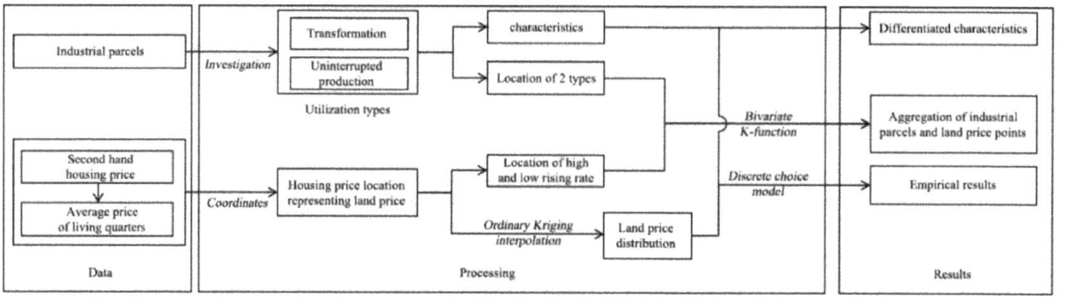

Figure 3. Framework of the methodology.

2.3.1. Method of Bivariate K-Function

This study employed the bivariate K-function to compare spatial distribution characteristics of L1 and L2. Ripley's K-function, coined by Byth and Ripley [25], is typically utilized to analyze the spatial pattern based on Euclidean distance [26]. The bivariate K-function is described as the anticipated point amount of pattern 2 within a given radius of an arbitrary position of pattern 1, divided by the point intensity of pattern 2 [27]. Essentially, the bivariate K-function aims to examine the mutual influence between two spatial distribution patterns. It is widely employed to measure the agglomeration of urban land use and the proximity to urban facilities [28,29]. This study adopted this method measuring the aggregation distance between industrial land parcels and living quarters in 2015 and 2019 within a 1000 m buffer of Minhang District using R programming, as adopted by [30].

2.3.2. Method of Discrete Choice Model

This study introduced econometric methods to estimate causal relationships between land prices rising expectations and industrial land transformation. To be specific, a series of discrete choice models (i.e., the LOGIT, PROBIT, and IVPROBIT models) were employed to estimate the influence of various factors on the binary decision-making for industrial land transformation [8,31]. The formula is shown in Equation (1):

$$\text{Transform}_i = a_1\text{Price}_i + a_2X_i + \varepsilon_0 \quad (1)$$

where Transform is a binary dependent variable representing whether an industrial land parcel is transformed (this study supposes that industrial land renewal is faced with only two choices: transformation and non-transformation); Price (land prices or its growth rate) is the independent variable; X refers to control variables; and $\varepsilon 0$ stands for the error.

It was noted that the LOGIT and PROBIT models inadequately addressed the endogeneity of land prices in industrial land transformation and may fail to yield convincing results. Extant literature suggested the externality of regeneration in the built environment, including industrial heritage and raised real estate values within a certain distance [32–35]. Accordingly, this study employed the IVPROBIT model to enhance the reliability of the results [36]. The distance from industrial land parcels to high schools or primary schools is selected as the instrumental variable because it is widely recognized that compulsory schools contribute to the rising value of surrounding real estate in China [37,38].

All in all, a total of sixteen discrete choice models were built in this study. Models (1), (3) and (5) were employed to examine the effect of land prices on industrial land transformation, while Models (2), (4) and (6) were employed to examine the effect of the expected land price growth rate on industrial land transformation. Models (7) to (10) were employed to investigate the impact of land-leasing policies on industrial land transformation by dividing the independent variable Price into two categories (i.e., before and after 2007). This is because the year 2007 is considered as a time watershed for the land-leasing policy change in China [8]. Models (11) and (12), using three dummy variables (i.e., dummy_district, dummy_city, and dummy_state) as substitutions for the control variable (i.e., Indpark), were used to explore the industrial park administrative levels' effect on industrial land transformation. Models (13) to (16) were used to examine the industrial land output's moderation effect on transformation by introducing a series of interaction variables (i.e., Area × Efficiency, Price × Efficiency, Price × Area, and Price × Efficiency × Area). All relevant variables are shown in Table 3.

Table 3. The description of selected variables.

Categories	Variables	Definition	Obs.	Mean	Std. Dev	Min	Max
Dependent variable	*Transform*	A binary variable to identify whether an industrial land parcel is transformed (0 = No; 1 = Yes)	5610	0.6230	0.4847	0	1
Independent variables	$lnPrice_{2015}$	The logarithm of mean land prices of the industrial land parcel in 2015 (CNY/m^2)	5610	10.1229	0.1852	9.6535	10.6863
	$Price_{Grow}$	Land price growth rate of the industrial land parcel from 2015 to 2019 (%)	5610	0.6476	0.0762	0.5023	0.9112
Control variables	*lnEfficiency*	The logarithm of output efficiency of the industrial land parcel (10^4 CNY/m^2)	5610	6.5351	1.9068	−0.1124	16.4058
	lnArea	The logarithm of the mean area of the industrial land parcel (m^2)	5610	8.7251	1.4084	2.4974	13.6191
	lnYear	The logarithm of the existing year (until 2015) of the industrial land parcel	5610	2.3769	0.7452	0	3.4657
	lnAgg500	The logarithm of the area of other industrial land within a 500 m buffer (not including itself, m^2)	5610	7.0587	4.6975	0	13.8146
	Tran500	The proportion of L1 in total industrial land areas within a 500 m buffer (not including itself)	5610	0.5466	0.3558	0	1

Table 3. *Cont.*

Categories	Variables	Definition	Obs.	Mean	Std. Dev	Min	Max
	lnMetro_{dist}	The logarithm of the nearest distance to a subway station from the industrial land parcel (m)	5610	6.9559	0.9477	0	8.6897
	lnCBD_{dist}	The logarithm of the distance from the industrial land parcel to Xujiahui (m)	5610	9.5188	0.3566	7.6424	10.1199
	CCA	A binary variable to identify whether the industrial land parcel locates in the CCA (0 = No; 1 = Yes)	5610	0.7642	0.4246	0	1
	Indpark	A binary variable to identify whether the industrial land parcel locates in industrial parks (0 = No; 1 = Yes)	5610	0.4193	0.4935	0	1
Instrumental variable	lnSchDist	The logarithm of the nearest distance to a high school or primary school (m)	5610	6.6813	0.8925	0	8.1565
Dummy variables	dummy_district	A binary variable to identify whether the industrial land parcel locates in district-level industrial parks (0 = No; 1 = Yes)	5610	0.2141	0.4102	0	1
	dummy_city	A binary variable to identify whether the industrial land parcel locates in city-level industrial parks (0 = No; 1 = Yes)	5610	0.1086	0.3111	0	1
	dummy state	A binary variable to identify whether the industrial land parcel locates in state-level industrial parks (0 = No; 1 = Yes)	5610	0.0966	0.2955	0	1
Interaction variables	*Area × Efficiency*	The product of the mean area and output efficiency of the industrial land parcel	5610	0.2811	1.3509	−15.8358	16.8490
	Price2015 × Efficiency	Interaction variables of Price and Efficiency, in which the price represents the expected land prices of land parcels in 2015	5610	−0.0103	0.2003	−3.7776	2.2318
	PriceGROW × Efficiency	Interaction variable of Price and Efficiency, in which the price represents the expected land prices of land parcels in 2015	5610	−0.0023	0.0763	−0.5436	0.9225
	Price2015 × Area	Interaction variable of Price and Area, in which the price represents the expected land prices of land parcels in 2015	5610	−0.0013	0.2510	−2.0604	1.6730
	PriceGROW × Area	Interaction variable of Price and Area, in which the price represents the land prices' expected growth rate of land parcels from 2015 to 2019	5610	0.0061	0.1062	−1.0430	0.7942
	Price2015 × Efficiency × Area	Interaction variable of Price, Efficiency and Area, in which the price represents the expected land prices of land parcels in 2015	5610	−0.0035	0.2447	−3.6971	3.6560
	PriceGROW × Efficiency × Area	Interaction variable of Price, Efficiency and Area, in which the price represents the land prices' expected growth rate of land parcels from 2015 to 2019	5610	−0.0012	0.0958	−1.0974	1.5482

Note: Xujiahui is the nearest urban sub-center of Shanghai city to Minhang district; 2. Interaction variable values are the product of involving variables after being centralized.

3. Results

3.1. *Differences in Characteristics and Locations for Industrial Land Transformation*

3.1.1. Differences in Characteristics between L1 and L2

Table 4 compares characteristic indicators between L1 and L2. There was a larger amount of L1 than L2 in Minhang District, Shanghai. The mean area and the surrounding industrial land area of L2 were slightly larger than those of L1. The land acquisition time of L1 was earlier than its counterpart. This is to say, the early conveyed land was more likely to have been in transformation at present. A lower efficiency in L1 indicated that the enterprises on L1 might be facing failure in manufacturing, or their original intention was to wait for real-estate-oriented redevelopment. This was similar between L1 and L2 in terms of the distance to a subway station and Xujiahui. There was a higher percentage of L2 locating in the CCA and industrial parks.

3.1.2. Spatial Distribution of Industrial Land and Living Quarters

Figure 4 presents the spatial distribution of industrial land (L1 and L2) and living quarters (P1 and P2). L1 was mostly distributed separately and located on the junction area between industrial clusters and other land-use functions. P1 was mainly located near L1 (Zone A in Figure 4) but relatively away from L2 (Zone B in Figure 4).

Table 4. Discrepancies between L1 and L2.

Indicators (Mean Values)	Obs.	*lnArea*	*lnYear*	*lnEfficiency*	*lnAgg500*	*Tran500*	*lnMetrodist*	*lnCBDdist*	*CCA Rate* (%)	*Indpark Rate* (%)
L1	3495	8.4218	2.5501	−2.4733	6.6544	0.6229	6.9802	9.4937	68.64	28.32
L2	2115	9.2264	2.0906	−1.7220	7.7267	0.4205	6.9158	9.5603	89.26	64.39

Note: The definitions and descriptions of relevant indicators are the same as those in Table 3.

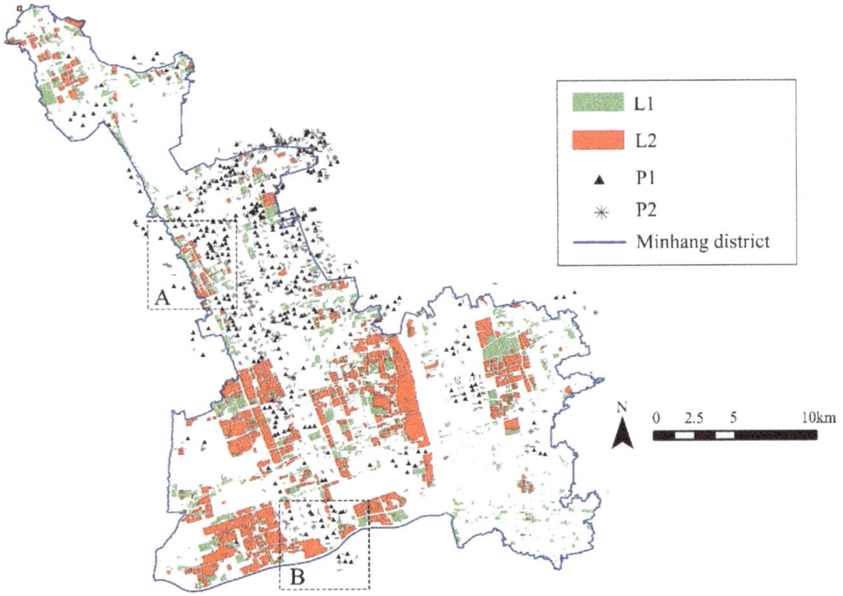

Figure 4. Spatial distribution of the industrial land and living quarters.

Figure 5 describes the clustering of industrial land and living quarters within a certain distance. Industrial land and living quarters appeared slightly discrete within a certain calculation radius. However, when the value of the K-function fell above the upper bound of the 95% confidence interval under the null hypothesis (grey zone in Figure 5), their agglomeration pattern reversed as the calculation radius increased.

There were evident differences in the clustering radius from industrial land (L1 and L2) to living quarters (P1 and P2). The minimum clustering radius from industrial land parcels to surrounding living quarters roughly fell by around 400 m (Figure 5a). The minimum clustering radius to surrounding living quarters from L1 (about 350 m) was shorter than that from L2 (about 600 m), as suggested in Figure 5b,c. The minimum clustering distance between L1 and P1 was as the shortest as approximately 350 m (Figure 5d), while the aggregation of L1 and P2 emerged from a radius of around 450 m (Figure 5e). The minimum clustering radius from L2 to P1 was over 600 m, which was significantly larger than the radius from L1 to P1, as suggested in Figure 5d,f. In a word, L1 was located nearer to living quarters, especially to P1.

To conclude, the results from Sections 3.1.1 and 3.1.2 presented significant differences in characteristics and locations between L1 and L2, verifying Hypothesis 1.

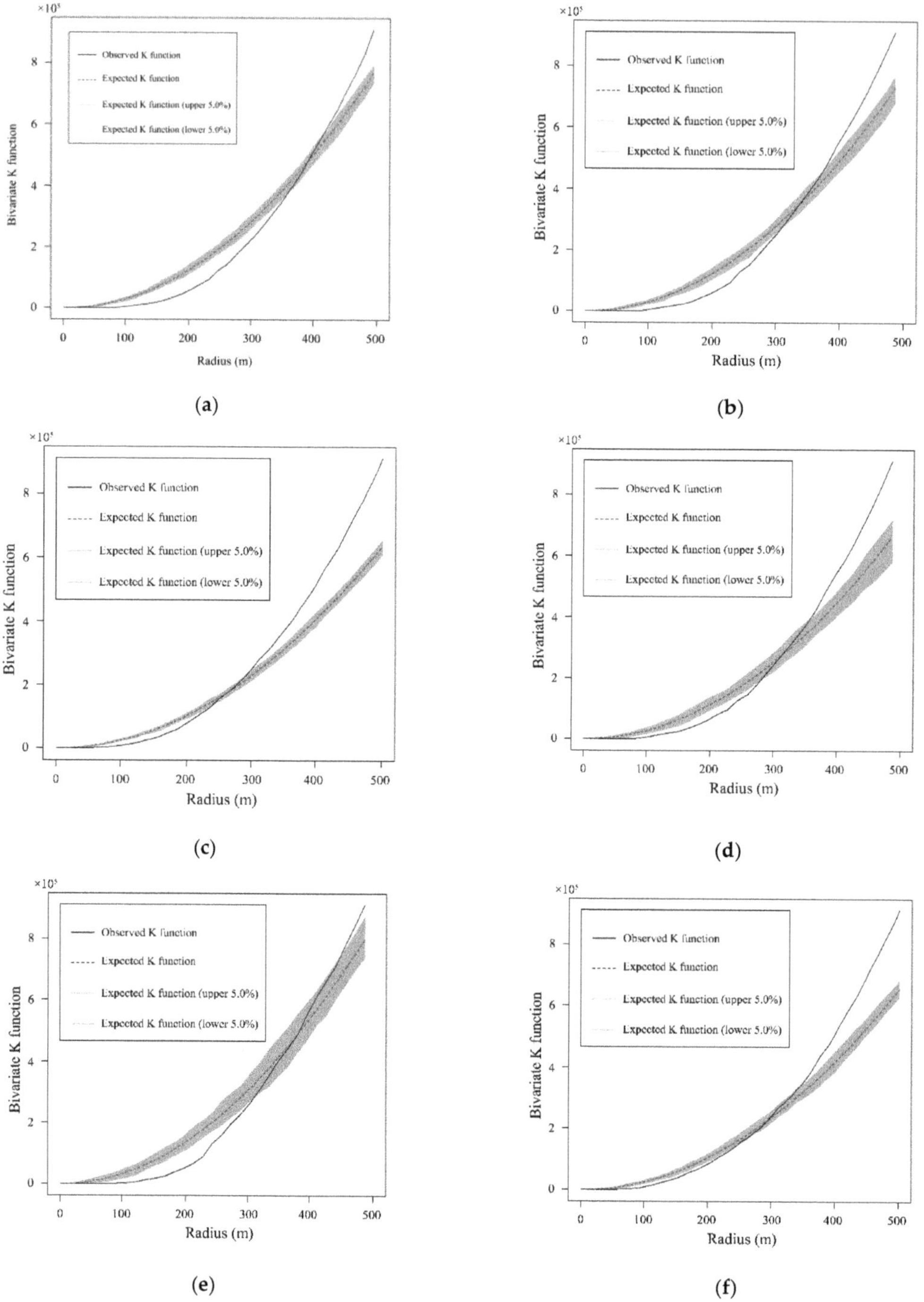

Figure 5. Clustering of industrial land (L1 and L2) and living quarters (P1 and P2): (**a**) Industrial parcels—Living quarters; (**b**) L1—Living quarters; (**c**) L2—Living quarters; (**d**) L1–P1; (**e**) L1–P2; (**f**) L2–P1.

3.2. Positive Impact of Land Price on Industrial Land Transformation

Table 5 shows the effect of a range of factors on industrial land transformation. The results of the LOGIT and PROBIT model are similar in the significance of the coefficients, and the IVPROBIT model strengthens the results. It is noticeable that both the land price and its growth rate yielded a significantly positive coefficient in Models (1) to (6), verifying Hypothesis 2.

Table 5. Results of the land price's effect on the industrial land transformation.

	Model (1)	Model (2)	Model (3)	Model (4)	Model (5)	Model (6)
VARIABLES	**LOGIT**	**LOGIT**	**PROBIT**	**PROBIT**	**IVPROBIT**	**IVPROBIT**
$\ln Price_{2015}$	1.199 ***		0.670 ***		10.32 ***	
	(0.311)		(0.182)		(2.635)	
$Price_{Grow}$		0.907 **		0.502 **		12.12 ***
		(0.425)		(0.249)		(2.931)
lnEfficiency	−0.608 ***	−0.610 ***	−0.346 ***	−0.347 ***	−0.300 ***	−0.315 ***
	(0.0437)	(0.0438)	(0.0250)	(0.0250)	(0.0269)	(0.0243)
lnArea	−0.274 ***	−0.272 ***	−0.163 ***	−0.163 ***	−0.197***	−0.207 ***
	(0.0260)	(0.0258)	(0.0151)	(0.0151)	(0.0207)	(0.0211)
lnYear	0.370 ***	0.357 ***	0.220 ***	0.213 ***	0.298 ***	0.184 ***
	(0.0490)	(0.0490)	(0.0288)	(0.0288)	(0.0413)	(0.0347)
lnAgg500	−0.0216 ***	−0.0232 ***	−0.0138 ***	−0.0147 ***	−0.00146	−0.0141 ***
	(0.00695)	(0.00695)	(0.00410)	(0.00409)	(0.00611)	(0.00487)
Tran500	1.076 ***	1.095 ***	0.649 ***	0.660 ***	0.354 ***	0.494 ***
	(0.0950)	(0.0947)	(0.0559)	(0.0557)	(0.106)	(0.0783)
$\ln Metro_{dist}$	−0.0101	0.00395	−0.00303	0.00467	−0.108 ***	0.0139
	(0.0341)	(0.0340)	(0.0205)	(0.0205)	(0.0390)	(0.0254)
$\ln CBD_{dist}$	0.319 *	−0.178 *	0.184 *	−0.0928	4.283 ***	−0.0276
	(0.171)	(0.106)	(0.0985)	(0.0606)	(1.120)	(0.0701)
CCA	−0.482 ***	−0.441 ***	−0.270 ***	−0.249 ***	−0.580 ***	−0.275 ***
	(0.104)	(0.102)	(0.0590)	(0.0583)	(0.110)	(0.0673)
Indpark	−0.718 ***	−0.712 ***	−0.433 ***	−0.430 ***	−0.540 ***	−0.517 ***
	(0.0799)	(0.0798)	(0.0475)	(0.0474)	(0.0661)	(0.0608)
Instrumental variable	No	No	No	No	Yes	Yes
Constant	−14.07 ***	2.086 *	−7.866 ***	1.161 *	−143.3 ***	−6.373 ***
	(4.521)	(1.102)	(2.629)	(0.631)	(36.97)	(2.014)
AR	-	-	-	-	0.000	0.000
Wald	-	-	-	-	0.000	0.000
Observations	5610	5610	5610	5610	5610	5610

Robust standard errors in parentheses; *** $p < 0.01$, ** $p < 0.05$, * $p < 0.1$.

Additionally, as shown in Table 5, the production efficiency, mean area, and locational factors (e.g., located in the CCA or industrial parks) played a significantly negative role in industrial land transformation. However, a range of factors significantly contributed to industrial land transformation, e.g., the earlier acquisition time and the higher proportion of surrounding L1. Moreover, it is interesting to find that the coefficient of lnAgg500 displayed great significance in models except for Model (5), while the coefficient of lnMetrodist only demonstrated great significance in Model (5).

3.3. Influence of Policy Factors and Industrial Parks on Industrial Land Transformation

Table 6 shows the effect of industrial park administration levels and land-leasing policies on industrial land transformation. According to results of Models (7) to (10), industrial land parcels acquired after 2007 had a lower coefficient of the land price and its growth rate than before 2007. Generally, industrial parks' higher administrative level discouraged industrial land parcels from transformation according to results of Models (11) and (12). However, the effect of the dummy variable for state-level industrial parks was not significant in Model (12).

Table 6. Results of the effects of industrial park administration level on industrial land transformation.

	Model (7)	Model (8)	Model (9)	Model (10)	Model (11)	Model (12)
VARIABLES	IVPROBIT	IVPROBIT	IVPROBIT	IVPROBIT	IVPROBIT	IVPROBIT
	Before 2007	After 2007	Before 2007	After 2007		
$lnPrice_{2015}$	22.65 **	4.230 **			10.61 ***	
	(10.04)	(1.814)			(2.705)	
$Price_{Grow}$			21.62 ***	6.054 **		13.46 ***
			(7.428)	(2.601)		(3.332)
lnEfficiency	−0.264 ***	−0.297 ***	−0.268 ***	−0.317 ***	−0.306 ***	−0.295 ***
	(0.0649)	(0.0330)	(0.0496)	(0.0315)	(0.0263)	(0.0269)
lnArea	−0.150 ***	−0.193 ***	−0.192 ***	−0.187 ***	−0.199 ***	−0.209 ***
	(0.0352)	(0.0332)	(0.0336)	(0.0324)	(0.0211)	(0.0220)
lnYear	1.002 ***	−0.0458	0.744 ***	−0.148 **	0.271 ***	0.188 ***
	(0.186)	(0.0620)	(0.110)	(0.0708)	(0.0385)	(0.0362)
lnAgg500	0.00616	−0.00464	−0.00450	−0.0182 **	−0.00218	−0.0140 ***
	(0.0118)	(0.00894)	(0.00768)	(0.00781)	(0.00602)	(0.00503)
Tran500	0.0963	0.394 ***	0.458 ***	0.399 ***	0.388 ***	0.425 ***
	(0.314)	(0.111)	(0.142)	(0.111)	(0.0986)	(0.0897)
$lnMetro_{dist}$	−0.300 **	−0.0563	−0.0260	−0.00183	−0.111 ***	0.0347
	(0.145)	(0.0386)	(0.0413)	(0.0399)	(0.0402)	(0.0273)
$lnCBD_{dist}$	9.765 **	1.626 **	0.00265	0.00327	4.357 ***	−0.0145
	(4.371)	(0.730)	(0.108)	(0.114)	(1.136)	(0.0731)
CCA	−1.057 ***	−0.246 *	−0.424 ***	−0.00588	−0.601 ***	−0.270 ***
	(0.392)	(0.133)	(0.119)	(0.158)	(0.115)	(0.0692)
Indpark	−0.388 ***	−0.662 ***	−0.428 ***	−0.650 ***		
	(0.114)	(0.105)	(0.0915)	(0.103)		
dummy_district					−0.344 ***	−0.578 ***
					(0.0756)	(0.0741)
dummy_city					−0.554 ***	−0.828 ***
					(0.0922)	(0.129)
dummy_state					−0.923 ***	−0.148
					(0.170)	(0.0992)
Instrumental Variables	Yes	Yes	Yes	Yes	Yes	Yes
Constant	−320.7 **	−56.35 **	−14.04 ***	−2.413	−146.8 ***	−7.416 ***
	(142.0)	(24.99)	(4.821)	(2.143)	(37.80)	(2.284)
AR	0.0000	0.0159	0.0000	0.0145	0.0000	0.0000
Wald	0.0240	0.0197	0.0036	0.0199	0.0001	0.0001
Observations	3934	1676	3934	1676	5610	5610

Standard errors in parentheses; *** $p < 0.01$, ** $p < 0.05$, * $p < 0.1$.

To conclude, the above results have shown that administrative grades of industrial parks and policy factors can significantly affect industrial land regeneration, thus verifying Hypothesis 3.

3.4. Moderation Effects of Output on Industrial Land Transformation

Figure 6 presents the spatial distribution of industrial land output in Minhang District. The industrial land with the highest output still kept manufacturing, although a large amount of industrial land with a comparatively higher output has been transformed. It was also found that large-scale and centrally distributed industrial land with low outputs was under transformation.

Table 7 shows the moderation effect of the industrial land output on its transformation based on a series of interaction variables. The coefficient of Area × Efficiency was significantly negative in Models (13) to (15). Considering the negative coefficients of lnArea and lnEfficiency, the land area and land use efficiency could co-function to prevent industrial regeneration. The significantly negative coefficients of Price × Efficiency × Area in Model (14) and Price × Efficiency in Model (16) revealed that the marginal effect of industrial land prices could be undermined when industrial land was efficient and productive.

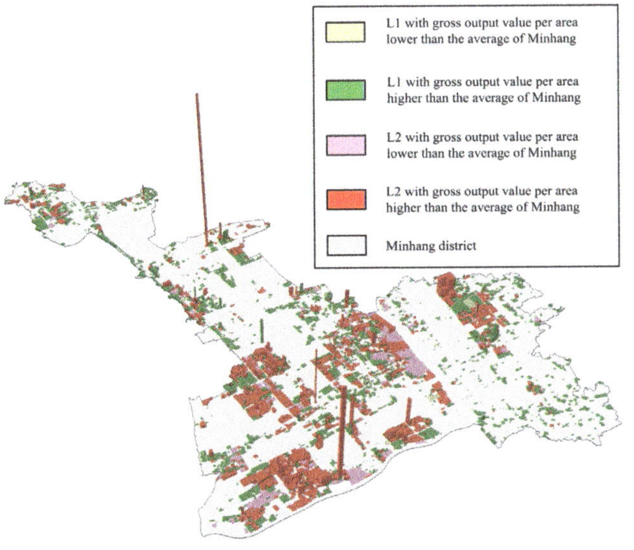

Figure 6. Distribution of industrial land output value in Minhang District.

Table 7. The results of moderation effects of output on industrial land transformation.

	Model (13)	Model (14)	Model (15)	Model (16)
VARIABLES	PROBIT	IVPROBIT	PROBIT	IVPROBIT
lnPrice$_{2015}$	0.650 *** (0.183)	11.41 *** (3.085)		
Price$_{Grow}$			0.549 ** (0.257)	17.00 *** (4.982)
lnEfficiency	−0.333 *** (0.0243)	−0.234 *** (0.0495)	−0.332 *** (0.0245)	−0.436 *** (0.0881)
lnArea	−0.173 *** (0.0155)	−0.238 *** (0.0297)	−0.175 *** (0.0155)	−0.239 *** (0.0381)
lnYear	0.221 *** (0.0290)	0.332 *** (0.0471)	0.212 *** (0.0289)	0.113 ** (0.0551)
lnAgg500	−0.0138 *** (0.00411)	−0.00332 (0.00664)	−0.0143 *** (0.00409)	−0.00771 (0.00706)
Tran500	0.644 *** (0.0559)	0.377 *** (0.115)	0.656 *** (0.0558)	0.374 *** (0.120)
lnMetro$_{dist}$	−0.000753 (0.0207)	−0.0577 (0.0455)	0.00435 (0.0205)	0.0213 (0.0367)
lnCBD$_{dist}$	0.177 * (0.0982)	4.407 *** (1.285)	−0.0975 (0.0605)	−0.117 (0.108)
CCA	−0.267 *** (0.0588)	−0.570 *** (0.129)	−0.246 *** (0.0580)	−0.420 *** (0.114)
Indpark	−0.433 *** (0.0474)	−0.555 *** (0.0744)	−0.433 *** (0.0474)	−0.549 *** (0.0862)
Area × *Efficiency*	−0.0525 ** (0.0215)	−0.136 *** (0.0442)	−0.0588 *** (0.0217)	−0.0114 (0.0582)
Price × *Efficiency*	−0.0255 (0.131)	−0.175 (0.930)	−0.351 (0.317)	−11.04 ** (5.079)
Price × *Area*	0.109 (0.0861)	0.536 (0.664)	−0.0374 (0.193)	−8.076 (5.016)
Price × *Efficiency* × *Area*	0.0967 (0.124)	−2.847 ** (1.117)	−0.380 (0.274)	2.490 (3.303)
Instrumental Variables	No	Yes	No	Yes
Constant	−7.499 *** (2.631)	−155.4 *** (43.00)	1.326 ** (0.633)	−8.384 *** (3.201)
AR	-	0.0000	-	0.0000
Wald	-	0.0001	-	0.0025
Observations	5610	5610	5610	5610

Robust standard errors in parentheses; *** $p < 0.01$, ** $p < 0.05$, * $p < 0.1$.

To conclude, output factors directly affect the choice of transformation model and moderate the influence of land price on transformation, thus verifying Hypothesis 4.

4. Discussion

4.1. The Game Framework for Industrial Land Transformation

China's urbanization brings with it the rapid expansion of the urban population and built-up areas. In urban growth areas, various forces to enlarge industrial land-leasing scales form a sophisticated situation, representing the total amount of development demands from different stakeholders. The conflict of interest in urbanization is caused by the divergence in the goals of governments at various administrative levels, the differential interests of involved groups, and multiple urban functions, etc. The goal determines transformation options because the scale and location entropy are different among industrial, commercial, and residential land. The objectives also influence the decision-making process because land occupants intend to fully obtain land values reflected in the different land use nature [39]. However, it is manifest and observable that the intervention from various stakeholders could lead to the multi-dimensional spatial fragmentation of urban areas [40].

The residential and commercial redevelopment of industrial land can cause a potential return from land value appreciation. The added value of real estate from industrial transformation attracts various participants to promote urban renewal [15], further pushing up prices of the surrounding land and housing [35]. This makes relevant stakeholders more willing to become involved in the transformation processes. Faced with this option, enterprises have to decide whether to keep on manufacturing. This study revealed that the commercial real estate redevelopment of industrial land was largely affected by residential property price rising, comparing the profit from manufacturing products. From this perspective, industrial land transformation is the process of land resources allocation under the land market economy mechanism. This process can cause a higher input–output efficiency and significantly contribute to GDP growth as well as local finance. Local governments can thus distribute more financial resources for public affairs. However, given that the income from real estate development is higher than that from industry and commerce, industrial land developers may be distracted from their intended purpose to restrain manufacturing and innovation research [14]. Therefore, this could lead to excessive commercial real estate development, a decline in the real economy, and social inequity, although industrial land transformation in commercial functions is economically rational.

At last, industrial enterprises' producing and profiting conditions affect revenue expectations from the industrial transformation, adding to the cost of redevelopment; that is to say, the output factors can be involved in the game process. According to Section 3.4, the industrial land efficiency was suggested as a mediator to regulate industrial land transformation. However, the driving force from land prices rising expectations is so intensive that some efficient industrial land has been transformed, even the redevelopment of this land calls for a higher rate of return to cover the loss from transformation.

4.2. The Role of Industrial Land Leasing for Local Governments

Industrial land leasing has become not only an essential part of the "land finance game" [41] but also a competitive weight of the "attracting investment" policy [3]. First, industrial land transformation driven by the land price increment is a process of urban reconstruction somewhat driven by anticipated financial revenue. The government dominates the primary land market by controlling the development region, transaction prices, and industrial land planning [42]. The existing fiscal and taxation system's constraints in China allow the local government to obtain financial revenue by controlling the land supply, leasing, and conveyance [43,44]. Industrial land is often allocated in the peri-urban area or suburb area with the attractive promise of low or even zero land prices when the local government attempts to attract external investment [45]. By simultaneously leasing a large amount of industrial land at a low price and pushing up the price of residential land, the local government makes full use of land composition balance theories [46]. Much of local governments' extrabudgetary income increase comes from land transactions, leading to an expanding metropolitan area [47]. The over-speed land urbanization strengthens the opinion that urban sprawl is promoted for economic growth in cities [48]. The massive industrial land allocated in this process generates fiscal revenue through both budget and extrabudgetary channels [1,2], which raises an overall fixed asset price. However, this land-centered development model [49,50] can only bring economic growth in the short run. It does not create much new social value or provide more opportunities for wealth accumulation.

Second, industrial land leasing has become a vital competition tool for regional and local governments after establishing a system of tax distribution in China [51]. The existing finance and taxation management system in China requires local governments to allocate industrial land within their administrative jurisdiction. Each jurisdiction has its own industrial park at different administrative grades. This will bring about competitive relationships among local governments because a higher-level industrial park can provide more attractive incentives for industrial enterprises. Local governments often use differentiated supply strategies to make full use of the economic and financial value of industrial and other functional land use [52,53]. This can be found more commonly in these areas (e.g., Shanghai) where long-term economic growth pressure exists [54,55].

4.3. The Regeneration Behavior of Industrial Enterprises

The diversity of enterprise types and industrial land ownerships results in different industrial land development patterns. On the one hand, the diversity of industrial enterprise identities may add to the complexity of regeneration approaches. Industrial land can be used by private companies, collective groups, state-owned enterprises, and various joint ventures in the peri-urbanization area. The differences in industrial enterprise identities lead to differential land-leasing prices, location choices, and priority in facility supports when industrial land is allocated. Considering the differences in property right structures of state-owned, private-owned, and joint-venture enterprises, the motive force and path of their transformation could be also different [56,57]. On the other hand, the land used by industrial enterprises is either state-owned or collectively owned. The state-owned industrial land is mostly in industrial parks, while collective land is located in rural areas, such as inner and outer suburbs, etc. Therefore, collective industrial land transformation involves the land expropriation process and the inevitable negotiation between governments and villagers, adding to the transition cost and uncertainty [13,39]. Moreover, according to Section 3.1.1, the coefficients of lnAgg500 and Tran500 indicated that the peer effect of industrial transformation existed and promoted the overall regeneration of industrial zones [58].

The relocation of industrial enterprises address also affects regeneration. Some enterprises are faced with the relocation of production workshops in urban regeneration. In most cases, industrial parks provide enterprises with tax compensation and other incentive policies to encourage them to move in their manufacturing base. However, industrial parks have a selection process for prospective enterprises based on productivity and industrial categories [11,59]. The entry criteria may exclude inefficient enterprises.

Additionally, the land prices rising expectations from industrial land developers can encourage industrial land redevelopment. To seek potential profit margins from land transactions, industrial enterprises intend to capture more industrial land than they need in manufacturing [60]. If these enterprises are unfortunately shut down due to the enterprise life cycle rule, they are willing to turn into real estate development instead of exploring new manufacturing opportunities in the industrial field. Thus, the 50-year term for industrial land tenure in China seems too long [61,62]. Some enterprises have even found such institutional loopholes. They tend to introduce nominal investment to stimulate the land price increase and wait for governments to Shouchu the land (repurchase and reserve of land resources by authorities), rather than manufacturing after obtaining industrial land.

The withdrawal system of inefficient land to some extent mitigates issues of the idle land and restricts the speculation behavior of industrial enterprises [60].

4.4. Targeted Guidance to Regulate Industrial Land Transformation

The positive effect of land price appreciation on regeneration exposes the risk of industrial land management. Industrial land transformation offers a chance for low-output or even shut-down enterprises to survive. On the contrary, the institutional arrangement to curb land speculation not only hinders the redevelopment of idle industrial land, but also forces the conveyance of new industrial land to achieve a quantitatively reasonable demand in the name of the development of the industrial economy. Over the past few decades, this process repeats, and inefficient industrial land continually emerges. The constant industrial land allocation provides hotbeds for the prevalence of land price-oriented regeneration. The sustainable industrial development associated with the urbanization process is thus threatened. This has become a core problem of urbanization in China [8].

The targeted guidance for regeneration is an effective approach to resolving spatial problems of industrial land. A series of urban space problems are caused due to excessive leasing scale and inefficient use of industrial land. For instance, urban spatial fragmentation is strengthened [39,56,57] because industrial land is mainly distributed in urban fringe and suburbs and mixed with other functional lands [51]. The spatial problem also intensifies the contradiction among various types of stakeholders [39] and limits a region's redevelopment [57,63]. According to Section 3.1.1, the transformation occurs on relatively small and independent industrial land parcels, leaving the distribution of L2 to be aggregated but the distribution of L1 to be fragmented. Moreover, according to Section 3.2, the significantly negative effect of the mean area and land efficiency on industrial land transformation indicated that the land parcels with a lower efficiency and smaller areas were more likely to be transformed. The redevelopment of fragmented land parcels can reduce the regional industrial fragmentation, and thus improve the urban spatial pattern. The government can guide inefficient industrial land and transform it into space for public facilities according to functional requirements. The policies of promoting production can bring together high-value enterprises and strengthen industrial aggregation. Therefore, this study insists that industrial land transformation can optimize the urban form and the industrial layout.

Improving inefficient land use and promoting its regeneration is related to the local government's philosophy of urbanization and urban renewal. Formal industrial land transformation cannot be completed without the approval of the local government in China. The desire for fiscal revenue and extrabudgetary income in land transactions urges local governments to enlarge the industrial land scale and promote industrial land redevelopment [44]. In turn, regeneration can be useful for governments to promote the transformation of substandard industrial enterprises. The management mechanism in land transactions and prices results in the deviation between industrial land's transfer cost and market price [8]. The parcels with advantageous locations can be easily transformed by market forces, while those with poorer performance and sites should be transformed into other infrastructures. Both of the two regeneration patterns can help relieve the financial pressure of local governments. Considering the positive impact of industrial land regeneration, it is necessary to implement supervision and withdraw measures on inefficient land from the perspective of benefit balance sharing. This can stimulate the full participation of market subjects in the regeneration of industrial land.

4.5. Highlights, Limitations, and Suggestions for Further Research

This study focuses on how land price appreciation affects industrial land transformation, enlightened by the large amount of inefficient and idle industrial land located in Minhang District, Shanghai. There are two main highlights of this study: first, econometric methods were employed to explore causal relationships between industrial land transformation and land prices rising expectations, which fills the methodological gap in industrial land transformation research. Another highlight is that a fine-grained industrial land data (i.e., at the parcel level) were adopted in this study. This can provide the potential to conduct a more in-depth and detailed analysis, compared to previous studies that used macro- or meso- level data.

However, this study still has several defects that need improving. First, this study did not consider industrial land parcels' influence outside of the study area in Shanghai due to unavailable industrial data in other districts adjacent to the Minhang District. Second, the industrial clusters' edge did not strictly follow the district boundary because of industrial land sprawl in southwestern Shanghai. This may conceal the real distribution of industrial land parcels. Third, there are a few limitations in the living quarter data. For instance, the location of second-hand housing transactions is unstable. The interpolation results thus cannot be directly compared. To calculate land prices' growth rate, this study only took the living quarter data appearing in both 2015 and 2019 into consideration, which actually narrowed the sample size. The built year and types of living quarters were omitted, resulting in certain errors in price estimation. The acquired housing price might not be the real trading price given the price fluctuations and artificial adjustments for advertising. Moreover, the Euclidean distance alone was considered when obtaining the spatial distribution of land prices using the Kriging interpolation method, which might weaken the result to some extent. Fourth, the potential influence of relevant factors (e.g., types of land ownership, industrial categories, and density of investment and labor) was not directly estimated due to unavailable data. The omitted variables in the empirical stage added to the difficulties of evaluating the effect of land price. We managed to employ relevant variables like efficiency, neighbor industrial land, industrial parks, and CCA to substitute the effect of unobserved factors, but the performance was still hard to evaluate. Fifth, the lack of data for enterprises below the designated scale (the annual enterprise income is no more than CNY 20 million) might lead to errors in the output efficiency analysis.

Finally, this study suggests a few directions for further research. First, the regeneration of industrial land in this study was illustrated as a static result instead of a dynamic process. In fact, the regeneration involves various stakeholders and lasts for quite a long time covering several stages. Future studies can focus on the dynamic process. For instance, based on multi-source data (e.g., remote sensing and electricity consumption data), the dynamics of industrial land from manufacturing to transformation can be estimated by figuring out the exact time of transition and the real condition of manufacturing. Second, the subject of industrial land transformation is complex, and the orientations of different interest groups do not concur with each other. This method of divergence affects the transformation process. This process's theoretical and practical value will play a significant role in the spatial planning of regeneration and urban management, deserving an in-depth analysis in the future. Third, the government introduced a series of regulations to control the transfer process and transaction value to curb land speculation [4,51]. Still, this problem has not effectively been solved [8]. Scholars can further focus on providing policymakers with theoretical and practical supports to mitigate bad effects from industrial land transformation in the game process.

5. Conclusions

This empirical study took the Minhang District of Shanghai as the study area and verified four hypotheses. First, this study showed distinct discrepancies in some features (e.g., the mean area, output efficiency, surrounding industrial land area, and location) between L1 and L2. Second, this study proved land prices rising expectations to be a driving force for promoting industrial land transformation. Third, it was found that two factors (e.g., land leasing policies and whether the industrial land is located in industrial parks) could influence the effect of land prices rising expectations on industrial land transformation. Finally, this study also found that the industrial land output could serve as a moderator to regulate the transformation process. These findings will provide policy makers (e.g., relevant authorities and urban planners) with significant references to coordinate the transformation process in a sustainable manner, given that industrial land transformation can bring both advantages and disadvantages.

It is usually believed that, after industrialization, industrial land vacancies will appear and wait for industry upgrading or the redevelopment of other types of land-use. However, China's situation is slightly different. Some industrial landowners do not really start production after acquiring the land, but rather wait for policy shifts and reap excess profits by adjusting to commercial housing development, because the industrial land acquisition cost is very low at the start of the projects. As a result, the development of commercial housing excessively relies on price increase expectations rather than the market supply and demand relationship, causing defects in the matching of urban functional space. This experience of Chinese cities in the process of rapid urbanization driven by industrialization can provide reference for later countries.

Author Contributions: Conceptualization, Methodology, Data curation, Supervision, Project administration, Validation, Resources, Formal analysis, Funding acquisition, F.Y.; Data curation, Investigation, Software, Writing—Original draft preparation, P.T.; Writing—Review and Editing, X.C. and J.W.; Visualization, J.W. All authors have read and agreed to the published version of the manuscript.

Funding: This research was funded by the National Natural Science Foundation of China, grant number 51778436.

Institutional Review Board Statement: Not applicable.

Informed Consent Statement: Not applicable.

Data Availability Statement: The data presented in this study are available on request from the first author.

Conflicts of Interest: The authors declare no conflict of interest.

References

1. Cai, M. Revenue, time horizon, and land allocation in China. *Land Use Policy* **2017**, *62*, 101–112. [CrossRef]
2. Tao, R.; Su, F.; Liu, M.; Cao, G. Land leasing and local public finance in China's regional development: Evidence from prefecture-level cities. *Urban Stud.* **2010**, *47*, 2217–2236. [CrossRef]
3. Zhang, L.; Nian, Y.; Pi, J.; Zhou, Y. Land policy, land supply structure and housing prices. *China J. Econ.* **2017**, *4*, 91–118. [CrossRef]
4. Lin, Y.; Qin, Y.; Yang, Y.; Zhu, H. Can price regulation increase land-use intensity? Evidence from China's industrial land market. *Reg. Sci. Urban Econ.* **2020**, *81*, 103501. [CrossRef]
5. Alder, S.; Shao, L.; Zilibotti, F. Economic reforms and industrial policy in a panel of Chinese cities. *J. Econ. Growth* **2016**, *21*, 305–349. [CrossRef]
6. Zhang, J. Interjurisdictional competition for FDI: The case of China's "development zone fever". *Reg. Sci. Urban Econ.* **2011**, *41*, 145–159. [CrossRef]
7. Schminke, A.; van Biesebroeck, J. Using export market performance to evaluate regional preferential policies in China. *Rev. World Econ.* **2013**, *149*, 343–367. [CrossRef]
8. Tu, F.; Yu, X.; Ruan, J. Industrial land use efficiency under government intervention: Evidence from Hangzhou, China. *Habitat Int.* **2014**, *43*, 1–10. [CrossRef]
9. Meng, Y.; Zhang, F.-R.; An, P.-L.; Dong, M.-L.; Wang, Z.-Y.; Zhao, T. Industrial land-use efficiency and planning in Shunyi, Beijing. *Landsc. Urban Plan.* **2008**, *85*, 40–48. [CrossRef]
10. Sun, Y.; Ma, A.; Su, H.; Su, S.; Chen, F.; Wang, W.; Weng, M. Does the establishment of development zones really improve industrial land use efficiency? Implications for China's high-quality development policy. *Land Use Policy* **2020**, *90*, 104265. [CrossRef]
11. Huang, D.; Wan, W.; Dai, T.; Liang, J. Assessment of industrial land use intensity: A case study of Beijing economic-technological development area. *Chin. Geogr. Sci.* **2011**, *21*, 222–229. [CrossRef]
12. Harvey, D. The geography of capitalist accumulation: A recpnstruction of the Marxian theory. *Antipode* **1975**, *7*, 9–21. [CrossRef]
13. Lai, Y.; Tang, B. Institutional barriers to redevelopment of urban villages in China: A transaction cost perspective. *Land Use Policy* **2016**, *58*, 482–490. [CrossRef]
14. Wang, W.; Rong, Z. Housing boom and firm innovation: Evidence from industrial firms in China. *China Econ. Q.* **2014**, *13*, 465–490. [CrossRef]
15. Gao, J.; Chen, W.; Yuan, F. Spatial restructuring and the logic of industrial land redevelopment in urban China: I. Theoretical considerations. *Land Use Policy* **2017**, *68*, 604–613. [CrossRef]

16. Tian, L.; Ma, W. Government intervention in city development of China: A tool of land supply. *Land Use Policy* **2009**, *26*, 599–609. [CrossRef]
17. Gao, J.; Chen, W. Spatial restructuring and the logic of industrial land redevelopment in urban China: III. A case study of the redevelopment of a central state-owned enterprise in Nanjing. *Cities* **2020**, *96*, 604–613. [CrossRef]
18. Gao, J.; Chen, W.; Liu, Y. Spatial restructuring and the logic of industrial land redevelopment in urban China: II. A case study of the redevelopment of a local state-owned enterprise in Nanjing. *Land Use Policy* **2018**, *72*, 372–380. [CrossRef]
19. Wang, S.; Wang, J.; Wang, Y. Effect of land prices on the spatial differentiation of housing prices: Evidence from cross-county analyses in China. *J. Geogr. Sci.* **2018**, *28*, 725–740. [CrossRef]
20. Wen, H.; Goodman, A.C. Relationship between urban land price and housing price: Evidence from 21 provincial capitals in China. *Habitat Int.* **2013**, *40*, 9–17. [CrossRef]
21. Dubin, R.A. Predicting house prices using multiple listings data. *J. Real Estate Financ. Econ.* **1998**, *17*, 35–59. [CrossRef]
22. Kuntz, M.; Helbich, M. Geostatistical mapping of real estate prices: An empirical comparison of kriging and cokriging. *Int. J. Geogr. Inf. Sci.* **2014**, *28*, 1904–1921. [CrossRef]
23. Li, H.; Wei, Y.D.; Wu, Y.; Tian, G. Analyzing housing prices in Shanghai with open data: Amenity, accessibility and urban structure. *Cities* **2019**, *91*, 165–179. [CrossRef]
24. Si, Z.; Shi, Y. The impacts of class 3A comprehensive hospitals on population distribution and housing price: Take children's hospital of Fudan University for example. *Econ. Geogr.* **2013**, *33*, 74–81.
25. Byth, K.; Ripley, B.D. On sampling spatial patterns by distance methods. *Biometrics* **1980**, *36*, 279–284. [CrossRef]
26. Lotwick, H.W.; Silverman, B.W. Methods for analyzing spatial processes of several types of points. *J. R. Stat. Soc. Ser. B* **1982**, *44*, 406–413.
27. Wiegand, T.; Moloney, K.A. Rings, circles, and null-models for point pattern analysis in ecology. *Oikos* **2004**, *104*, 209–229. [CrossRef]
28. Austin, S.B.; Melly, S.J.; Sanchez, B.N.; Patel, A.; Buka, S.; Gortmaker, S.L. Clustering of fast-food restaurants around schools: A novel application of spatial statistics to the study of food environments. *Am. J. Public Health* **2005**, *95*, 1575–1581. [CrossRef]
29. Cuthbert, A.L.; Anderson, W.P. Using spatial statistics to examine the pattern of urban land development in Halifax-Dartmouth. *Prof. Geogr.* **2002**, *54*, 521–532. [CrossRef]
30. Baddeley, A.; Turner, R. spatstat: An R package for analyzing spatial point patterns. *J. Stat. Softw.* **2005**, *12*, 1–42. [CrossRef]
31. Itoh, R. Estimating the willingness to pay of industrial firms for Japanese industrial parks. *Urban Stud.* **2013**, *50*, 2753–2765. [CrossRef]
32. Chau, K.W.; Wong, S.K. Externalities of urban renewal: A real option perspective. *J. Real Estate Financ. Econ.* **2014**, *48*, 546–560. [CrossRef]
33. Lee, C.-C.; Liang, C.-M.; Chen, C.-Y. The impact of urban renewal on neighborhood housing prices in Taipei: An application of the difference-in-difference method. *J. Hous. Built Environ.* **2017**, *32*, 407–428. [CrossRef]
34. Rossi-Hansberg, E.; Sarte, P.D.; Owens, R., III. Housing externalities. *J. Political Econ.* **2010**, *118*, 485–535. [CrossRef]
35. van Duijn, M.; Rouwendal, J.; Boersema, R. Redevelopment of industrial heritage: Insights into external effects on house prices. *Reg. Sci. Urban Econ.* **2016**, *57*, 91–107. [CrossRef]
36. Newey, W.K. Efficient estimation of limited dependent variable models with endogenous explanatory variables. *J. Econom.* **1987**, *36*, 231–250. [CrossRef]
37. Li, H.; Wei, Y.D.; Yu, Z.; Tian, G. Amenity, accessibility and housing values in metropontan USA: A study of salt Lake county, utah. *Cities* **2016**, *59*, 113–125. [CrossRef]
38. Zheng, S.; Kahn, M.E. Land and residential property markets in a booming economy: New evidence from Beijing. *J. Urban Econ.* **2008**, *63*, 743–757. [CrossRef]
39. Wang, B.; Li, T.; Yao, Z. Institutional uncertainty, fragmented urbanization and spatial lock-in of the pen-urban area of China: A case of industrial land redevelopment in Panyu. *Land Use Policy* **2018**, *72*, 241–249. [CrossRef]
40. Guo, Y.; Yuan, Q.; Tan, S.; Xiang, Z. Urban renewal to integrate fragmented rural industrialized areas: A case study of Nanhai District, Foshan City. *City Plan. Rev.* **2020**, *44*, 53–61, 89.
41. Jiang, Z. Industrialization, local government efforts and land finance: A perspective of land finance in China. *China Ind. Econ.* **2014**, *10*, 33–45. [CrossRef]
42. Lin, S.-W.; Ben, T.-M. Impact of government and industrial agglomeration on industrial land prices: A Taiwanese case study. *Habitat Int.* **2009**, *33*, 412–418. [CrossRef]
43. Cao, G.; Feng, C.; Tao, R. Local "Land finance" in China's urban expansion: Challenges and solutions. *China World Econ.* **2008**, *16*, 19–30. [CrossRef]
44. Xu, N. What gave rise to China's land finance? *Land Use Policy* **2019**, *87*, 104015. [CrossRef]
45. Peng, S.; Wang, S.; Chen, C.; Wei, H.; Wang, J.; Wu, X. Analysis of the economic rationality of local government's downtick of the industrial land: Evidence from the municipalities in Guangdong Province. *J. Nat. Resour.* **2015**, *30*, 1078–1091. [CrossRef]
46. Zhang, X.; Lin, Y.; Wu, Y.; Skitmore, M. Industrial land price between China's Pearl River Delta and Southeast Asian regions: Competition or coopetition? *Land Use Policy* **2017**, *61*, 575–586. [CrossRef]
47. Gomez-Antonio, M.; Hortas-Rico, M.; Li, L. The causes of urban sprawl in spanish urban areas: A apatial approach. *Spat. Econ. Anal.* **2016**, *11*, 219–247. [CrossRef]

48. Gao, J.; Wei, Y.D.; Chen, W.; Chen, J. Economic transition and urban land expansion in provincial China. *Habitat Int.* **2014**, *44*, 461–473. [CrossRef]
49. Lin, G.C.S. Reproducing spaces of chinese urbanisation: New city-based and land-centred urban transformation. *Urban Stud.* **2007**, *44*, 1827–1855. [CrossRef]
50. Ye, L.; Wu, A.M. Urbanization, land development, and land financing: Evidence from Chinese cities. *J. Urban Aff.* **2014**, *36*, 354–368. [CrossRef]
51. Zhou, L.; Tian, L.; Gao, Y.; Ling, Y.; Fan, C.; Hou, D.; Shen, T.; Zhou, W. How did industrial land supply respond to transitions in state strategy? An analysis of prefecture-level cities in China from 2007 to 2016. *Land Use Policy* **2019**, *87*, 104009. [CrossRef]
52. Huang, Z.; Du, X. Government intervention and land misallocation: Evidence from China. *Cities* **2017**, *60*, 323–332. [CrossRef]
53. Shen, T.; Ai, G.; Cui, N.; Gu, H. Interactive behavior of local government industrial land transfer strategy in the Yangtze River Delta. *J. Ind. Technol. Econ.* **2019**, *38*, 47–53. [CrossRef]
54. Fang, L.; Tian, C.; Yin, X.; Song, Y. Political cycles and the mix of industrial and residential land leasing. *Sustainability* **2018**, *10*, 3077. [CrossRef]
55. Huang, Z.; Du, X. Strategic interaction in local governments' industrial land supply: Evidence from China. *Urban Stud.* **2017**, *54*, 1328–1346. [CrossRef]
56. Tian, L. Land use dynamics driven by rural industrialization and land finance in the pen-urban areas of China: "The examples of Jiangyin and Shunde". *Land Use Policy* **2015**, *45*, 117–127. [CrossRef]
57. Zhu, J.; Guo, Y. Fragmented peri-urbanisation led by autonomous village development under informal institution in high-density regions: The case of Nanhai, China. *Urban Stud.* **2014**, *51*, 1120–1145. [CrossRef]
58. Lieberman, M.B.; Asaba, S. Why do firms imitate each other? *Acad. Manag. Rev.* **2006**, *31*, 366–385. [CrossRef]
59. Huang, Z.; He, C.; Wei, Y.H.D. A comparative study of land efficiency of electronics firms located within and outside development zones in Shanghai. *Habitat Int.* **2016**, *56*, 63–73. [CrossRef]
60. Wang, K.; Xiong, K.; Gao, W. Research on the relationship between the conveyance mode of industrial land use rights and the output elasticity of land in the development zones. *China Land Sci.* **2013**, *27*, 4–9.
61. Li, J.; Dai, S. Retrospect and prospect of China's urban land use system reform in the past 60 years. *Rev. Econ. Res.* **2009**, *63*, 2–10. [CrossRef]
62. Liu, J. Analysis on reform of industrial land transfer and utilization institution. *J. Party Sch. Cent. Comm. C. P. C.* **2015**, *19*, 65–70. [CrossRef]
63. Wu, F.; Zhang, F.; Webster, C. Informality and the development and demolition of urban villages in the Chinese peri-urban area. *Urban Stud.* **2013**, *50*, 1919–1934. [CrossRef]

sustainability

Article

Market-Driven Rural Construction—A Case Study of Fuhong Town, Chengdu

Yujun Zhou *, Jingming Liu and Xiang Kang

School of Public Policy and Administration, Xi'an Jiaotong University, Xi'an 710049, China; jimmyliu1993@stu.xjtu.edu.cn (J.L.); kangxiang9523@stu.xjtu.edu.cn (X.K.)
* Correspondence: zhou20191202@stu.xjtu.edu.cn

Abstract: Although local government has played an important role in rural China's development, some disadvantages of government-led rural construction have gradually emerged with changes in socioeconomic structure, which have negative impacts on rural development. To solve the problems of the traditional rural construction pattern, the introduction of market mechanisms into rural construction became the consensus in theory and in practice. Extant studies emphasize the importance of a market-driven rural construction pattern; however, they do not discuss how to practice this pattern in detail. Thus, this paper uses a case study and comparative analysis to illustrate the background, implementation process and outcomes of the market-driven pattern, aiming to identify the intrinsic dynamics among the local government, market capital and villagers in the market-driven pattern. We argue that although the transformation from a government-led to market-driven pattern is a gradual process, the market-driven pattern is an alternative to the traditional pattern and can better fulfill villagers' interests and enhance sustainable rural development.

Keywords: rural construction; the market-driven pattern; rural construction land consolidation (RCLC); rural China

Citation: Zhou, Y.; Liu, J.; Kang, X. Market-Driven Rural Construction—A Case Study of Fuhong Town, Chengdu. *Sustainability* **2022**, *14*, 6014. https://doi.org/10.3390/su14106014

Academic Editors: Wenwu Tang, Dianfeng Liu and Jianxin Yang

Received: 7 March 2022
Accepted: 10 May 2022
Published: 16 May 2022

1. Introduction

Rural decline is a worldwide problem in the context of urbanization and industrialization [1] (Gao and Wu, 2017) and is expressed as rural outmigration, inefficient rural land use, declining rural industry and ultimately insufficient endogenous development capability in rural areas [2]. Among them, land-related issues are considered to be a major factor in rural decline [3,4] because land is important for rural industry development and villagers' livelihoods. To revitalize rural areas, western countries began comprehensive rural reconstruction in the middle or late stages of industrialization and urbanization, which focused on encouraging urban residents migrate to rural areas to relieve crowded cities [5–8]. Contrarily, the outflow of production factors from rural areas and the ensuing widening urban–rural gap are the main problems in rural East Asia [9,10]. Both Japan and South Korea have paid attention to the promotion of self-help and cooperation among rural communities through the "new village movement" campaigns [11,12]. However, these related experiences are insufficient to explain the complexities of China's rural development under the distinct rural construction land regime and uneven urban–rural relationship. With a long-term planned economy since 1949, China has gradually formed a dual system to separate urban and rural areas. As the representative of public interests, local governments can use their "extra-territorial power" [4,13] to acquire land from the rural collective and deprive villagers of land development rights [14], while the rural collective, as a legal owner of rural land, is merely informed that the decision has been made to acquire their land for development [15]. Meanwhile, the availability of rural construction land for residential or public purposes is limited, and transferring this land to commercial end-users is prohibited. Contrastingly, state-owned land can be used for business opportunities, especially for real estate. In addition to the limited land use rights, the particularity of rural construction land also lies in its allocation principle, causing less efficient land use. As a major composition of rural construction land, rural residential land is allocated to collective members for free and for an unlimited duration. Similarly, residential land cannot be transferred to users outside of the rural collective.

This dual land-ownership system reflects an asymmetric urban–rural relationship. To some degree, government-led urbanization results in rapid urban growth at the expense of villagers' interests. Under this pattern, the so-called rural construction is just a supporting tool for land finance and urban expansion, while the rural collective's endogenous development capability and sustainability are ignored. On the one hand, the GDP-oriented performance evaluation system and administrative resource allocation result in urbanization charactered with "rural land urbanization" [16–18]. The local government depends highly on land finance, and some enterprises prefer land speculation than industrial development, resulting in the wasting of land resources. On the other hand, the local government pays little attention to villagers' sustainable livelihood after land requisition. Landless farmers, especially those who are middle-aged and elderly, find it difficult to find non-agricultural employment [19,20]. This, coupled with the higher cost of living after concentrated resettlement, means that many villagers actually face greater economic pressure than before. Considering these negative impacts and indirect costs, villagers may lack enough economic incentive to participate in land requisition [21,22]; however, they have little bargaining power to decide whether and how to participate when facing the "coercive force of the state". As a result, villagers are forced to become involved in the government-led urbanization scheme and are excluded from sharing economic development outcomes. This passive rural construction led by the government results in many serious socioeconomic issues such as "hollowing villages", abandoned arable land, idle homestead land, poor rural living environments and weak rural governance [23,24].

To solve the problem of the "thriving cities, declining villages" and "strong government and weak market" [25–27] associated with the government-led rural construction pattern, the introduction of a market mechanism into rural construction became the consensus in theory and in practice [23,27–29].

In theory, both urban and rural areas are spatial expressional forms resulting from the interaction of production factor aggregation, market transactions and public goods provision [30]. Economies of scale and transaction efficiency are driving forces for their emergence, development and maturity. Just as enterprises are explained as an organizational form of saving transaction costs by Coase, cities and villages have the same nature as enterprises for realizing economies of scale and decreasing transaction costs. The concentration of transactions in a certain space results from humans' spontaneous choice to minimize their market transaction costs. Subsequently, there will be a virtuous circle between population agglomeration, public goods provision and more market transactions, boosting the development of cities and villages. Hence, urbanization or rural development should come back to its essence of concentration and seek for a co-ordination between the market and government. In practice, the Chinese Central Government opts to use the market mechanism instead of coercive forces [10,31] to promote rural revitalization and new-type urbanization, emphasizing the importance of "people, land, capital and industry" for rural revitalization and an equal urban–rural relationship [32–34]. Since the Central Government pointed out "the decisive role of the market mechanism in resource allocation" in 2013, China continuously deepens the reform of the economic system with a focus on a market-oriented resource allocation system. In February 2022, the Central Government further emphasized the importance of the market mechanism in realizing efficient mobility and the reasonable allocation of production factors, as well as a balanced allocation of public resources between urban and rural areas.

Under these guidelines, many local governments began to introduce the market mechanism to rural construction based on context-based innovative institutions, such as "land tickets" in Chongqing, "land coupons" in Yiwu and "land reclamation coupons" in Henan. These local experiments, based on rural construction land consolidation (RCLC), have attracted much attention from academics. RCLC is widely considered to be a spatial problem-solving instrument for land management [35] to enhance efficient land use [23], as well as to improve rural production and living conditions [36], community building [37] and endogenous development capability [38]. However, it also has a profound impact on the villagers' production and lifestyle. The introduction of the market mechanism and private capitals to RCLC further complicate the relationship between the government, the market capital and villagers. Zhou et al. [39,40] pointed out that the resettlement of villagers to high-rise apartments is essentially a way to use rural residential land for urban expansion, and the local government and market capital have become the residual claimant of differential land rent.

Extant research points out the importance of the market mechanism for rural construction; however, such studies do not discuss how to integrate it into rural construction in detail. Thus, this paper establishes a framework of "institutional environment–governance structure–performance" based on the New Institutional Economics, using a case study and comparative analysis to illustrate the background, implementation and outcomes of the market-driven pattern. By taking a company-dominated RCLC project in Chengdu as a case study, we aim to identify the intrinsic dynamics among the local government, market capital and villagers in rural revitalization via asking two questions: how is the market-driven pattern implemented, and what are the differences between the market-driven and government-led patterns? It is hoped that this paper will improve our knowledge of the market-driven rural construction pattern and offer new guidance for the design and implementation of resource governance policies, and finally for rural revitalization.

In the following part, this paper first introduces some basic information of the study area and data collection process. Then, the detailed implementation process and outcomes of the market-driven rural construction scheme are demonstrated. Based on the discussion above, the next section compares the market-driven pattern with the government-led one from the perspective of "people, land, capital and industry". The last section presents the conclusion and suggestions.

2. Case Selection and Data Collection

2.1. Study Area

Fuhong town is 25 km away from Chengdu Municipality and has convenient access for transportation. In 2012, the town covered an area of 39.36 square kilometers, including arable land of 38,346 mu (about 25.564 square kilometers) and consisted of nine administrative villages. (Because the Fuhong town began coordinating the new-type urbanization and agricultural modernization through the comprehensive rural construction land consolidation in 2012, this information mainly reflects the situation circa 2012.) In particular, Fuhong town is located within the scope of the Longquan Mountain Tourist Area in Chengdu and owns a national-level scenic spot with more than 13,000 mou of apricot tree stands. In addition, it is adjacent to several Ancient Town tourism attractions. Despite better resource endowments and location, Fuhong town was still a traditional agricultural town with an agricultural population of 26,419 and about 3000 non-agricultural residents, and was one of the poorest towns in the Qingbaijiang District by 2012. Farmers' annual net income per capita was only RMB 4000. As Chengdu is a national pilot for Urban–Rural Coordination Reform, the town government grasped this political opportunity and cooperated with an external private investor on rural construction.

Fuhong town was chosen as a case example for following reasons. Firstly, all of the nine administrative villages were incorporated into a comprehensive RCLC project, while most similar projects in China were carried out within a single administrative village. Considering the potential for construction land consolidation in each village, the economies of scale of the project implementation and the town's development planning, the local government decided to optimize the layout of the rural space, especially the rural construction land through RCLC, ultimately to coordinate new-type urbanization and rural construction. Secondly, the local government introduced a private company into the RCLC project. As a new actor, this company was involved in the whole implementation processes from providing funds, preparing township development planning and constructing concentrated residence areas to developing rural industry. This means that the company participated fully in the rural and town construction, which is traditionally undertaken by the local government.

2.2. Data Collection Process

To understand the detailed transformation process of the rural construction pattern and the invisible relationships between the different stakeholders, we employed a case study method to elucidate the implementation and impacts of the market-driven pattern using qualitative and quantitative analyses. A group of eight researchers conducted an in-depth field investigation for 7 days during October 2016 and June 2017. Quantitative data were collected from semi-structured interviews and a questionnaire survey to validate the qualitative description and arguments. The interviewees included township-level officers, villager cadres, a private investor and some villagers.

For the township-level officers from the related departments and the villager cadres, the questions in the semi-structured interview concerned the general implementation process, including decision making, project management and their personal comments on the project. Specifically, the interview questions included the following: (1) Why did you decide to implement the rural construction through rural construction land consolidation? (2) Why did you decide to introduce market capitals? (3) How did you introduce this project to the villagers and what kind of policies did you explain to the villagers? (4) What roles do you think you or your department/organization played? (5) How did you protect villagers' rights and interests during the whole process?

To understand the investment incentives of the private company, questions in the semi-structured interview referred to how the company designed, planned and implemented the project and how the villagers were negotiated with. A questionnaire was also used to obtain some specific data about the project, including basic information regarding the investor, the quantity of demolition and reclamation, the construction standards of the concentrated residential areas, investment costs and returns, ways in which construction land quotas were saved and encountered problems.

For the affected villagers, 137 questionnaires were administered to a random sample. The questionnaire included seven sections: (1) interviewees' household demographic characteristics and changes in employment; (2) changes in household income and expenditure; (3) functions, areas and location of residential land and farmhouse, areas of resettlement house; (4) types and quantities of compensation obtained and costs undertaken; (5) changes in living environment; (6) the ownership and use of reclaimed homestead; (7) reasons for participation and satisfaction levels.

The age of the interviewees was mainly between 35 and 64 years old, with an average of about 60 years old. This is because we rarely found interviewees under 35 years old in the field investigation. Female interviewees accounted for 45.99%. To some extent, this reflects the general phenomenon of hollowing out and aging of the village, and the information provided by these people who lived in the village for a long time also better reflects the impacts of a RCLC project on their welfare. In addition, the majority of the participants were from agricultural households and were ordinary villagers, which avoids the impact of certain special factors, such as social capital and elite capture, on the participation willingness and project outcomes. The detailed characteristics of the interviewees are shown in Table 1.

Table 1. Basic characteristics of interviewees.

Indicator	Classification	Proportion (%)
Age	Under 34 years old	1.46%
	35–64 years old	59.85%
	Over 65 years old	37.96%
Gender	Female	45.99%
	Male	53.28%
Level of education	Illiteracy	29.93%
	Primary school	51.82%
	Junior high school	15.33%
	Senior high school	1.46%
	Higher education	0.73%
Type of household registration	Agriculture	93.43%
	Non-agriculture	5.84%
Marital status	Unmarried	2.92%
	Married	84.67%
	Divorced	0.73%
	Widowed	10.95%
Village cadre	Yes	5.84%
	No	93.43%
Member of CPC	Yes	5.84%
	No	93.43%

3. Institutional Background of Market-Driven Rural Construction

As a national pilot for Urban–Rural Coordination Reform, Chengdu has had rich experience in using rural construction land consolidation (RCLC) for rural construction since 2006. An RCLC project is composed of three areas: demolition and reclamation areas (chaijiu qu), resettlement areas (anzhi qu) and construction areas (jianxin qu). To ensure that the quantity of arable land does not decrease and the quantity of construction land does not increase, a RCLC project promotes rural space reconfiguration through demolition and reclamation, resettlement, construction quota transactions and utilization. Figure 1 shows the spatial mechanism of an RCLC. The first area refers to the reclamation of scattered rural construction land, which mainly consists of residential land, as arable land. By demolishing the villagers' scattered old farmhouses and reclaiming them as arable land, more arable land is created, which is more than the quantity the local government is required to preserve. This amount of arable land is registered as "newly-created construction land quotas". Then, the affected villagers are resettled into higher-density concentrated residential areas that occupy a smaller area of rural construction land than before. Thus, some construction land quotas are saved and can be used in the construction areas (jianxin qu). The saved construction land quotas can be used to convert some arable land in a desirable location near the town into rural commercial construction land for industrial development. They can also be sold to end-users of urban land through the Chengdu Agricultural Equity Exchange. During the whole process, the quantities of arable land and construction land remain unchanged within the town, but the rural construction land is more concentrated.

More importantly, different from being resettled to high-rise apartments under the traditional pattern and other villages' RCLC practices, there are often three kinds of resettlement options for villagers to choose from voluntarily in Chengdu: high-rise apartments in the township, high-rise apartments or relatively concentrated single houses in the village and monetized resettlement. Different resettlement options mean different benefits and costs for the villagers, the investor and the local government.

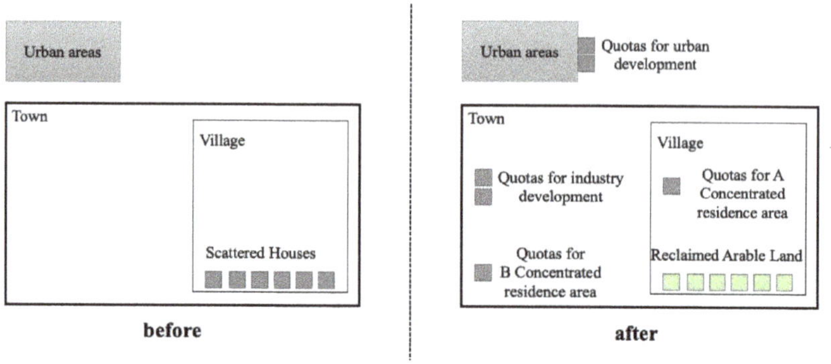

Figure 1. Spatial mechanism of rural construction land consolidation.

When villagers are resettled into high-rise apartments with a higher density than before, most find it difficult to adapt to this "upstairs" lifestyle because of the higher living costs and the inconvenience in terms of farming. Villagers not only have to pay for water, electricity and gas, but also cannot find space to store their farming tools and plant vegetable for self-sufficiency, and the scattered old farmhouses have now disappeared. However, higher residential density often means more saved quotas for transaction, higher returns and ensuing higher compensation. Villagers pay less to buy their new resettlement houses because of the lower construction costs and receive more monetary compensation because of more saved quotas. For the investor, more saved quotas mean more investment returns as a result of rural industry development indirectly, or from quota transactions directly. For the local government, it is much more convenient and cost-efficient to offer infrastructure and public services in a large-scale concentrated residence area. If villagers are resettled into single houses that occupy more rural construction land, fewer quotas are saved and less compensation is paid to villagers; living costs and farming costs are also lower than the first way, because villagers can be self-sufficient by growing vegetables near the homestead plots. What is more, the resettlement houses can be built by the villagers themselves or by the investor, since the construction costs of the former is relatively low. As for monetized resettlement, this is the simplest way. As long as villagers have a fixed house and stable income, they can choose monetized resettlement. Table 2 gives a brief comparison of these resettlement options.

Table 2. Comparison of different resettlement options.

	Resettlement in the Township	Resettlement in the Village		Monetized Resettlement
Type of resettlement house	High-rise apartments	High-rise apartments	single houses	No need for resettlement
Quantity of saved quotas	+++	++	+	++++
Farming radius and costs	++	+	+	No farming costs
Living costs	+++	++	+	Unchanged living costs
Construction costs of resettlement houses	+	++	Depending on construction standard	None
Monetized compensation	+++	++	+	++++
Degree of lifestyle change	+++	++	+	Unchanged lifestyle

Note: the greater the number of "+", the higher the degree.

4. Case Study of Market-Driven Rural Construction

4.1. Implementation Procedures of Market-Driven Pattern

Through semi-structured interviews with government officers, project managers and villager cadres, the implementation processes of the market-driven pattern, in which a private company plays a leading role, are summarized in Figure 2.

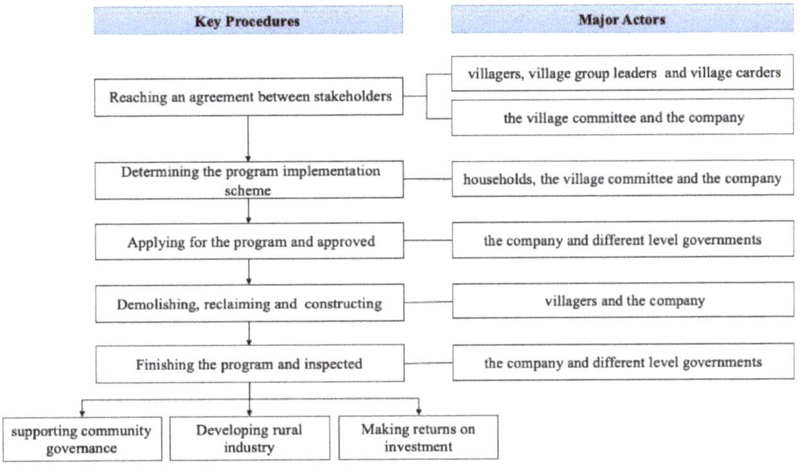

Figure 2. Project's key implementation steps.

(1) Reaching an agreement between the Village Committee and the private investor

With support and guidance from the local government, the company drafted a project scheme based on the land use situation and potential for land consolidation in each village. Then it introduced the implementation process, benefits and costs of the project to the villagers and village cadres, as well as communicating directly with the villagers to determine their willingness to participate and interest demands. After rounds of negotiation with villagers and democratic discussions within the village, the Village Committee (VC), representing all of the villagers, reached an agreement with the company regarding how to implement the RCLC project.

At this step, whether the villagers can reach a consensus on the implementation details is critical for a smooth implementation. Before contracting with the company, the village cadres took the village group as a basic unit for propaganda and mobilization. (In China, a villager cadre refers to a leader in an administrative village that usually consists of several natural villages.) The village cadres explained to every village group leader about key questions, such as "what is RCLC?" and "what are their costs and benefits, rights and obligations?", and then the group leaders delivered the information to every household. In the meantime, a formal announcement on these issues was publicized by the VC. After seven working days, the VC held a Villager Meeting to collect villagers' opinions on the project and discuss whether to participate in this project. If more than 80% of the households agreed to participate, the Village Council (cunmin yishihui, YSH) would subsequently organize a public discussion with every household representative about a range of issues including participation qualification, compensation standards, resettlement house construction planning and allocation procedures. Following this, a "Consultation Form" containing the discussion outcomes was publicized and delivered to every household to collect their opinions and confirmation in writing. When the final implementation rules had been agreed by more than two-thirds of the villagers or household representatives after repeated amendments, the villagers submitted some materials to the VC including a letter of attorney, application and commitment. Once these materials had been verified, the villagers and the VC reached a written agreement to formally confirm their participation. Importantly, villagers who refused to participate also had to submit a letter of commitment to avoid possible contradictions in the future. Figure 3 shows the democratic decision-making procedures in the village in brief.

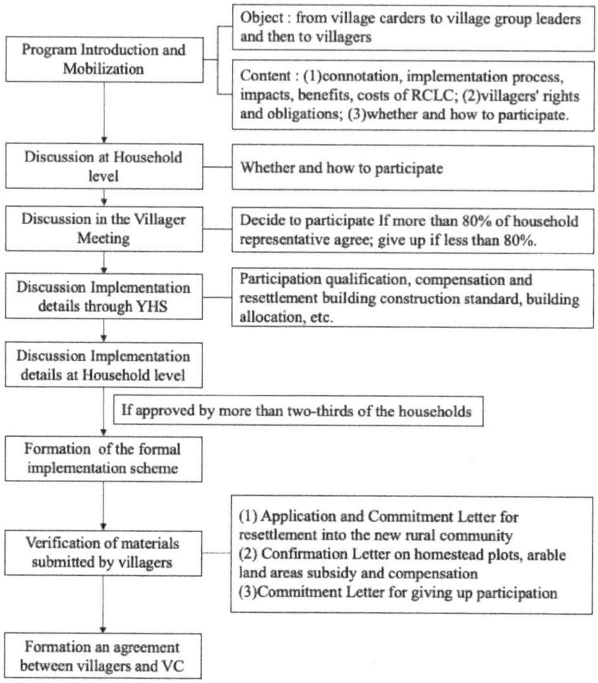

Figure 3. Democratic decision-making procedures in the village.

(2) Determining the implementation scheme

Based on the potential for land consolidation and villagers' willingness, the company proposed an implementation scheme focusing on resettlement and compensation standards. Three resettlement options were offered: ① six-story apartments in the newly-built township. Every villager was offered 30 square meters of floor space for living and 5 square meters of floor space for commercial purposes, in exchange for their old residential land. Taking a three-person family, for example, this family could get a resettlement house of 90 square meters and a commercial house of 15 square meters in principle. In order to use commercial houses as a whole and achieve a long-term income, the villagers do not use the 5 square meters of floor space by themselves but transfer the use rights to the company. The rights for use in this respect were transferred to the company to build and operate large-scale commercial buildings near the newly planned town; the villagers could receive an annual dividend according to their shares converted from the rights they relinquished, and the VC representing the rural collective obtained the ownership. ② Three-story apartments in the village. Every participant was entitled to a floor space of 35 square meters per capita for living. ③ Monetized resettlement. Villagers could voluntarily give up the 35-square meter resettlement houses and get RMB 45,500 per capita at the price of RMB 1300/m^2. In addition, all affected villagers could receive subsidies and compensation according to the standard of RMB 20,000 per mou (1 mou is equal to 667 square meters) for their demolished residential land.

(3) Applying for project approval, construction and inspection

From negotiating with villagers in Xianfeng Village (Xianfeng village is one of the nine administrative villages; we used it as an example because the implementation procedures and standards are similar in every village, and it participated early in the RCLC), 150 households (about 500 people) refused to participate. Finally, 66 households (204 people) chose to resettle in the village, and 651 households (2134 people) chose to resettle in the town. Since more than 80% households agreed to participate, the company prepared the final project scheme and submitted it for approval to different government departments. Once approved, the company began demolition, reclamation, resettlement and compensation. After the construction was complete, the company applied to relevant government departments to verify the quality of the new resettlement buildings and reclamations, meanwhile formulating a resettlement houses allocation scheme and helping to prepare the villagers to move.

(4) Making investment returns

According to the previous agreement, the company would bear all of the project costs including resettlement houses, infrastructure and public services, villagers' compensation, demolition and reclamation. In exchange, it would own all of the saved construction land quotas. For the saved quotas of 585.91 mu, 488.58 mu were bought back by the Chengdu Agricultural Equity Exchange and the remaining 97.33 mu were mainly used for industrial development in the town. Specifically, there were four ways to recover the investment costs and make profits. Firstly, the company built shops totaling 48,843.3 square meters on the ground floor of the high-rise resettlement apartments in the town, and these shops could be sold or rented to anyone, even non-collective members. Secondly, some arable land near the newly-built township was converted into rural commercial construction land using saved quotas to construct an industrial park and standardized factory buildings. According to the agreement with the VC, the company owns all of the saved quotas and associated rights to use the rural commercial construction land, thus, the company could transfer the land use rights in the industrial park for transfer fees or lease the factory buildings for rent. Thirdly, the company could directly sell some of the saved quotas to enterprises engaging in leisure agriculture, rural tourism, etc. Lastly, the saved quotas could be bought back by the Chengdu Agricultural Equity Exchange at a fixed price.

(5) Supporting community governance and rural industry development

The company was responsible for not only engineering the construction, but also community governance and industry development in cooperation with the local government. During the project implementation, the local government invested special financial funds in improving the infrastructure and public services of the newly planned township, and organized different community activities to help the villagers to adapt to their new lifestyles sooner. In addition, the company, in cooperation with the local government, successively introduced a number of enterprises engaging in modern agriculture, rural tourism and labor-intensive processing factories.

4.2. Impacts of Market-Driven Pattern on Rural Revitalization

By introducing private capital into rural construction, Fuhong town used a company-dominated RCLC as a platform to integrate rural revitalization and urbanization in situ. At present, Fuhong town has an established modern agricultural system led by planting roses and apricots, as well as a rural tourism service system based on leisure agriculture. In 2019, the town received more than 2.15 million tourists and RMB 165 million in tourism revenue. In addition, the town introduced some manual processing factories, providing non-agricultural employment opportunities for villagers. With the industry development, farmers' annual net income per capita also increased from RMB 4000 in 2012 to RMB 20,572 in 2017. In this section, we will focus on the villagers' welfare changes and their perceptions about the project, to reflect on the impacts of the market-driven pattern on rural revitalization.

(1) Impacts on rural industrial development

Our villager questionnaire contained two questions about rural industrial development. The first was: "Are there any measures taken to promote industrial development during the project implementation?". Most interviewees said that some measures had been taken, such as a newly established collective economic organization, an introduction of industrial enterprises, professional skills training for villagers and arable land circulation.

However, 37.74% of the interviewees said that they never heard of any measures being taken. The second question was: "How is your reclaimed residential land used?". After the villagers' scattered residential land was reclaimed as arable land, most arable land (41.94%) was idle, 25% of the villagers chose to cultivate by themselves, 12.90% of the villagers gave the land to their relatives or neighborhoods for free, and only 8.06% and 12.10% of arable land, respectively, was transferred to agricultural businesses and the collective economic organization. This shows that the arable land use efficiency decreased, although the quantity of arable land had not changed.

(2) Impacts on villagers' wealth

After the project implementation, the villagers' employment structure changed gradually and there was more non-agricultural employment available, which helped to improve their income. Reflected by the questionnaire, the proportion of the population engaging in agriculture decreased from 34.07% to 9.90%, and the proportion of those working in non-agricultural employment outside the town decreased slightly from 22.54% to 20.14%. Those in non-agricultural employment working in the town increased from 18.98% to 23.44%, and the proportion of self-employed villagers engaging in restaurant and agritainment work, etc., in the town increased from 0.68% to 3.82% (Table 3). These data demonstrate that the project slowed down the rural outmigration to some extent.

Table 3. Changes in household employment.

Employment Type	Before the Project	Percentage	After the Project	Percentage
Agriculture	201	34.07%	57	9.90%
Non-agricultural rural employment in town	112	18.98%	135	23.44%
Non-agricultural rural employment outside town	133	22.54%	116	20.14%
Self-employment	4	0.68%	22	3.82%
Others (students, full-time mothers, etc.)	140	23.73%	246	42.71%
Total	590		576	

Note: "Before the project" reflects the information before 2012, while "After the project" reflects the information in 2017, so the total number decreased from 590 to 576 because some people died during this period. Moreover, the total number is not 137 because these data contain employment situation information of interviewees' family members.

To further understand the reasons for the employment structure changes, we also asked two questions: "Do you think your employment changes relate to the project?" and "What do you think led to your employment changes? (Multiple choices are allowed)". N total, 55.12% of the interviewees thought that the project led to their employment changes. As for influential factors, 77.78% of the interviewees thought the changes to employment were not directly related to project. Table 4 shows the villagers' views on the reasons for the employment changes, reflecting that the impact of the project on promoting non-agricultural employment was not that obvious.

Table 4. Reasons for employment changes.

Question	Answers	Proportion
Reasons for your and your family members' employment changes	Changes in macroeconomic conditions	17.78%
	Changes in education level	0.00%
	Changes in professional skills	2.22%
	Changes in age	57.78%
	Secondary industry introduced after project	6.67%
	Development of tourist industry	11.11%
	Other reasons	4.44%

What is more, changes in living style and employment further affected household income and expenditure (Table 5). After the project, the average annual total income and expenditure of the households all increased, and the net income increased by RMB 74,238.94. From the perspective of income structure, except from agricultural income, other types of income increased. However, agriculture expenditure decreased in parallel with lower agricultural income, while other types increased significantly. As for reasons of income–expenditure change, only 43.73% of the villagers related the income increase with the project, while 65.97% thought the project increased their expenditure, especially daily consumption costs.

Table 5. Changes in household income and expenditure.

Income (per Capita)/RMB	Before Project	After Project	Income Gap between and after Project
1. Total income	62,719.58	136,958.52	74,238.94
1-1 Agricultural income	5969.56	2216.09	−3753.48
1-2 Operational income	4989.05	20,336.23	15,347.18
1-3 Wage	44,725.55	99,731.88	55,006.34
1-4 Transfer income	4427.99	6512.71	2084.72
1-5 Property income	248.91	5351.38	5102.47
1-5-1 House rental income	21.90	221.74	199.84
1-5-2 Dividend income from collective economic organizations or cooperatives	0.00	28.99	28.99
1-5-3 Rent for arable land circulation	193.43	2936.74	2743.31
1-5-4 Insurance income	820.12	3812.25	2992.13
2. Total expenditure	39,874.58	79,619.51	39,744.93
2-1 Agricultural production expenditure	4079.71	724.64	−3355.07
2-2 Operational expenditure	2153.62	12,443.48	10,289.86
2-3 Consumption expenditure	33,589.57	66,982.17	33,392.61
2-3-1 For daily food	13,911.45	30,691.30	16,779.86
2-3-2 For water, electricity, gas, etc.	1806.38	5533.62	3727.25
2-3-3 For medicine	4484.49	10,748.84	6264.35
2-3-4 For education	4300.00	5063.04	763.04
2-3-5 For social communication (marriage, funeral, friends, etc.)	8972.46	10,730.43	1757.97
2-3-6 For estate management	0.00	11.30	11.30
2-4 Insurance expenditure	285.30	341.91	56.60

Note: this table only lists some key income and expenditure types in the leftmost column, and ignores some types that only 1–2 people had.

(3) Impacts on living environment

For villagers, both house quality and residential safety were enhanced when they were resettled to concentrated residential areas. Their living environment was significantly improved and they enjoyed similar public facilities to urban residents (Table 6). The concentrated residential areas were equipped with not only water, electricity, gas, etc., but also basic service facilities such as garbage and sewage treatment stations and a kindergarten. There were also cleaning and security personnel responsible for the community's environment and safeguarding. However, the villagers' residential land areas decreased significantly from 368.97 square meters per household to 98.85. We argue that the decrease in residential land is just a phenomenon and trend during the rural reconstruction, and is not a problem. The key issue behind it refers to how to balance the relationship between residential land demand and land use efficiency.

(4) Impacts on protection of villagers' rights and interests

The questionnaire survey also provided a way to understand the impacts of the project on the degree of protection of villagers' rights. In general, over 80% of the interviewees felt satisfied with the living environment, employment and economic conditions, land readjustment and other factors (see Table 7). Democratic negotiation and decision making were practiced in the project: some key issues were highlighted and discussed with all participants, and then the outcomes were agreed by more than two-thirds of the participants (see Tables 8 and 9). However, villagers' rights to know about and make decisions regarding these issues were not quite protected enough. In particular, many interviewees pointed out problems such as unreasonable compensation, monetary compensation in arrears and longer farming radius. Thus, some villagers thought that the company and local government encroached on their interests and that the village cadres did not protect their rights very well (see Table 10).

Table 6. Changes to infrastructure and public services.

	Before Project		After Project	
	Yes	No	Yes	No
Are there street lamps on the village's main road?	14.60%	85.40%	98.54%	1.46%
Is there a waste disposal station in the village?	8.03%	91.97%	100.00%	0.00%
Is there a sewage treatment plant in the village?	0.00%	100.00%	96.35%	3.65%
Are there shops in village?	37.96%	62.04%	94.89%	5.11%
Is there a kindergarten in the village?	5.11%	94.89%	88.32%	11.68%
Are there entertainment or fitness facilities in the village?	0.73%	99.27%	88.32%	11.68%
Are there medical and health facilities in the village?	31.39%	68.61%	96.35%	3.65%
Is there a bus station in the village?	8.76%	91.24%	99.27%	0.73%
Are there cleaners employed in the village?	2.19%	97.81%	99.27%	0.73%
Are there security staff employed in village?	17.52%	82.48%	99.27%	0.73%

Table 7. Interviewees' general satisfaction with the project.

Degree of Satisfaction with the Whole Implementation Process	Frequency	Ratio of Frequency (%)
−2 (not satisfied)	4	2.92
−1 (below average)	23	16.79
0 (average)	68	49.64
1 (above average)	29	21.17
2 (completely satisfied)	13	9.49
Total	137	100.00

Note: From −2 to 2, the degree of satisfaction increases gradually.

Table 8. Interviewees' decision-making rights.

		Standard for Compensation (Ratio of Frequency%)	Construction Standard of Resettlement House (Ratio of Frequency%)	Allocation Standard of Resettlement House (Ratio of Frequency%)
Whether to ask for your family's opinion	No	8.03%	21.90%	13.97%
	Yes, but disagree	4.38%	2.92%	2.21%
	Yes, and somewhat agree	61.31%	48.91%	49.26%
	Yes, and fully agree	26.28%	26.28%	34.56%
Decision-making method	Through the VC	59.85%	58.99%	63.57%
	Through the Village Meeting	7.30%	7.91%	15.71%
	Through the government or company	11.68%	15.83%	7.14%
	Do not know how to make decisions	21.17%	17.27%	13.57%

Table 9. Interviewees' right to know.

Key Issues	Ratio of Frequency (%): Which Rules did your Household Know About? (Multiple Choice)	Ratio of Frequency (%): Which Rules Were You Notified or Informed About? (Multiple Choice)
Standard for resettlement and compensation	87.59%	24.09%
Construction standard of resettlement house	68.61%	22.63%
Land property rights adjustment	35.04%	8.76%
Land reclamation	56.20%	9.49%
Allocation standard of resettlement house	74.45%	13.87%
Funding use	0.00%	2.19%
Project supervision	5.84%	3.65%
Know little about these	10.22%	75.18%

Table 10. Interviewees' evaluation on the company, the local government and village cadres.

		Frequency	Ratio of Frequency (%)
Does the company encroach on your rights?	Yes	72	52.55
	No	65	47.45
Does the local government (or investor) encroach on your rights?	Yes	51	37.23
	No	86	62.77
Do the village cadres protect your rights?	Yes	61	44.53
	No	76	55.47

One interviewee told us that for the villagers resettled in the newly planned township, the resettlement standards were a floor space of 30 square meters per capita for living and 5 square meters per capita for commercial purposes. However, these were not the original standards. At first, the company told the villagers that the 5 square meters would be compensated with a payment of RMB 500 per square meter (that is, RMB 2500 per capita), but the villagers did not receive the money for a long time. To pacify them, the company said that the 5 square meters would be converted into shares instead and used in the town-level industry park, and then the villagers would receive dividends every year. Another other problem highlighted by many participants concerned house allocation. Due to the existence of shops on the ground floor of the resettlement apartments, villagers who were allocated homes on the second floor by drawing lots actually lived on the third floor. This meant that they needed to pay more to buy their resettlement houses because the construction costs varied across the different floors. For example, the construction costs of the second and third floor were RMB 280/m^2 and RMB 240/m^2, respectively; thus, the households living on the third floor needed to pay an extra RMB 40/m^2 to the investor.

In sum, the production, living and ecological spaces in the countryside were optimized after the project. By reconfiguring the layout of rural construction land, the land demands for rural industrial development were guaranteed. Although the impacts of the project on changes to employment structure were not that obvious at first, the levels of agricultural modernization and industrial integration were higher than before, which established a foundation for future development. As reflected by the field investigation, the villagers' job opportunities and income sources became diversified with the upgrading of the industrial structure, and some migrant workers even returned for entrepreneurship and employment in the town. For the local government, it was more convenient to provide infrastructure and public services in more concentrated residential areas.

5. Discussion

Drawing on the field investigation results, this section aims to identify the key issues related to rural construction and elucidate how an alternative approach to the government-led pattern functions. Further, we attempt to analyze the relationships among the main actors under the different patterns.

5.1. Comparison of Two Patterns

(1) A traditional pattern of government-led rural construction

Rural construction led by the government is, in essence, a tool to deal with urban sprawl; that is, local governments convert rural land into state-owned land through land expropriation, and then transfer it to end-users by bidding, auction, or listing in the primary land market at different prices. The logic behind this is that the local government depends on monopoly power for land conversion and coercive power to acquire the land from rural collectives at a low price, and then to transfer the land to those who require it at a high price. During this process, the local government obtains high land-transfer fees while the villagers receive a lower amount of compensation, and the government normally does not pay much attention to rural construction once the landless peasants are resettled and compensated. What is worse, the landless peasants cannot find non-agricultural employment from urban industry development under this pattern, because the local government prefers the introduction of capital- or technology-intensive enterprises and foreign-owned businesses rather than small and medium-sized factories. However, the former is far beyond the villagers' skills, and the latter could offer job opportunities for villagers. As a result, the urban–rural gap is likely to be widened further, and rural outmigration cannot be reversed due to the neglect of rural industry development, creating a "hollowing-out" of villages and inefficient land use [23,24]. It does not mean that the local government should not introduce these kinds of enterprises, but should pay attention to the affected villagers' sustainable livelihoods.

Considering the sustainability of this pattern, one important thing is the funding source for rural construction. Often, rural construction funds come from the budgetary fiscal revenues and land transfer fees. To attract industrial enterprises, which are main sources of government tax revenues, local governments tend to lower the price of industrial land. This approach not only makes some enterprises care little about land use efficiency and even engage in land speculation, but also increases the government's financial burden. The extra financial burden caused by unreasonably low industrial land prices is always compensated by higher commercial land transfer fees from real estate developers. This means that rural construction funds are highly dependent on the sustainable prosperity of the real estate market. However, increasingly high house prices caused by land transfer fees has depressed housing demand, and coupled with the regulation from central government, the real estate market is gradually depressed, which makes this funding source unsustainable.

Briefly speaking, this pattern itself and following villagers' livelihoods are unsustainable. With the aim of coordinating the urban–rural relationship and rural revitalization, it is urgently necessary to transform government-led rural construction patterns.

(2) Market-driven rural construction

To relieve the financial burden on rural construction, local governments turn towards encouraging market capitals to invest in rural areas. Typically, there are two fund sources in the market-driven pattern: the rural collective's own funds, or external funds from the private sector. The former means that the rural collective mortgages the use right of rural construction land or saved construction land quotas to directly receive loans from local banks. The latter means that a private company pays the necessary funds in advance in exchange for the ownership of saved quotas or the right to utilize rural construction land. Obviously, land is key for rural construction funds under both patterns, but the difference from the traditional pattern is that these fundraising methods would not burden the local government as much and would raise resource use efficiency. In the market-driven pattern, investors aiming to maximize profits will take market demands, local conditions and benefit–cost analysis into consideration to choose an appropriate land use pattern and industry type. For example, rural areas with distinct natural resources are suitable for sightseeing agriculture; rural areas in a desirable location may develop industrial parks to attract small and medium-sized enterprises.

Another important difference is that the land is still owned collectively in the market-driven pattern and land use rights are transferred to the end-users through land circulation, lease or shareholding [41]. There are two ways to use rural construction land for commercial purposes: one is that the rural collective develops non-agricultural industry by itself, an approach exemplified by Zhenggezhuang Village in Beijing. The other is that a private investor rents the land with the right to develop industry, and the rural collective is passively involved in industrialization, a method represented by the Jiaolong Industrial Park in Chengdu. However, both of these approaches depend highly on the land's location and scale, because not all land parcels are suitable for industrial development. Now, the case village offers a third way to combine rural construction and new-type urbanization by implementing RCLC. The rural construction land is abstracted into construction land quotas, which represent the right to convert arable land into construction land. After deducting the quotas for building concentrated residential areas, most quotas are used to convert arable land for rural industrial development. Land supply in this way is more flexible in scale and location, as well as more cost-friendly for the end-users, and ultimately can better contribute to rural sustainable development due to a virtuous interaction between industrial development and population agglomeration. On the one hand, enterprises entering into the village- or town-level industrial parks are often labor-intensive small- or medium-sized enterprises that are difficult to find room for in urban industrial parks. However, these enterprises can provide many job opportunities for the low-skilled labor force in rural areas, which would improve the possibility of population agglomeration. On the other hand, resettling villagers from different villages into concentrated residential areas in the town will also promote population agglomeration, which in turn supports rural industrial development.

In sum, an intuitive difference between government-led and market-driven rural construction is an institutional arrangement, that is, land expropriation and RCLC. This difference subsequently effects the main actors, funding sources, land use efficiency and, ultimately, the rural development approach. Table 11 compares some key features of the two patterns discussed above. Land expropriation, as an urban-oriented institutional arrangement, aims to fulfill land demands for urban expansion. To realize "urban modernization" and industrialization as local governments prefer, they tend to monopolize land conversion rights. As land is an important funding source, this monopoly power, which excludes other market entities from land access, inevitably leads to the local government playing a leading role in the whole implementation procedure including fundraising, planning, negotiation and construction. When the local government relaxes this monopoly power and gives other actors access to land, the market actor will have economic incentives to invest in rural areas. Thus, to some extent, RCLC is an innovative institutional arrangement that is jointly chosen by the government and the market forces [42–44]. In the RCLC project, land users do not wish to obtain state-owned land, since rural construction land can also fulfill their demands due to the removal of controls and restrictions on rural land management. Hence, RCLC is a more flexible and responsive way to meet the land demand from small and medium-sized or labor-intensive enterprises, since these enterprises are normally overlooked by the local government. In next section, we will further discuss how these different institutional arrangements impact the rural construction pattern and the relationships among the main actors.

Table 11. Comparisons between government-led and market-driven patterns.

Key Issues	Government-Led Pattern	Market-Driven Pattern
Industry development	Neglects rural industry development	Emphasizes introduction of small and medium-sized or labor-intensive enterprises
Population agglomeration	Has little impact on rural population outmigration	Makes villagers more concentrated and attracts migrant workers back
Land acquisition	Compulsory land expropriation and rural land is changed to state-owned land	Land use rights are transferred and land is still owned collectively
Fundraising	Land transfer fees from estate developers and land mortgage loans	Market capital

5.2. *Characteristics of Market-Driven Rural Construction*

Our empirical research reveals that rural construction in China is now more complicated than a top-down land expropriation led by the local government. Based on the experiences of the village used in this case study, new development patterns and relationships among the main actors in the transformative process of rural construction are identified. Market-driven rural construction, which is supported by the local government and organized by private companies, has the following characteristics.

Firstly, the decision-making mechanism is transformed from administrative control to a market mechanism. Due to the top-down land control in China, it is impossible to form a completely free land market. "Market mechanism" here means that some market tools are introduced into land governance and resources allocation, allowing private companies to participate in rural construction through RCLC or construction land quota transactions. The two patterns are characterized by different decision-making mechanisms. In the government-led approach, key issues such as construction land quota price, villagers' compensation standards and resettlement options are decided by administrative orders based on bureaucratic hierarchy [45] and villagers have little room for negotiation. In the market-driven pattern, all decisions are guided by the price mechanism and benefit–cost analysis, and key issues are negotiated by the company and rural collectives. Due to the lack of coercive power such as that held by the government, the company must depend on economic incentives to persuade villagers to participate in an RCLC project. For the investor, direct negotiations help to understand villagers' actual demands and expectations, to avoid possible contradictions and struggles in the future. For the villagers, the market-driven pattern with little government intervention endows them with more bargaining power to express their true expectations and protect their interests. Villagers then have more choice about whether to participate or not, how to be resettled, etc.

Secondly, the local governments' role is transformed from being a direct participant to an indirect supervisor and supporter. The introduction of market capital into rural construction not only reduces the local governments' financial burden, but also relieves them from specific affairs such as scheme design, project implementation and communication with villagers. By delivering these responsibilities to the company, the local government avoids direct interest contradictions with villagers, and can better mediate any conflict between villagers, village cadres and the company [12]. This transformation reflects thew fact that governments' rural development strategies become rural-oriented instead of urban-biased. Through cooperation with a private company, the local government can offer better infrastructure and public services at a relatively low cost, to enhance villagers' social welfare.

Thirdly, better protection of the rural collective's land development rights and interests takes place. In the government-led pattern, rural land is expropriated as state-owned land for urban development, meaning that the rural collective permanently loses land development rights and potential development space. However, villagers only receive one-time monetary compensation, which cannot make up for their indirect losses or opportunity costs of losing future development opportunities. Although the villagers also lose their land development rights under the market-driven pattern, most saved construction land quotas are used for rural industry development in the village or town. Hence, the employment opportunities provided by industrial development could compensate villagers for the loss of opportunity costs associated with losing their land development rights. In addition, the rural collective does not lose land completely, because the land is still owned collectively and the company just obtains land use rights for a period of time.

Lastly, there is a balance of villagers' willingness and implementation costs using an appropriate resettlement radius. An economy of scale in terms of industrial development and public services can only be achieved with a population concentration of a certain scale. Concentrated residential areas for resettling landless peasants are usually located near their original village or near the city. For the former, a small-scale population concentration is obviously not conducive to industrial development. Without industrial development, some key problems such as rural outmigration are still not resolved, meaning that rural construction may become an action of "building a new hollowed village". For the latter, huge differences in living habits and ideas will make it difficult for villagers to adapt to an urban lifestyle. Thus, it is important to choose an appropriate resettlement radius. The innovation of the market-driven pattern lies in its flexibility in resettlement options. By offering villagers different resettlement options and economic incentives, most villagers from different villages can be resettled into the more concentrated residential areas in the town, which solves both of the abovementioned resettlement problems.

6. Conclusions

The market-driven rural construction approach mentioned in this study differs from the traditional government-led pattern, whereby a private company dominates the key construction procedures and integrates population agglomeration, rural space restructuring and industry development through RCLC. The contributions of this pattern to rural sustainability are follows: (1) it makes up for insufficient government funding for rural construction by encouraging market capital. It is also conducive to attracting more market capital to engage in rural construction with the support of government funds. (2) It provides land for rural industry development in a responsive and flexible way, and the industry development can, in turn, offer diversified employment opportunities for villagers to increase their operational or property income. (3) It enhances villagers' governance ability by involving them in the decision-making process.

Under this pattern, different actors realize their interests in a cooperative way. For villagers, their living environment and employment opportunities are improved after the project. They are included in the decision-making processes of design, construction and supervision, and thus have more choice of options to better secure their interests and rights. For private investors, they can gain investment returns through construction land quota transactions, or obtain desirable land use rights in a relative low-cost and flexible way. For local governments, a private company dominates the main implementation procedures as an investor and a developer, which reduces their financial burden and political risks.

However, the path dependency of the institutional change implies that it is not easy to transform from government-led to market-driven rural construction. Both local governments and villagers need time to adapt to this newly emerging pattern, and not all governments embrace this transformation. The fact that some affected villagers felt dissatisfied with the project in our field investigation suggests that the transformation to a market-driven pattern is a gradual process. Yet, the experience in the case study demonstrates that, irrespective of the diversity and complexity of the local circumstances, it is possible and significant to explore a market-driven rural construction that underlines institutional change and government transformation. Accordingly, this paper provides two suggestions to better implement the market-driven pattern. Firstly, the upper-level government should encourage context-based innovative institutions and experiments to diversify participants and financial sources for rural construction. For the rural construction land system, a more liberal transaction mechanism of construction land quotas is necessary. Secondly, institutional barriers hampering the cross-village flow of production factors within the country should be relaxed. Local governments should enhance the open degree of rural collectives, allowing rural labor force and land flow between different villages based on the principle of "Linking Land to People", that is, construction land quotas are allocated to places with population inflow.

Obviously, whether this local experiment in our study can be adapted to other rural areas requires further discussion with more empirical cases. In addition, this paper only focuses on the relationship between the local government, the private investor and villagers in this new pattern, and thus, two interesting questions arise that deserve further study: the first is how the transformation of the rural construction pattern influences the grassroots governance structure in the village, and the second is about the reconfiguration of rural land property rights during this transformation.

Author Contributions: Conceptualization, Y.Z. and J.L.; methodology, Y.Z.; validation, Y.Z., J.L. and X.K.; formal analysis, Y.Z.; investigation, Y.Z.; resources, Y.Z. and X.K.; writing—original draft preparation, Y.Z.; writing—review and editing, Y.Z. All authors have read and agreed to the published version of the manuscript.

Funding: This research received no external funding.

Conflicts of Interest: The authors declare no conflict of interest.

References

1. Gao, J.; Wu, B. Revitalizing traditional villages through rural tourism: A case study of Yuanjia Village, Shanxi Province, China. *Tour. Manag.* **2017**, *63*, 223–233. [CrossRef]
2. Mohabir, N.; Jiang, Y.; Ma, R. Chinese floating migrants: Rural-urban migrant laborer's intentions to stay or return. *Habitat Int.* **2017**, *60*, 101–110. [CrossRef]

3. Tan, S.; Heerink, N.; Kruseman, G.; Qu, F. Do fragmented landholdings have higher production costs? Evidence from rice farmers in Northeastern Jiangxi province, P.R. China. *China Econ. Rev.* **2008**, *19*, 347–358. [CrossRef]
4. Gomes, E.; Banos, A.; Abrantes, P.; Rocha, J.; Kristensen, S.B.P.; Busck, A. Agricultural land fragmentation analysis in a peri-urban context: From the past into the future. *Ecol. Indic.* **2018**, *97*, 380–388. [CrossRef]
5. Pašakarnis, G.; Maliene, V. Towards sustainable rural development in Central and Eastern Europe: Applying land consolidation. *Land Use Policy* **2010**, *27*, 545–549. [CrossRef]
6. Chigbu, U.E. Village renewal as an instrument of rural development: Evidence from Weyarn, Germany. *Community Dev.* **2012**, *43*, 209–224. [CrossRef]
7. Mao, D.; Wang, P. The Evolution of Western Restructuring Research. *Sociol. Stud.* **2014**, *1*, 194–216. (In Chinese)
8. Patrick, K.; Kundolf, S.; Mettenberger, T.; Tuitjer, G. Rural regeneration strategies for declining regions: Trade-off between novelty and practicability. *Eur. Plan. Stud.* **2018**, *26*, 229–255.
9. Jiao, B. Experience and Enlightenment of modern rural construction in the suburbs of Tokyo, Japan—Taking Sanying city and Huiyuan village as examples. *Shanghai J. Econ.* **2007**, *10*, 107–112. (In Chinese)
10. Chen, X.; Xin, X. The core of China's rural revitalization: Exerting the functions of rural area. *China Agric. Econ. Rev.* **2019**, *12*, 1–13. [CrossRef]
11. Long, H.; Liu, Y.; Li, X.; Chen, Y. Building new countryside in China: A geographical perspective. *Land Use Policy* **2010**, *27*, 457–470. [CrossRef]
12. Wong, S.; Dai, Y.; Tang, B.; Liu, J. A new model of village urbanization? Coordinative governance of state-village relations in Guangzhou City, China. *Land Use Policy* **2021**, *109*, 105500. [CrossRef]
13. Liu, R.; Jiang, J.; Yu, C.; Rodenbiker, J.; Jiang, Y. The endowment effect accompanying villagers' withdrawal from rural homesteads: Field evidence from Chengdu, China. *Land Use Policy* **2021**, *101*, 105107. [CrossRef]
14. Lai, Y.; Wang, J.; Lok, W. Redefining property rights over collective land in the urban redevelopment of Shenzhen, China. *Land Use Policy* **2017**, *69*, 485–493. [CrossRef]
15. Song, J. Official relocation and self-help development: Three housing strategies under ambiguous property rights in China's rural land development. *Urban Stud.* **2008**, *52*, 121–137. [CrossRef]
16. Long, H. Land consolidation: An indispensable way of spatial restructuring in rural China. *J. Geogr. Sci.* **2014**, *24*, 211–225. [CrossRef]
17. Yan, S.; Wu, Q. Incoordination between urban expansion and urban population growth and their interaction. *China Popul. Resour. Environ.* **2016**, *26*, 9. (In Chinese)
18. Yang, X.; Lu, X. Driving factors of urban land urbanization in China from the perspective of spatial effects. *China Popul. Resour. Environ.* **2021**, *31*, 9. (In Chinese)
19. Wang, H. Political and economic consequences of government-led urbanization in central and Western China. *J. Northwest AF Univ.* **2016**, *2*, 103–109. (In Chinese)
20. Kan, K. Creating land markets for rural revitalization: Land transfer, property rights and gentrification in China—ScienceDirect. *J. Rural. Stud.* **2021**, *81*, 68–77. [CrossRef]
21. Liu, P.; Ravenscroft, N. Collective action in implementing top-down land policy: The case of Chengdu, China. *Land Use Policy* **2017**, *65*, 45–52. [CrossRef]
22. Ma, Y.; Chiu, R.L. Governing rural redevelopment and re-distributing land rights: The case of Tianjin. *Land Use Policy* **2018**, *72*, 533–546. [CrossRef]
23. Zhou, Y.; Li, Y.; Xu, C. Land consolidation and rural revitalization in China: Mechanisms and paths. *Land Use Policy* **2020**, *91*, 104379. [CrossRef]
24. Liu, Y. Introduction to land use and rural sustainability in China. *Land Use Policy* **2018**, *74*, 1–4. [CrossRef]
25. Liu, Y.S. Research on the urban-rural integration and rural revitalization in the new era in China. *Acta Geogr. Sin.* **2018**, *73*, 637–650. (In Chinese)
26. Yang, Y.; Liu, Y.; Li, Y.; Li, J. Measure of urban-rural transformation in Beijing-Tianjin-Hebei region in the new millennium: Population-land-industry perspective. *Land Use Policy* **2018**, *79*, 595–608. [CrossRef]
27. Liu, Y.; Liu, Y.; Chen, Y.; Long, H. The process and driving forces of rural hollowing in China under rapid urbanization. *J. Geogr. Sci.* **2010**, *20*, 876–888. [CrossRef]
28. Cai, Y.; Selod, H.; Steinbuks, J. Urbanization and land property rights. *Reg. Sci. Urban Econ.* **2018**, *70*, 246–257. [CrossRef]
29. Li, Y.; Jia, L.; Wu, W.; Yan, J.; Liu, Y. Urbanization for rural sustainability-Rethinking China's urbanization strategy. *J. Clean. Prod.* **2018**, *178*, 580–586. [CrossRef]
30. Mumford, L. The city in history: Its origins, its transformations, and its prospects. *J. Aesthet. Art Crit.* **1961**, *67*, 5.
31. Huang, Y.; Hui, E.C.; Zhou, J.; Lang, W.; Chen, T.; Li, X. Rural Revitalization in China: Land-Use Optimization through the Practice of Place-making. *Land Use Policy* **2020**, *97*, 104788. [CrossRef]
32. Li, Y.; Westlund, H.; Liu, Y. Why some rural areas decline while some others not: An overview of rural evolution in the world. *J. Rural. Stud.* **2019**, *68*, 135–143. [CrossRef]
33. Han, J. How to promote rural revitalization via introducing skilled labor, deepening land reform and facilitating investment? *China Agric. Econ. Rev.* 2020; *ahead-of-print*. [CrossRef]

34. Guo, Y.; Liu, Y. Poverty alleviation through land assetization and its implications for rural revitalization in China. *Land Use Policy* **2021**, *105*, 105418. [CrossRef]
35. Rao, J. Comprehensive land consolidation as a development policy for rural vitalisation: Rural In Situ Urbanisation through semi socioeconomic restructuring in Huai Town. *J. Rural. Stud.* **2020**. [CrossRef]
36. Li, Y.; Fan, P.; Liu, Y. What makes better village development in traditional agricultural areas of China? Evidence from long-term observation of typical villages. *Habitat Int.* **2019**, *83*, 111–124. [CrossRef]
37. Wu, B.; Liu, L. Social capital for rural revitalization in China: A critical evaluation on the government's new countryside programme in Chengdu. *Land Use Policy* **2019**, *91*, 104268. [CrossRef]
38. Zhou, L.; Zhang, W.; Fang, C.; Sun, H.; Lin, J. Actors and network in the marketization of rural collectively-owned commercial construction land (RCOCCL) in China: A pilot case of Langfa, Beijing. *Land Use Policy* **2020**, *99*, 104990. [CrossRef]
39. Zhou, F.; Wang, S. Upstairs farmers and capital to the countryside: A sociological study of Urbanization. *Soc. Sci. China* **2015**, *1*, 18. (In Chinese)
40. Jiao, C.; Zhou, F. "Capital to the countryside" and the reconstruction of villages. *Soc. Sci. China* **2016**, *1*, 100–116. (In Chinese)
41. Liu, S. Land capitalization and rural urbanization: A survey of Beijing city's three mode. *China Open. J.* **2011**, *2*, 17–22. (In Chinese)
42. Gu, S.; Yang, W. Reflection on the five biases in the current development of Urbanization. *Chin. J. Popul. Sci.* **2012**, *3*, 2–8. (In Chinese)
43. Tu, S.; Long, H.; Zhang, Y.; Ge, D.; Qu, Y. Rural restructuring at village level under rapid urbanization in metropolitan suburbs of China and its implications for innovations in land use policy. *Habitat Int.* **2018**, *77*, 143–152. [CrossRef]
44. Tian, Y.; Qian, J.; Wang, L. Village classification in metropolitan suburbs from the perspective of urban-rural integration and improvement strategies: A case study of Wuhan, central China. *Land Use Policy* **2021**, *111*, 105748. [CrossRef]
45. Zhang, B.; Niu, W.; Ma, L.; Zuo, X.; Kong, X.; Chen, H.; Zhang, Y.; Chen, W.; Zhao, M.; Xia, X. A company-dominated pattern of land consolidation to solve land fragmentation problem and its effectiveness evaluation: A case study in a hilly region of Guangxi Autonomous Region, Southwest China. *Land Use Policy* **2019**, *88*, 104115. [CrossRef]

sustainability

Article

A Study on the Factors Influencing Farmers' Intention to Revitalize Idle Homesteads Based on Improved TPB Framework—Analysis of the Moderating Effect of Farmer Differentiation

Mengyuan Lu [1,*], Bin Guo [2], Guohua Chen [1], Lei Yuan [2], Runnan Xing [2] and Yu Huang [2]

1 School of Management, Xi'an University of Architecture and Technology, Xi'an 710055, China
2 School of Public Administration, Xi'an University of Architecture and Technology, Xi'an 710055, China
* Correspondence: lumengyuan@xauat.edu.cn

Abstract: Under the dual structure of urban and rural lands, revitalizing rural idle homesteads is an effective measure for reducing resource waste and improving the efficiency of rural land use. Therefore, the intention of farmers, as the primary decision-makers in revitalizing rural idle homesteads, is the key to the success of revitalizing idle homesteads. With an analytical framework based on the theory of planned behavior (TPB), this study used multiple linear regressions to analyze the survey data (N = 680). The results showed that attitude toward the behavior (AB), subjective norms (SN), and perceived behavioral control (PBC) had significant positive effects on the farmers' intention to revitalize, with SN, PBC, and AB in descending order of influence. It confirmed that the TPB applies to the study of farmers' intention to revitalize idle homesteads in the context of China. In addition, this study focused on the social phenomenon of farmer differentiation, which is prominent in the urbanization process. It emphasizes the moderating effect of farmer differentiation on the relationships of "attitude toward the behavior–intention to revitalize," "subjective norm–intention to revitalize," and "perceived behavioral control–intention to revitalize," and further improves TPB. The present empirical study using hierarchical regression found that the deeper the differentiation of farmers, the stronger the effective influence of AB, SN, and PBC on farmers' intention to revitalize idle homesteads. Therefore, it is suggested that the Chinese government should enhance farmers' intention to revitalize by cultivating a positive attitude toward the behavior, strengthening the positive influence of subjective norms on farmers, and enhancing farmers perceived behavioral control. Furthermore, it is more important to pay full attention to the phenomenon of farmer differentiation and design a revitalization policy according to the differences in sensitivity of different types of farmers to attitudes toward the behavior, subjective norms, and perceived behavioral control.

Keywords: farmers' intention to revitalize idle homesteads; improved theory of planned behavior (TPB); farmer differentiation; moderating effect

Citation: Lu, M.; Guo, B.; Chen, G.; Yuan, L.; Xing, R.; Huang, Y. A Study on the Factors Influencing Farmers' Intention to Revitalize Idle Homesteads Based on Improved TPB Framework—Analysis of the Moderating Effect of Farmer Differentiation. *Sustainability* **2022**, *14*, 15759. https://doi.org/10.3390/su142315759

Academic Editors: Dianfeng Liu, Wenwu Tang and Jianxin Yang

Received: 26 October 2022
Accepted: 22 November 2022
Published: 26 November 2022

1. Introduction

Population and land are the essential elements that constitute the national situation and power, which restrict and influence social and economic development to a considerable extent. The rapid development of urbanization has brought about social and economic development and has changed the relationship between people and land in rural areas—the most prominent of which is the use of homesteads. Homesteads belong to rural (collective) construction land, a type of land exclusively used by farmers to build residential houses and ancillary facilities, which are owned by farmers and village collectives. Before 2019, rural homesteads were not legally allowed to house migrants. Therefore, with the migration of the agricultural population to cities and further urbanization, rural homesteads have not decreased but increased. A large number of homesteads are also shown to be idle [1,2].

According to China's 2017 National Land Consolidation Plan (2016–2020), as of the end of 2017, the area of rural homesteads in China was approximately 191,333 km^2. With the migration of nearly 15 million agricultural populations and urbanization every year, the idle area has reached 20,000 km^2 [3,4]. As a developing country, it is necessary for China to revitalize rural idle homesteads to promote urbanization and rural revitalization and increase the property income of farmers under the constraints of limited total urban construction land and the dual structure of the land. The specific approach is that rural collective economic organizations and farmers can reuse idle homesteads and houses on the ground through self-employment, leasing, shareholding, and cooperation [5,6]. Since 2015, the Chinese government has introduced a series of measures to encourage the revitalization of idle homesteads. The 2018 Central Government Document No. 1 clearly proposes to explore the "separation of ownership, qualification and use" of the homestead and appropriately release the right to use homesteads and farmers' houses. As the main body and direct stakeholder, the intention of farmers to revitalize plays a crucial role in the revitalization of rural idle homesteads.

At the same time, farmer differentiation in the process of urbanization has become one of the most important social phenomena worthy of our attention [7,8]. Since the 1980s, urbanization has gradually broken down the barriers of institutional mechanisms that restricted the flow of various factors between urban and rural areas under the previous urban–rural dual structure; good public infrastructure and more employment opportunities in cities have enabled labor factors to flow between urban and rural areas and across different industries. The migration of the agricultural populations have caused differences in their livelihoods, sources of income, and land dependence [9,10]. These differences have led to the differentiation of farmers through a self-accumulation cycle. This differentiation is essentially a decline in the share of the farmer's income from the farm. The liberalization of institutional policies and the development of agricultural transformation are essential factors in the differentiation of farmers, driving their evolution from pure, to migrant, to urbanized farmers [11]. At present, the farmer differentiation is clear, the proportion of pure farmers is decreasing, and the proportion of urbanized farmers is increasing. Moreover, rural homesteads, which are used to guarantee the production and living of farmers, have gradually lost their effectiveness. A large number of homesteads are idle and inefficiently used, which is a tremendous waste of land resources in China. Therefore, the phenomenon of farmer differentiation is worthy of attention. Based on this, it is of great theoretical and practical significance to explore the influencing factors and mechanisms of farmers' intention to revitalize and analyze the moderating effect of farmers' differentiation, then propose a strategy to enhance the intention of the farmer to revitalize the homestead.

As a unique product of China's dual structure, the rural homestead is the legacy and evolution of the land system reform since the founding of the People's Republic of China. In Western countries, where rural land property rights are clear and predominantly private, there is no concept similar to the homestead. However, at present, the United States and most of the countries undergoing urbanization are also undergoing major social and economic changes in their rural areas. The problems of the migrant rural population and low utilization of rural homesteads are gradually becoming prominent. The concept of rural residential land utilization has become the focus of research by foreign scholars [12,13]. Geographical location [14], ecological changes [15], and accessibility to public services [16] are essential factors in the reuse of rural residential land. In line with the reform process in China, there are few studies on the revitalization of farmers' idle homesteads, but studies around the withdrawal of idle homesteads have yielded some results. It is well known that the property rights of rural homesteads in China are collectively owned by the village and have the function of guaranteeing the production and livelihood of farmers, so the reform of idle rural homesteads in China is unique and complicated compared with Western countries [17]. China's policies on the withdrawal and revitalization of idle homesteads repeatedly emphasize respecting farmers' intentions. Farmers, as "rational economic people," are the main subjects and direct stakeholders in the revitalization of the idle homestead, and their intention to revitalize plays a vital role in the revitalization and utilization of idle homesteads in rural areas. On the subjective level, policy cognition [18,19] and generational differences [20], and on the objective level, factors such as the housing environment [21], social security policy [17], and land income [22], have an impact on farmers' intention to withdraw. Only a few scholars have focused on the influence of the three capital attributes and policy advocacy's effects on farmers' intention to recycle [23,24]. Finally, some scholars have focused on the influence of farmer differentiation on the conservation input behavior of arable land quality and the adoption behavior of new agricultural technologies [25,26]. In summary, first, most studies on idle homesteads in the context of China focus on farmers' intention to withdraw from idle homesteads. However, with the further deepening of the reform of the homestead system, the focus of the reform has gradually shifted from farmers' withdrawal from idle homesteads to the revitalization of idle homesteads [17–22]. Second, fewer studies systematically analyze the factors influencing farmers' intention to revitalize using mature theory [23,24]. Lastly, fewer scholars have paid attention to the social phenomenon of farmer differentiation [25,26].

Therefore, this paper builds a theoretical model based on the theory of planned behavior, which is a mature theory for studying individual intention behaviors and enables the exploration of the influencing factors and mechanisms of farmers' intention to revitalize idle homesteads. The paper also emphasizes the influence of farmers' differentiation on the relationships of "attitude toward the behavior–intention to revitalize," "subjective norm–intention to revitalize," and "perceived behavioral control–intention to revitalize." In this study, we try to answer three questions. First, what factors influence farmers' intention to revitalize their idle homesteads? Second, to what extent do these factors influence farmers' intention to revitalize? Third, is there a moderating effect of farmer differentiation? The contributions of this paper are as follows: first, this paper creates an analytical framework to improve the theory of planned behavior by using farmer differentiation as a moderating variable. Second, it discusses the factors influencing farmers' intention to revitalize their homesteads. Third, the moderating role of farmer differentiation is emphasized.

The remainder of the article is organized as follows: Section 2 builds the theoretical framework and presents the research hypotheses. Section 3 introduces the data sources and empirical methods. Section 4 presents the empirical results. Section 5 discusses the findings. Section 6 presents the research conclusions and policy recommendations.

2. Theoretical Framework and Research Hypotheses

2.1. Analysis of the Influencing Mechanism of Farmers' Intention to Revitalize Idle Homesteads Based on the Theory of Planned Behavior

The theory of planned behavior (TPB), as an extension and refinement of the theory of rational behavior, has become one of the classic theories for predicting and explaining individual intention and behavior. Ajzen proposes that attitude toward the behavior (AB), subjective norms (SN), and perceived behavioral control (PBC) can help predict and explain behavioral intention (BI) and thus influence individual behavior (behavior) [27].

Attitude toward the behavior refers to an individual's overall evaluation of a certain behavior, and subjective norms refer to the influence from the social group that an individual feels when performing a behavior, especially whether the individual should follow the preferences of significant others. Perceived behavioral control, as an advancement of the theory of rational behavior, refers to an individual's perception of the ease of performing a behavior and the resources under individual control. The more positive the attitude of farmers, the more subjective norms, the stronger the perceived behavior control, the greater the behavior intention will be, and vice versa [28]. This theory has been applied by scholars in various countries to explain the farmers' intentions to reform [29], manage pastures sustainably [30], conserve land [31], and the intention of urban residents to separate waste [32]. However, due to the heterogeneity within the context of the individual, it is unknown whether the theory of planned behavior applies to idle homestead revitalization intention in China.

Considering that the revitalization of idle homesteads in China is still in the initial stages, intention and behavior are always considered as one, but this cannot be consistent. In order to avoid confusion between intention and behavior, this study focused on the motivation stage of planned behavior theory to conduct in-depth research on farmers' intention to revitalize idle homesteads [33] (Figure 1).

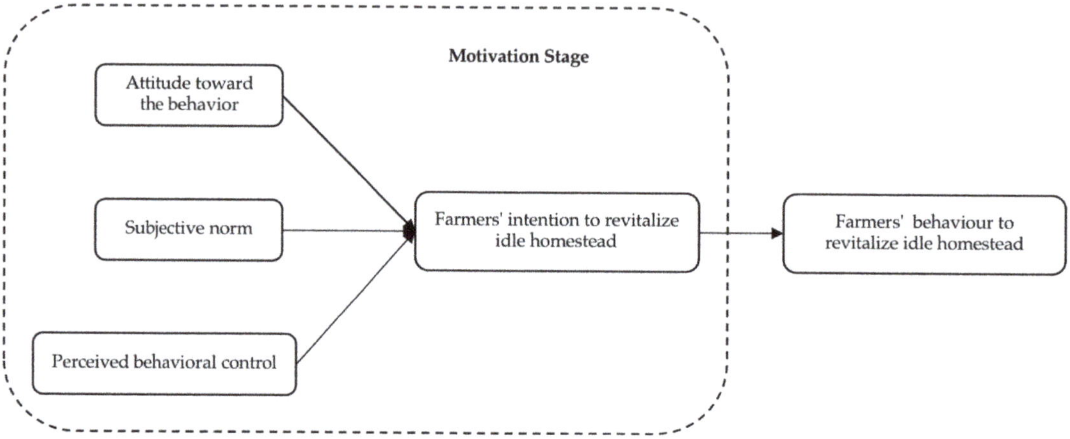

Figure 1. Simplified model of the theory of planned behavior.

1. Farmers' attitude toward the behavior;

Attitude toward the behavior refers to the degree of farmers' agreement with the act of revitalizing idle homesteads. The more positive the attitude is towards the behavior, the stronger the farmers' intention to revitalize. This leads to the first hypothesis:

Hypothesis 1 (H1). *Farmers'attitude toward the behavior has a significant positive influence on their intention to revitalize.*

2. Subjective norms of farmers' feelings;

Subjective norms refer to the influence of social networks that farmers perceive when they engage in revitalizing behavior. When the subjective norms are more positive, they help to enhance farmers' intention to revitalize. This leads to the second hypothesis:

Hypothesis 2 (H2). *Subjective norms perceived by farmers have a significant positive effect on their intention to revitalize.*

3. Farmers' perceived behavioral control.

Perceived behavioral control refers to farmers' perception of the difficulty level and their own controllable resources when revitalizing idle homesteads. The stronger the perceived behavioral control of farmers, the more likely they will be willing to revitalize. This leads to the third hypothesis:

Hypothesis 3 (H3). *The perceived behavioral control of farmers has a significant positive effect on their intention to revitalize.*

2.2. The Moderating Effect of Farmer Differentiation

The higher the farmer's agricultural income proportion, the shallower the degree of differentiation; the lower the farmer's agricultural income proportion, the deeper the degree of differentiation (referred to as "shallowly differentiated" and "deeply differentiated" in the following) [34].

For deeply differentiated farmers, first, they are less dependent on homesteads and more receptive to new concepts, so they have fewer doubts about the income after the revitalization of homesteads and pay more attention to the property functions of homesteads. Therefore, deeply differentiated farmers recognize the revitalization of idle homesteads more, and their attitude toward the behavior is more positive than shallowly differentiated farmers. Second, deeply differentiated farmers are more susceptible to the influence of subjective norms, mainly because the differentiation of farmers is manifested in the migration of the agricultural population to urban and rural areas and the civilization of the agricultural migrant population. This makes farmers not only migrate into a local society based on geographical kinship but also obliges farmers to abide by modern social norms based on business relationships. The two are intertwined and interact to varying degrees, presenting a certain complementary relationship. Finally, the livelihood capacity and control over resources of deeply differentiated farmers are stronger; therefore, their perceived behavioral control is stronger. This will further enhance the attitude towards behavior, subjective norms, and perceived behavioral control of deeply differentiated farmers, which ultimately has a stronger impact on the relationships of "attitude toward behavior–intention to revitalize," "subjective norm–intention to revitalize," and "perceived behavioral control–intention to revitalize."

However, for the shallowly differentiated farmers, first, since their primary source of income is agriculture, they are highly dependent on homesteads and have a low acceptance of new things, such as idle homesteads, which leads to less positive attitudes towards behavior. Second, the shallowly differentiated farmers have lived in rural societies for a long time, and their informal social network is more developed than deeply differentiated farmers'; thus, it is not easy to receive the directive norms from the government in the formal social network, so they experience less subjective norms. Finally, shallowly differentiated farmers have a lower sense of self-efficacy and weaker control over resources, which makes their perceived behavioral control weaker. This will lead to the further weakening of farmers' attitude toward behavior, subjective norms, and perceived behavioral control by shallowly differentiated farmers, and ultimately have a dampening effect on the relationships of "attitude toward behavior–intention to revitalize," "subjective norm–intention to revitalize," and "perceived behavioral control–intention to revitalize."

Based on this, farmer differentiation is used as a moderating variable to improve the theory of planned behavior (Figure 2); Moreover, the following hypotheses are proposed:

Hypothesis 4 (H4). *The influence of attitude toward behavior on farmers' intention to revitalize is moderated by the differentiation of farmers. The differentiation of farmers strengthens the influence of behavioral attitudes on farmers' intention to revitalize, and the higher the degree of differentiation, the stronger the intention of farmers to revitalize.*

Hypothesis 5 (H5). *The influence of subjective norms on farmers' intention to revitalize is moderated by farmers' differentiation. Farmers' differentiation strengthens the effect of subjective norms on farmers' intention to revitalize, and the higher the degree of differentiation, the stronger the farmers' intention to revitalize.*

Hypothesis 6 (H6). *The influence of perceived behavioral control on farmers' intention to revitalize is moderated by farmers' differentiation. Farmers' differentiation strengthens the effect of perceived behavioral control on farmers' intention to revitalize. The higher the degree of differentiation, the stronger the farmers' intention to revitalize.*

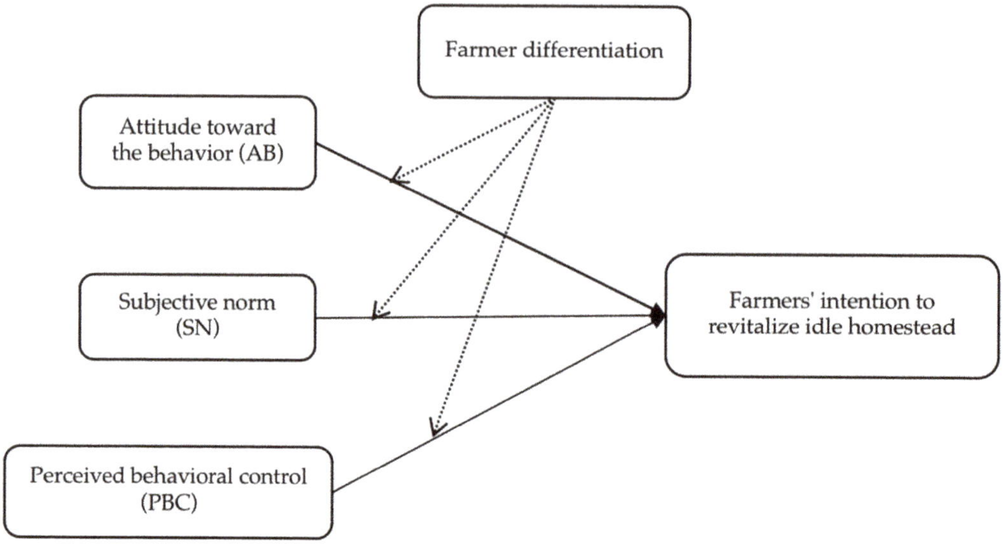

Figure 2. Improving the theoretical model of planned behavior.

3. Materials and Methods

3.1. Research Region

The Shaanxi Province is one of the core provinces in Northwest China and the bridgehead of the "Belt and Road" initiative. From north to south, Shaanxi Province is differentiated into three natural regions, namely the Northern Shaanxi Plateau, Guanzhong Plain, and Southern Shaanxi Qinling Mountain. It has been pointed out in this study that research on the extent of idle homesteads in 140 sample villages is in four major regions of China, namely, East, Central, West, and Northeast. The average rate of idle homesteads was 10.7%, with the rate of idle homesteads in the western region being 11.4%, ranking second among the four major regions and higher than the national average. The reason for the idle homestead in the western region is mainly due to the relatively backward economic development of the region and the lack of industrial support in the rural areas. So, a large number of the agricultural population has moved to the cities and tertiary industries, thus causing a large number of homesteads to be idle [35]. According to the data of the seventh census of China, as of November 2020, the urbanization rate of China's resident population was 63.89%. Among them, the urbanization rate of the Shaanxi Province reached 62.66%, which is 16.96 percentage points higher than the urbanization rate of the Shaanxi Province 10 years ago and 2.75 percentage points higher than the national increase. The total number of agricultural population migrants from urbanization reached 10 million, with the characteristics of a large total and high age range. Rapid urbanization has brought about a change in the relationship between rural people and land. Many rural homesteads have been idle and used inefficiently due to the migration and urbanization of a large part of the agricultural population. To cope with the problem of wasted rural land resources and to facilitate the urbanization of the migrating agricultural population, the Shaanxi Province has selected 12 counties (districts), including national-level pilot counties, for the homestead system reform. The characteristics of the Shaanxi Province, such as the widespread differentiation of farmers and the prominent problem of the idle homestead, meet the needs of this study (Figure 3).

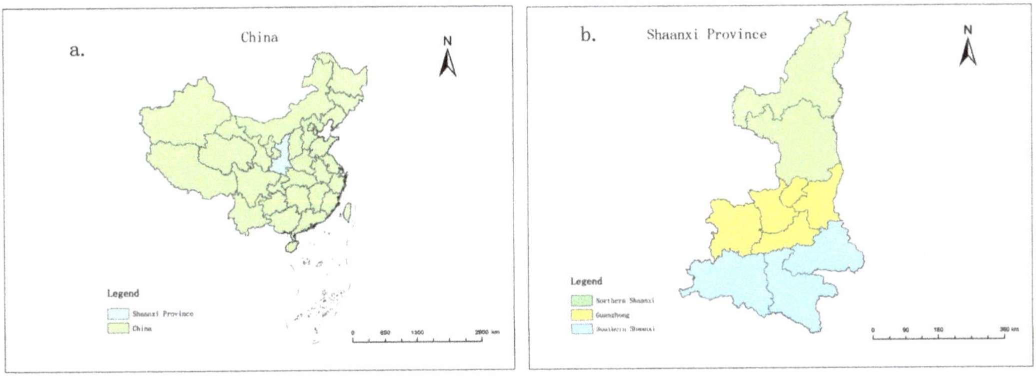

Figure 3. Overview of Shaanxi Province: (**a**) The location of Shaanxi Province in China, and (**b**) the distribution of the three natural regions in Shaanxi Province.

3.2. Data Source

To truly grasp the real intentions of different types of farmers, the research team conducted a field survey in February 2022, during the first month of the Chinese Lunar New Year before farmers went out to work, and in August, during the summer vacation of Chinese schools. Before the formal survey, all survey team members received expert training and conducted a pre-survey in the Gaoling District, Xi'an, to refine the questionnaire based on the survey, and the pre-survey data were not included in the final data analysis. The survey used a multi-stage sampling method to select the sample. In the first stage, ten counties (districts) were selected from three regions in the Shaanxi Province, which included the Yuyang District, Zichang County, Fu County, Dali County, Gaoling District, Yangling District, Liquan County, Pingli County, Zhashui County, and Chenggu County, as selected by using the stratified sampling method (Figure 4). Two towns were selected in each county (district) in the second stage using the random sampling method.

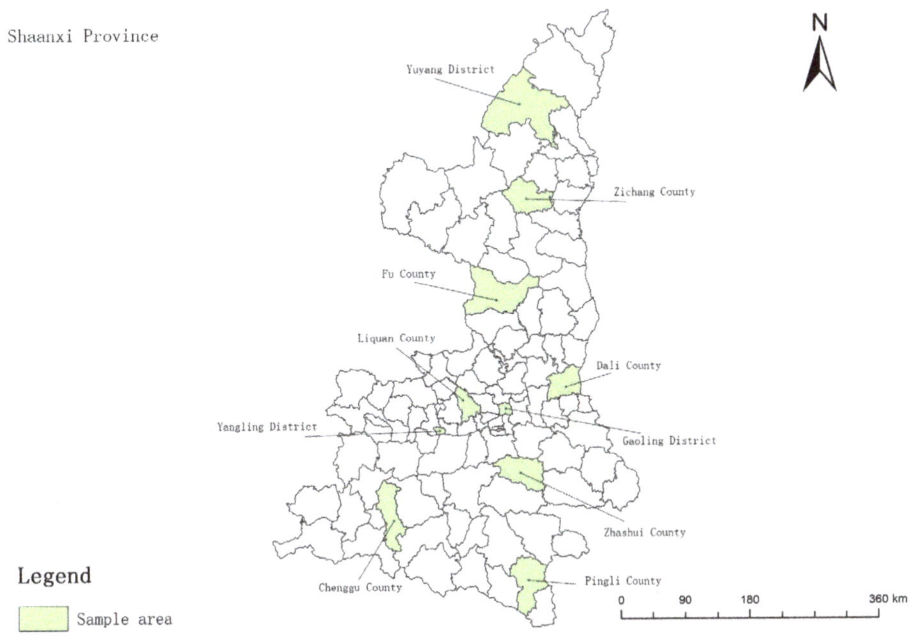

Figure 4. Locations of sampled counties.

In the third stage, two villages were selected in each town by systematic random sampling method. In the fourth stage, the snowball method was applied to find 20 farmers in each village, and a total of 800 farmers were targeted to conduct the survey questionnaire. The questionnaire mainly collected data involving personal, family, and homestead characteristics, intention, behavior attitude, subjective norms, and the perceived behavior control of farmers. A total of 800 questionnaires were distributed, of which 680 were valid, with a valid return rate of 85%.

3.3. Variable Selection

(1) Dependent variable:

The dependent variable was the farmers' intention to revitalize idle homesteads. In this study, whether a farmer was willing to revitalize the idle homestead was used to measure the farmer's intention to revitalize. Furthermore, the Likert 5-point scale method was used (very unwilling = 1; unwilling = 2; generally = 3; willing = 4; very willing = 5), where the larger the value, the stronger the farmers' intention.

(2) Independent variables:

The attitude toward the behavior dimension can be measured via the experiential and instrumental attitudes, and we further selected the "increase family income and employment opportunities" to measure "experiential attitude" and "promote industrial scale and rural development" to measure "instrumental attitude" [36].

Directive and descriptive norms can be used to measure the subjective norm dimension, so in this study, we chose "government and village collective encouragement" to measure "directive norms" and "support from family and neighbors" to measure "descriptive norms" [33].

The perceived behavioral control dimension can be measured by self-efficacy and control force. Therefore, in this study, we chose "overcoming the difficulties and taking risks of the revitalize" to measure "self-efficacy" and "access to relevant resources and familiarity with the revitalize model" to measure "control force" [37].

These three variables were used as core independent variables in this study. Each dimension was measured by a scale consisting of four question items, using a 5-point Likert scale method (strongly disagree = 1; disagree = 2; average = 3; agree = 4; strongly agree = 5). Estimating the internal consistency of the scale is necessary to ensure scale reliability [38]. Therefore, we used SPSS 26.0 to test the reliability of the scales. After testing, Cronbach's α values of 0.897 for the Attitude toward Behavior Scale, 0.875 for the Subjective Normative Scale, and 0.918 for the Perceived Behavioral Control Scale were obtained, all of which were above 0.700, indicating that the core independent variables had high internal consistency with the scales and excellent reliability [19,39,40]. Therefore, it was reasonable to use the mean values of the four-question items to represent their corresponding core independent variables in the model validation phase.

(3) Moderating variables:

Farmer differentiation referred to the proportion of agricultural income in the total household income of farmers, based on a criterion by the Institute of Rural Development, Chinese Academy of Social Sciences in 2002. Farmers with an agricultural income of 90% or more were defined as pure farmers and assigned a value of 1. Farmers with an agricultural income of 10–90% were defined as the agricultural migrant population and assigned a value of 2. Farmers with less than 10% agricultural income were defined as urbanized farmers and assigned a value of 3. The larger the value, the higher the proportion of non-agricultural income and the deeper the differentiation of farmers.

(4) Control variables:

To exclude interference with the independent variables, age, gender, occupation, and education level were controlled in the variables of individual farmers' characteristics. The number of laborers, annual income, and whether they have bought or plan to buy houses in towns were controlled by the variables of family characteristics. The idle status, quantity, and area were controlled by the variables of homestead characteristics.

Table 1 reports the definition, assignment, and descriptive statistics of all variables.

Table 1. Variable selection, assignment, and descriptive statistics.

Categories	Variable	Variable Definition and Assignment	Mean	Std
Dependent variable	Intention	Farmers' intention to revitalize idle homesteads. Very reluctant = 1; Reluctant = 2; Average = 3; Willing = 4; Very willing = 5.	4.06	1.053
Independent variables	Behavioral attitude	Increase family income. Strongly disagree=1; Disagree = 2; Indifferent = 3; Agree = 4; Strongly agree = 5.	3.82	1.160
		Increase employment opportunities. Strongly disagree = 1; Disagree = 2; Indifferent = 3; Agree = 4; Strongly agree = 5.	3.80	1.178
		Promote industrial scale. Strongly disagree = 1; Disagree = 2; Indifferent = 3; Agree = 4; Strongly agree = 5.	3.85	1.132
		Promote rural development. Strongly disagree = 1; Disagree = 2; Indifferent = 3; Agree = 4; Strongly agree = 5.	3.91	1.155
	Subjective norm	Government encouragement. Strongly disagree = 1; Disagree = 2; Indifferent = 3; agree = 4; Strongly agree = 5.	3.97	1.058
		Village collective encouragement. Strongly disagree = 1; Disagree = 2; Indifferent = 3; agree = 4; Strongly agree = 5.	4.06	1.005
		Support from friends and relatives. Strongly disagree = 1; Disagree = 2; Indifferent = 3; agree = 4; Strongly agree = 5.	4.04	1.045
		Neighborhood support. Strongly disagree = 1; Disagree = 2; Indifferent = 3; Agree = 4; Strongly agree = 5.	4.04	1.055
	Perceived behavioral control	Overcoming the difficulties of revitalization. Strongly disagree = 1; disagree = 2; Indifferent = 3; Agree = 4; Strongly agree = 5.	3.53	1.254
		Assume the risk of revitalization. Strongly disagree = 1; Disagree = 2; Indifferent = 3; agree = 4; Strongly agree = 5.	3.59	1.308
		Obtain relevant resources. Strongly disagree = 1; Disagree = 2; Indifferent = 3; agree = 4; Strongly agree = 5.	3.54	1.321
		Familiarize oneself with the revitalization model. Strongly disagree = 1; Disagree = 2; Indifferent = 3; Agree = 4; Strongly agree = 5.	3.62	1.285
Moderating variable	Farmer differentiation	Farmers' type. Pure farmer = 1; Agricultural migrant population = 2; Farmer that has been urbanized = 3.	1.95	0.779

Table 1. *Cont.*

Categories	Variable	Variable Definition and Assignment	Mean	Std
Control variables	Personal characteristics	Age. Youth (18–45 years old) = 1; Middle age (46–69 years old) = 2; Elderly (69 years old and above) = 3.	1.82	0.753
		Gender. Male = 1; Female = 2.	1.42	0.494
		Occupation. Non-farming = 0; Farming = 1.	0.48	0.500
		Education level. Elementary school and below = 1; Middle school to high school/junior college = 2; College and above = 3.	1.88	0.782
	Family characteristics	The number of the labor force (people,	2.03	0.741
		annual income). CNY 50,000 and below = 1; CNY 50,001–100,000 = 2; CNY 100,001–150,000 = 3; CNY 150,000 or more = 4.	1.65	0.940
		Have purchased or plan to purchase a house in town. No = 0; Yes = 1.	0.64	0.480
	Homestead characteristics	Idle status. Non-idle (less than 3 months) = 0; Seasonally idle (3 to 6 months) = 1; Year-round inactivity (more than 6 months) = 2	1.04	0.857
		The number of owned properties. One place = 0; Two or more places = 1.	1.38	0.485
		Area (mu).	286.7	0.226

Note: 1 mu = 1/15 hectare; CNY, or Yuan, is the Chinese currency: USD 1 = CNY 7.1082 in 2022.

3.4. Research Methods

In this study, the dependent variable "farmers' intention to revitalize idle homestead" was differentiated into five levels (very unwilling = 1; unwilling = 2; average = 3; willing = 4; very willing = 5), so the dependent variable can be treated as a continuous variable [41]. Moreover, the data met the prerequisites for using multiple linear regression. Therefore, a multiple linear regression model was used for data analysis with SPSS 26.0. The basic form of Model 1 (Formula (1)) was:

$$Y = \beta_0 + \beta_1 X_1 + \beta_2 X_2 + \beta_3 X_3 + \mu \quad (1)$$

where Y is the dependent variable, "farmers' intention to revitalize idle homestead," X_1 is the attitude toward behavior, X_2 is the subjective norm, and X_3 is the perceived behavioral control. β_i is the regression coefficient of the independent variable, which indicates the degree of influence of the independent variable on farmers' intention to revitalize idle homesteads, where the larger the parameter, the greater the influence. μ is the random error term. Before performing multiple linear regression, the independent variables were tested for independence, and no multiple covariances existed between the independent variables after testing.

This study was concerned not only with the direct relationship between the independent variables and the dependent variable but also with whether the moderating variables influence the effect of the independent variables on the dependent variable. Since the independent variables belong to continuous variables and the moderating variables belong to categorical variables, the method of hierarchical regression analysis was chosen to test the moderating effect of farmer differentiation. The basic form of Formula (2) was:

$$Y = \alpha_1 X + \alpha_2 M + \alpha_3 XM + \varepsilon \quad (2)$$

where Y is the dependent variable, "farmers' intention to revitalize idle homestead," and X is the independent variable, "attitude toward behavior, subjective norm, and perceived behavioral control." M is the moderating variable "farmers' differentiation." XM is the product of the independent variable and the moderating variable "attitude toward behavior × farmer differentiation, subjective norm × farmer differentiation, perceived behavioral control × farmer differentiation"; α_i is the estimated coefficient, and ε is the residual error. Based on Formula (2), using hierarchical regression, the regressions of the independent variables "attitude toward behavior, subjective norms, and perceived behavioral control" (Model 2), the moderating variable "farmer differentiation" (Model 3), and the product of the independent variables and the moderating variables (Model 4) was obtained. They were used to test the effects of farmer differentiation on the three relationships "attitude toward behavior–farmers' willingness to revitalize," "subjective norms–farmers' willingness to revitalize," and "perceived behavioral control–farmers' willingness to revitalize."

If the determination coefficient R^2 of Model 4 is greater than the R^2 of Models 2 and 3, or if the XM regression coefficient passes the significance test, it indicates that the differentiation of farmers plays a moderating role in the relationships of "attitude toward behavior–farmers' intention to revitalize," "subjective norms–farmers' intention to revitalize," and "perceived behavioral control–farmers' intention to revitalize."

4. Results

4.1. Demographic Analysis

Regarding the personal characteristics of farmers, males accounted for the majority of the sample at 57.6% (392 households), and females accounted for only 42.4% (288 households). Education was mostly elementary school and below. Junior high school to high school education accounted for 37.5% (255 households) and 37.4% (254 households), respectively, and only 25.1% (171 households) of farmers had a college education or above. The middle-aged and elderly accounted for 61.3% (417 households), and the occupational distribution was more balanced between farming and non-farming, accounting for 48.4% (329 households) and 51.6% (351 households), respectively. In terms of family characteristics, the average number of laborers in the sample family was 2.03, and 64.1% (436 households) had purchased or planned to purchase houses in town areas, which reflects the trend of migrating peasant populations to work in urban and rural areas with the further acceleration of urbanization. From the characteristics of the homestead, 65.3% (444 households) of the homesteads had different degrees of idleness, 37.6% (256 households) of the families had multiple houses in one family, the average area of the homestead was 286.7 m^2, and 52.9% (360 households) exceeded the standard of 200 m^2 per family. The phenomenon of multiple houses in one family and exceeding the standard area for one family exists widely, indicating that the homestead idleness in the Shaanxi Province is serious. In terms of the types of farmers, 33.1% (225 households) were pure farmers, 39.1% (266 households) were of the agricultural migrant population, and 27.8% (189 households) were urbanized farmers, which shows a clear differentiation of farmers.

Regarding the intention to revitalize idle homesteads among the respondents, 34.6% (235 households) and 42.1% (286 households) of the farmers indicated that they were willing and very willing to revitalize idle homesteads, respectively, together accounting for 76.7% (521 households) of the total sample. Meanwhile, 4.3% (29 households) and 4.1% (28 households) indicated that they were unwilling and very unwilling to revitalize idle homesteads, accounting for 8.4% (57 households) of the total sample, while 15% (102 households) were undecided. Most farmers were willing to revitalize their idle homesteads (Table 2).

Table 2. Basic characteristics of sampled farmers.

	N	%		N	%
Gender			**Idle status of homestead**		
Male	392	57.6	Year-round idle	264	38.8
Female	288	42.4	Seasonally idle	180	26.5
			Non-idle	236	34.7
Age			**Homestead area**		
Youth 18–45 years old	263	38.7	200 m^2 and below	320	47.1
Middle-aged 46–69 years old	274	40.3	More than 200–333 m^2	232	34.1
Elderly 69 years old and above	143	21	More than 333–667 m^2	114	16.8
			More than 667 m^2	14	2.0
Occupation			**Number of homesteads owned**		
Farming	329	48.4	One	424	62.4
Non-farming	351	51.6	Two or more	256	37.6
Educational level			**Degree of farmer differentiation**		
Elementary school and below	255	37.5	Pure farmers	225	33.1
Middle school to high school	254	37.4	Agricultural migrant population	266	39.1
College and above	171	25.1	Urbanized farmers	189	27.8
Number of family laborers					
One person	178	26.2	**Intention to revitalize**		
Two persons	307	45.1	Very willing	286	42.1
Three persons and more	195	28.7	Willing	235	34.6
Have purchased or intend to purchase a home in town			Do not care	102	15.0
Yes	436	64.1	Not willing	29	4.3
No	244	35.9	Very reluctant	28	4.1

Note: In the actual research, the homestead area item is measured in mu. In the paper, international units are used for conversion to facilitate readers' understanding.

4.2. Model Estimation Results

Based on the theoretical analysis, attitude toward behavior, subjective norms, and perceived behavioral control affect farmers' intention to revitalize idle homesteads, and the effects of attitude toward behavior, subjective norms, and perceived behavioral control on farmers' intention to revitalize varies depending on the degree of farmers' differentiation. Therefore, this study first examined the effects of attitude toward behavior, subjective norms, and perceived behavioral control on the intention to revitalize. Secondly, we examined how farmer differentiation moderates the effects of attitude toward behavior, subjective norms, and perceived behavioral control on the farmers' intention to revitalize.

1. Effects of attitude toward behavior, subjective norms, and perceived behavioral control on farmers' intention to revitalize;

The effects of attitude toward behavior, subjective norms, and perceived behavioral control on farmers' intention to revitalize idle homesteads were estimated using multiple linear regression equations (Model 1) without considering the moderating effects. The estimated results are shown in Table 3.

After controlling for a series of variables, such as farmers' personal, family, and homestead characteristics, attitude toward behavior, subjective norms, and perceived behavioral control, all passed the significance test, and the regression coefficients were significant at the 0.01 level in the positive direction. This points out that the more positive farmers' attitudes are toward idle homestead revitalization behaviors, the more positive the perceived subjective norms, and the stronger the perceived behavioral control, the stronger farmers' intention to revitalize, verifying H1, H2, and H3. This also indicates that the theory of planned behavior is applicable to the scenario of idle homestead revitalization behavior.

Table 3. Regression analysis of independent variables on farm households' intention to revitalize.

Variable	B	SD	T-Value
Behavioral attitude	0.08 ***	0.026	3.062
Subjective norm	0.39 ***	0.024	16.009
Perceived behavioral control	0.127 ***	0.024	5.318
Age	−0.162 ***	0.031	−5.244
Gender	−0.006	0.034	−0.186
Occupation	0.052	0.034	1.533
Educational level	0.062 ***	0.022	2.851
Number of family laborers	0.037	0.024	1.552
Annual income	−0.018	0.019	−0.966
Have purchased or intend to purchase a home in town	0.036	0.046	0.788
Idle status of homestead	0.574 ***	0.034	17.033
Number of homesteads owned	0.03	0.035	0.856
Homestead area	−0.076	0.074	−1.03
Constant	1.22 ***	0.16	7.621
R^2		0.835	
Observations		680	

Note: *** indicates significance at the 1% level.

2. Testing the moderating effect based on farmer differentiation;

In this study, we centralized the three core independent variables. We then tested the moderating effects of farmer differentiation on the relationships of "attitude toward behavior–intention to revitalize idle homestead," "subjective norms–farmers' intention to revitalize idle homestead," and "perceived behavioral control–farmers' intention to revitalize idle homestead," through hierarchical regressions (Table 4). The R^2 of each variable in Model 4 was larger than that of Models 2 and 3 in the three relationships of "attitude toward behavior–farmers' intention to revitalize idle homestead," "subjective norm–farmers' intention to revitalize idle homestead," and "perceived behavioral control–farmers' intention to revitalize idle homestead," The interaction coefficients of attitude toward behavior, subjective norms, perceived behavioral control, and farmer differentiation in Model 4 each passed the significance test.

Table 4. A test of the moderating effect of farmer differentiation.

Variable	Behavioral Attitudes–Farmers' Intention to Revitalize			Subjective Norm–Farmers' Intention to Revitalize			Perceived Behavioral Control–Farmers' Intention to Revitalize		
	Model 2	Model 3	Model 4	Model 2	Model 3	Model 4	Model 2	Model 3	Model 4
Behavioral attitudes	0.317 ***	0.308 ***	0.305 ***						
Subjective norm				0.457	0.435 ***	0.419 ***			
Perceived behavioral control							0.251 ***	0.233 ***	0.235 ***
Farmer differentiation		−0.187 ***	−0.187 ***		−0.106 ***	−0.11 ***		−0.176 ***	−0.176 ***
Behavioral attitudes × Farmer differentiation			0.118 ***						
Subjective norm × Farmer differentiation						0.061 **			
Perceived behavioral control × Farmer differentiation									0.081 ***
Age	−0.225 ***	−0.209 ***	−0.187 ***	−0.2 ***	−0.194 ***	−0.195 ***	−0.2 ***	−0.188 ***	−0.161 ***
Gender	−0.037	−0.039	−0.04	−0.002	−0.006	−0.003	−0.055	−0.057	−0.052
Occupation	0.034	−0.005	0.003	0.024	0.002	0.003	0.063	0.023	0.029
Educational level	0.048 *	0.022	0.027	0.067 ***	0.051 **	0.052 **	0.035	0.011	0.014
Number of family laborers	0.04	0.05 *	0.045 *	0.05 **	0.055 **	0.055 **	0.055 *	0.065 **	0.059 *
Annual income	−0.029	−0.031	−0.025	−0.01	−0.012	−0.011	−0.027	−0.028	−0.022
Have purchased or intend to purchase a home in town	0.089	0.039	0.051	0.071	0.046	0.056	0.099 *	0.055	0.07
Idle status of homesteads	0.694 ***	0.694 ***	0.701 ***	0.718 ***	0.725 ***	0.725 ***	0.696 ***	0.707 ***	0.715 ***
Number of homesteads owned	0.016	−0.018	−0.017	0.025	0.005	0.006	0.004	−0.028	−0.028
Homestead area	−0.064	−0.02	−0.033	−0.065	−0.038	−0.04	0.017	0.057	0.053
Constant	2.392 ***	2.872 ***	2.823 ***	1.569 ***	1.911 ***	1.978 ***	2.643 ***	3.125 ***	3.042 ***
R^2	0.761	0.779	0.787	0.822	0.827	0.828	0.741	0.756	0.761
Observations					680				

Note: *, **, and *** indicate significance at the 10%, 5%, and 1% levels, respectively; × indicates the interaction between items.

This indicates that, first, there are significant moderating effects of farmer differentiation in the relationships of "attitude toward behavior–farmers' intention to revitalize idle homestead," "subjective norms–farmers' intention to revitalize idle homestead," and "perceived behavioral control–farmers' intention to revitalize idle homestead." Second, the direction of the interaction term coefficient is consistent with the direction of the main effect (attitude toward behavior, subjective norms, and perceived behavioral control all have positive effects on farmers' intention to revitalize idle homesteads), indicating that the moderating effect is enhanced.

The effects of attitude toward behavior, subjective norms, and perceived behavioral control on farmers' intention to revitalize idle homesteads are stronger for deeply differentiated farmers. On the contrary, the effects of attitude toward behavior, subjective norms, and perceived behavioral control on farmers' intention to revitalize idle homesteads are smaller for shallowly differentiated farmers, which verifies H4, H5, and H6.

3. Effects of other control variables;

The two control variables (age and idle status) had a significant effect on the intention of farmers to revitalize idle homesteads, in which age had a negative effect but idle status had a positive effect.

This indicates that, first, the older the farmers are, the lower their intentions to revitalize idle homesteads. This may be because, on the one hand, older people have been living in a rural society for a long time and are influenced by the traditional concept of "private house and ancestral property." They believe that the homestead is their "root." On the other hand, elderly adults have a single channel to obtain information and are slow to accept new concepts, so they do not see the benefits of revitalizing their homesteads for the time being. Second, farmers are more willing to revitalize their idle homesteads throughout the year. This is largely due to the desire of farmers to realize the property function of their homesteads after urbanization or migration to the city, thus enhancing their livelihood in the city.

5. Discussion

In the survey area, 33.1% (225 households), 39.1% (266 households), and 27.8% (189 households) of the survey area were represented by pure farmers, the agricultural migrant population, and urbanized farmers, respectively, which once again confirms that the division of farmers has become a widespread phenomenon in the urbanization process [20]. Moreover, 65.3% of the homesteads were idle to varying degrees, indicating that the problem of idle or inefficient use of homesteads in China is severe, and the Chinese government must take appropriate measures to revitalize idle homesteads [42]. Therefore, it is necessary to conduct a study on farmers' intention to revitalize idle homesteads [23,24].

5.1. Effects of Attitude toward Behavior, Subjective Norms, and Perceived Behavioral Control on Farmers' Intention to Revitalize

As a mature theory for studying individual intention and behavior, TPB can fully explain the formation process of an individual's intention and behavior [43]. Therefore, in this study, we chose the TPB theory to explain the formation process of farmers' intention to revitalize idle homesteads [41]. The results of the study indicate that farmers' attitudes toward behavior, subjective norms, and perceived behavioral control of revitalizing idle homesteads all have positive effects on the intention to revitalize at the 1% significance level, which is consistent with theoretical expectations and also with previous empirical results on farmers' intention to withdraw idle homesteads [19,44]. This indicates that the theory of planned behavior also applies to the study of idle homestead revitalization intention in the Chinese context, further extending the explanatory scope of TPB. However, in the practical application, we found that behavioral experience, individual needs and emotions, and national culture influences farmers' intentions to revitalize. TPB theory needs to add other manifestations that can explain the explanatory power of behavior and intention, of which behavioral experience is considered the most.

First, the coefficient of attitude toward behavior was 0.08. For every unit increase in farmers' behavioral attitude toward revitalizing idle homesteads, farmers' intention to revitalize increased by 0.08 units, indicating that the more positive farmers' attitude toward revitalizing idle homesteads, the higher is farmers' intention to revitalize. The results also go hand-in-hand with Willock et al.'s (1999) argument [45]. The highest degree of influence of attitude toward behavior on farmers' intention in the context of green fertilizer adoption technology. In the context of farmers' green fertilizer adoption technology, attitude toward behavior had the highest degree of influence on farmers' intentions [46]. However, in the context of idle homestead revitalization, although farmers' attitude toward a behavior is a more critical influencing factor in the formation of farmers' intention to revitalize, it is less influential than the two factors of subjective norms and perceived behavioral control. This is because China has long restricted the free flow of land elements between urban and rural areas, coupled with the fact that farmers have stayed in rural areas for a long time and are less receptive to new concepts, which has led to fewer typical cases of successful revitalization in China, as farmers do not see the benefits of revitalization for themselves and their villages. In the field research, we found that the more inconvenient the traffic is, the less developed the network is, and the more the farmers do not believe that revitalization will bring any income. However, according to the four questions of the attitude toward behavior dimension, it can be seen that the current attitude toward farmers' behavior regarding the revitalization of the idle homestead is based on four aspects: whether the revitalization of the idle homestead will increase the family income, increase employment opportunities for individuals, promote the development of village industries on a large scale, and promote rural development. This is mainly because the property function of the idle homestead is becoming increasingly prominent at present. If the effect brought to farmers and villages after revitalization is not attractive enough to them, their attitude toward behavior will not be positive, thus affecting the farmers' intention to revitalize idle homesteads.

Second, the coefficient of subjective norms was 0.39, which significantly influenced farmers' intention to revitalize. Specifically, for every unit increase in farmers' subjective norms of revitalizing idle homesteads, farmers' intention to revitalize increased by 0.39 units, indicating that the social network influence positively influenced farmers' intention to revitalize, which is consistent with the study on farmers' intention to manage their farms sustainably [29]. As social beings, farmers are influenced by cultural norms and social expectations in their conscious and subconscious and feel pressure from society that affects their intention to recycle. Chinese law stipulates that village collectives are mass grassroots organizations that are self-managed, self-educated, and self-serving by farmers and are also vital to promoting public policies that can be effectively implemented. In the context of urbanization and land system reform, full attention should be paid to the village collectives' role in carrying on the top and bottom. Moreover, government guidance and support are highly directive and organizational. Friends and neighbors are the most important social resources for farmers, and the geographical proximity makes farmers' intentions more convergent, and other people's inventory behavior also has a strong demonstration effect on farmers [47]. Therefore, their advice and support are the primary reference and drivers of farmers' decisions. Furthermore, based on the four items of the subjective norm dimension, farmers' intention to revitalize is mainly influenced by the directive norms from the government and village collectives and the descriptive norms from friends and relatives. Farmers not only care about the influence of the credible government and village collectives but are also easily influenced by friends and relatives. The more positive the influence of the government, village collectives, and friends and neighbors, the more motivated they are to follow the subjective norms, thus increasing farmers' intention to revitalize.

Third, the coefficient of perceived behavioral control was 0.127, which was between the influence of attitude toward behavior and subjective norms on farmers' intention to revitalize. For every unit increase in farmers' perceived behavioral control in revitalizing idle homesteads, farmers' intention to revitalize increased by 0.127 units, indicating that the stronger farmers' self-efficacy and control, the stronger their intention to revitalize. This finding was in discordance with the findings of Armitage and Conner (2001) and Bijani et al. (2017) [48,49]. It is worth noting that China deeply recognizes that farmers are the main body of the idle homestead. Therefore, to improve farmers' intention to revitalize, the Chinese government has taken a series of measures to improve the control of farmers' perception behavior, such as policy publicity and explanation, linking social capital, and improving the security system. When farmers have the ability and opportunity to revitalize the idle homestead, they will participate in the idle homestead. According to the four items of the subjective norm dimension, the more farmers believe that they can overcome difficulties and bear the risks of revitalization, the more they can obtain relevant resources and become familiar with the revitalization model, and the higher their intention to revitalize.

5.2. The Moderating Effect Based on Farmer Differentiation

In addition to the direct effects of attitude toward behavior, subjective norms, and perceived behavioral control on farmers' intention to revitalize, this study took into account the current realistic context of farmers' differentiation in China and examined the effects of farmers' differentiation in the relationships of "attitude toward behavior–farmers' intention to revitalize," "subjective norms–farmers' intention to revitalize," and "perceived behavioral control–farmers' intention to revitalize idle homestead." While the deeper level of farmer differentiation suggested by Liu et al. (2020) has an inhibitory effect on farmers' behavior of exiting idle homesteads [20], the empirical results of this study found that the interaction item between farmer differentiation and attitudes toward behavior, subjective norms, and perceived behavioral control was significant and positive. These results indicate that farmer differentiation strengthened the positive relationship between attitude toward behavior, subjective norms, and perceived behavioral control on farmers' intention to revitalize, which was consistent with the theoretical analysis. The reason for the inconsistent findings is that this study focuses on the revitalization of the idle homestead, that is, the reuse of idle homesteads without changing the property rights relationship, rather than the withdrawal of the right to use an idle homestead. This verifies the necessity and importance of incorporating the variable of farmer differentiation into the theory of planned behavior and increases the explanatory power of the theory of planned behavior.

Actually, deeply differentiated farmers have stronger livelihood capacity and control, and their dependence on the rural homestead is much lower [50]. With the gradual disappearance of institutional barriers to free flow and the equal exchange of urban and rural factors, farmers gradually see the property value of their homesteads and hope to realize this [51]. Moreover, deeply differentiated farmers are positively influenced by descriptive norms from informal social networks based on the Consanguineous Relationship and Geographical Relationship and by directive norms from formal social networks based on the Business Relationship. Therefore, deeply differentiated farmers have a more positive attitude toward behavior, stronger perceived positive subjective norms, and stronger perceived behavioral control over the idle homestead. Thus, attitudes toward behavior, subjective norms, and perceived behavioral control strongly influence farmers' intentions to revitalize. The shallowly differentiated farmers are less receptive to new ideas, have narrower access to information, and are more susceptible to descriptive norms from informal social networks, so they have a more negative view of the revitalization of the idle homestead.

Moreover, the shallowly differentiated farmers have a weaker sense of self-efficacy and control due to a low-quality and less thorough understanding of national policies. The homestead assumes more of a security function, and shallowly differentiated farmers have stronger emotional attachments to it. Therefore, the influence of attitude toward behavior, subjective norms, and perceived behavioral control on farmers' intention to revitalize is relatively weak [50,52]. In short, the effects of attitude towards behavior, subjective norms, and perceived behavioral control on farmers' intention to revitalize are stronger in farmers with deeper differentiation.

The differentiation of farmers has led to differences in the sensitivity of different types of farmers' attitudes toward behavior, subjective norms, and perceived behavioral control. Therefore, policymakers should no longer treat farmers as a homogeneous group but should fully explore their heterogeneity and formulate land use policies according to their categories.

This study focused on the prominent social phenomenon of farmer differentiation. It used the theory of planned behavior to explore the influencing factors of farmers' intention to revitalize idle homesteads, which to a certain extent, makes up for the shortcomings of previous studies. However, there are still some potential limitations: First, the analysis of factors influencing farmers' intention to revitalize idle homesteads is based on the theory of planned behavior, and other mature theories, such as the TAM model, can be further considered in the future to clarify the influence of factors, such as perceived usefulness and perceived ease of use on a farmers' intention to revitalize. Second, this study only focused on farmers' intention to revitalize idle homesteads. However, intention does not necessarily lead to behavior, and the conversion process from intention to behavior should be paid attention to in future research. Regarding farmers' revitalized behavior, since the revitalization of the idle homestead is based on the Chinese scenario, and different cultures represent independent preferences for a state of affairs, the study of farmers' revitalized behavior can further consider factors such as national culture. Third, this study only selected the moderating variable of farmer differentiation to improve the theory of planned behavior. According to the idea of the inductive method, there may be other moderating variables that can be comprehensively studied in the future. Fourth, this study mainly took samples from the Shaanxi Province in Northwest China and did not consider other provinces. The results may differ due to the differences in the degree of idleness of homesteads and the degree of farmer differentiation among provinces, and the scope of the study can be further expanded in the future.

6. Conclusions and Implications

6.1. Conclusions

This study aims to explore the influencing factors of farmers' intention to revitalize idle homesteads and further consider the moderating role of farmer differentiation, which to a certain extent, enriches relevant research.

The main conclusion of this article is that, first, there are significant positive effects of attitude toward behavior, subjective norms, and perceived behavioral control on farmers' intention to revitalize. In particular, the influence of subjective norms is stronger, followed by perceived behavioral control and attitude toward behavior. The influence of subjective norms on farmers' intention to revitalize is the strongest. This is since Chinese society is vernacular, and farmers live around villages, which creates "local" constraints, making people from different villages isolated from each other and people from the same village familiar with each other. In a society of acquaintances without strangers, farmers are more vulnerable to the influence of significant others. Second, farmer differentiation has an enhancing effect in the relationships of "attitude toward behavior–intention to revitalize," "subjective norms–intention to revitalize," and "perceived behavioral control–intention to revitalize." Third, we should pay full attention to the phenomenon of farmers' differentiation and design the revitalization policy according to different types of farmers' sensitivity to differences in attitude toward behavior, subjective norms, and perceived behavioral control, and fully consider farmers' different demands.

This study is unique to China, and its contributions are worth acknowledging. First, this study expands the scope of what can be explained by the theory of planned behavior and improves the contribution of the theory. Second, this study improves the theory of planned behavior by incorporating farmer differentiation variables as moderating variables based on the phenomenon of farmer differentiation which is prominent in the current urbanization process. Third, this study provides some references for the formulation of rural land use policies in many countries that are experiencing urbanization and, thus, has potential application value. Therefore, as farmer differentiation has become a common social phenomenon in urbanization, national governments need to pay attention to farmer heterogeneity and formulate rural land use policies according to their categories.

6.2. Implications

Based on the above findings, to enhance farmers' intention to revitalize idle homesteads, we should not only pay attention to farmers' attitudes toward behavior, subjective norms, and perceived behavioral control but also pay attention to the influence of the moderating effect of farmers' differentiation. Therefore, this paper presents the following four suggestions to enhance farmers' revitalization intention.

First, to cultivate farmers' positive attitudes toward the behavior by enhancing their experiential and instrumental attitudes, the government should explore successful cases and benchmark farmers and give full play to the role of the media in guiding public opinion to establish farmers' correct and positive perceptions of revitalizing idle homesteads, thus improving farmers' attitude toward the behavior.

Second, the positive influence of subjective norms is strengthened through the implementation of directives and descriptive norms. The government should continuously improve the management system of rural homesteads and the supervision and management system of revitalizing idle homesteads. Additionally, they should open the supervision and reporting mechanism to break the "trust barrier" between farmers, the government, and village collectives. They should also actively collect public opinions so that various policies and systems can more effectively reflect the farmers' needs and interests, improve the guiding and driving effect of the directives and descriptive norms on farmers, create a strong social atmosphere of actively revitalizing idle homesteads, and thus strengthen the positive influence of subjective norms.

Third, it is necessary to enhance the perceived behavior control of farmers by improving their sense of self-efficacy and control force. On the one hand, the government should strengthen farmers' skills training and broaden their employment channels. On the other hand, it needs to continuously strengthen the institutional construction of idle homestead revitalization, standardize the revitalization procedures, and establish dispute resolution mechanisms to enhance farmers' perceived behavioral control.

Fourth, the government needs to pay great attention to the social phenomenon of farmer differentiation and promote the urbanization of the agricultural population. By improving the social security system in rural areas and promoting the equalization of public services in urban and rural areas, the government can narrow the gap between urban and rural areas at the root, reduce farmers' dependence on the homestead, and promote the urbanization of the agricultural population. When designing the policy mechanism for revitalizing idle homesteads, the government should respond to the differences in the sensitivity of different types of farmers to attitudes toward behavior, subjective norms, and perceived behavioral control in the context of rapid urbanization and design the revitalization policy according to the category.

Author Contributions: Conceptualization, M.L. and B.G.; methodology, M.L. and B.G.; software, M.L.; validation, M.L. and G.C.; investigation, M.L., L.Y., G.C., R.X. and Y.H.; resources, B.G.; data curation, M.L.; writing—original draft preparation, M.L.; writing—review and editing, M.L. and B.G.; visualization, G.C.; supervision, B.G.; project administration, B.G.; funding acquisition, B.G. All authors have read and agreed to the published version of the manuscript.

Funding: This research was funded by the Social Science Foundation of Shaanxi Provincial, China (2022F004) and the Research Project on Major Theoretical and Practical Issues in Philosophy and Social Sciences of Shaanxi Province, China (2021ZD1017).

Institutional Review Board Statement: Not applicable.

Informed Consent Statement: Not applicable.

Data Availability Statement: All the data will be available from the corresponding author upon reasonable request.

Conflicts of Interest: The authors declare no conflict of interest.

References

1. Liu, Y.; Li, J.; Yang, Y. Strategic adjustment of land use policy under the economic transformation. *Land Use Policy* **2018**, *74*, 5–14. [CrossRef]
2. Wei, H.; Yan, K.; Yu, W.; Cui, H.; Tan, Q. *China's Rural Development Report (2018): The Road to Overall Revitalization of Rural Areas in the New Era*; China Social Science Press: Beijing, China, 2018.
3. Ministry of Land and Resources (MLR); National Development and Reform Commission (NDRC) of China. *National Land Consolidation Plan (2016–2020)*; Ministry of Land and Resource and National Development and Reform Commission: Beijing, China, 2017. Available online: http://www.mlr.gov.cn/zwgk/zytz/201702/t20170215_1440315.htm (accessed on 25 October 2022).
4. Long, H.; Li, Y.; Liu, Y. Evolutionary characteristics of hollowed-out villages in China and their dynamic mechanisms. *J. Geogr.* **2009**, *64*, 53–63.
5. Zhang, Y.; Zhou, J.; Lu, P. The theory and practice of revitalizing and utilizing idle rural homestead in the context of rural revitalization–a study based on two cases in Anhui Province. *Agric. Econ. Issues* **2022**, *4*, 96–106.
6. Shan, Z.; Feng, C. The redundancy of homestead in rural China: The evolution process, status, and policy implications. *Land Use Policy* **2018**, *74*, 179–186. [CrossRef]
7. Zhao, L.; Liu, S.; Zhang, W. New trends in internal migration in China: Profiles of the new-generation migrants. *China World Econ.* **2018**, *26*, 18–41. [CrossRef]
8. Long, H.; Tu, S.; Ge, D.; Li, T.; Liu, Y. The allocation and management of critical resources in rural China under restructuring: Problems and prospects. *J. Rural. Stud.* **2016**, *47*, 392–412. [CrossRef]
9. Li, X.; Gao, Q. Behavioral logic, differentiation results and development prospects-an examination of farmer differentiation behavior in China since 1978. *Agric. Econ. Issues* **2013**, *34*, 56–65+111.
10. Chang, X.; Liu, L. Characterizing rural household differentiation from the perspective of farmland transfer in eastern China using an agent-based model. *Hum. Ecol.* **2018**, *46*, 875–886. [CrossRef]
11. Zhang, C.; Peng, C.; Kong, X. The evolutionary logic, historical evolution and future outlook of farmer differentiation. *Reform* **2019**, *2*, 5–16.
12. Short, A.G. Governing change: Land-use change and the prevention of nonpoint source pollution in the north coastal basin of california. *Environ. Manag.* **2013**, *51*, 108–125. [CrossRef]
13. Andreas, H.; Anka, L. Land consolidation for large-scale infrastructure projects in Germany. *Geod. Vestn.* **2014**, *58*, 46–68.
14. Wynne, L.; Ruoso, L.E.; Cordell, D.; Jacobs, B. 'locationally disadvantaged': Planning governmentalities and peri-urban agricultural futures. *Aust. Geogr.* **2020**, *51*, 377–397. [CrossRef]
15. Ghosh, S.; Chifos, C. The 1985 siting of a Toyota manufacturing plant in rural Kentucky, USA: The ensuing land use change and implications for planning. *Landsc. Urban Plan.* **2017**, *167*, 288–301. [CrossRef]
16. Strbele, M.; Hunziker, M. Are suburbs perceived as rural villages? landscape-related residential preferences in switzerland. *Landsc. Urban Plan.* **2017**, *163*, 67–79. [CrossRef]
17. Chen, H.; Zhao, L.; Zhao, Z. Influencing factors of farmers intention to withdraw from rural homesteads: A survey in zhejiang, China. *Land Use Policy* **2017**, *68*, 524–530. [CrossRef]
18. Fan, W.; Zhang, L. Does cognition matter? Applying the push-pull-mooring model to Chinese farmers' intention to withdraw from rural homesteads. *Pap. Reg. Sci.* **2019**, *98*, 2355–2369. [CrossRef]
19. Shi, R.; Hou, L.; Jia, B.; Jin, Y.; Zheng, W.; Wang, X.; Hou, X. Effect of Policy Cognition on the Intention of Villagers Withdrawal from Rural Homesteads. *Land* **2022**, *11*, 1356. [CrossRef]
20. Liu, R.; Yu, C.; Jiang, J.; Huang, Z.; Jiang, Y. Farmer differentiation, generational differences and farmers' behaviors to withdraw from rural homesteads: Evidence from Chengdu, China. *Habitat Int.* **2020**, *103*, 102231. [CrossRef]
21. Luo, Z.; Chunjie, L.I.; Zhanping, Y.U.; Zhang, H. Farmers' intention to withdraw from rural homesteads. *Asian Agric. Res.* **2022**, *14*, 8.
22. Zhang, X.; Han, L. Which Factors Affect Farmers intention for rural community remediation? A tale of three rural villages in China. *Land Use Policy* **2018**, *74*, 195–203. [CrossRef]
23. Feng, Y.; Wang, W.; Ding, Y. The impact of the "three capital" attribute on the viability of rural idle homestead sites. *Resour. Environ. Arid. Areas* **2021**, *35*, 58–64.
24. Hou, Y.; Pu, C.; Yan, Z. A study on the intention to choose the utilization mode of farmers' home base inventory based on SEM-Logit model. *Hubei Agric. Sci.* **2021**, *60*, 145–151.
25. Yang, Z.; Wang, Y.; Maidan, T. Analysis of Farmers' Conservation Input Behavior of Cropland Quality and Its Influencing Factors-Based on Part-Time Divergence Perspective. *China Popul.-Resour. Environ.* **2015**, *25*, 105–112.
26. Zhu, M.F.; Li, N.T. Research on behavioral differences and countermeasures of farmers' adoption of new agricultural technologies. *Agric. Technol. Econ.* **2001**, *2*, 26–29.

27. Ajzen, I.; Madden, T.J. Prediction of goal-directed behavior: Attitudes, intentions, and perceived behavioral control. *J. Exp. Soc. Psychol.* **1986**, *22*, 453–474. [CrossRef]
28. Ajzen, I. Theory of planned behavior. *Organ. Behav. Hum. Decis. Process.* **1991**, *50*, 179–211. [CrossRef]
29. Menozzi, D.; Fioravanzi, M.; Donati, M. Understanding Farmers' Responses to Cap Reform. In Proceedings of the 2014 International Congress, Ljubljana, Slovenia, 26–29 August 2014; European Association of Agricultural Economists: Wageningen, The Netherlands, 2014.
30. Hassan, A.; Mohammad, B.; Mansour, Z.; Fatemeh, N. Incorporating Land Use Changes and Pastoralists' Behavior in Sustainable Rangeland Management: Evidence from Iran. *Rangeland Ecol. Manag.* **2022**, *80*, 48–60.
31. Simmons, B.A.; Wilson, K.A.; Dean, A.J. Psychosocial drivers of land management behavior: How threats, norms, and context influence deforestation intentions. *Ambio* **2021**, *50*, 1364–1377. [CrossRef]
32. Razali, F.; Daud, D.; Wai, C.W. Waste separation at source behavior among Malaysian households: The Theory of Planned Behavior with moral norm. *J. Clean. Prod.* **2020**, *271*, 122025. [CrossRef]
33. Elahi, E.; Zhang, H.; Lirong, X.; Khalid, Z.; Xu, H. Understanding cognitive and socio-psychological factors determining farmers' intentions to use improved grassland: Implications of land use policy for sustainable pasture production. *Land Use Policy* **2021**, *102*, 105250. [CrossRef]
34. Chen, M.; Yuan, D.; Kuang, F.; Wu, Q.; Xie, X. The impact of farmers' differentiation and intergenerational differences on the adoption of ecological farming. *Chinas Popul. Resour. Environ.* **2019**, *29*, 79–86.
35. Li, T.; Long, H.; Wang, Y. Analysis on the idle degree of rural homestead in China and its causes. *China Land Sci.* **2019**, *33*, 64–71.
36. Davies, J.; Foxall, G.R.; Pallister, J. Beyond the intention-behaviour mythology: An integrated model of recycling. *Mark Theory* **2002**, *2*, 29–113. [CrossRef]
37. Zhang, G.; Zhang, L.; Qiu, X.; Zhu, W. Mechanism of fishermen's participation in professional cooperative organizations based on the theory of planned behavior. *Agric. Econ. Issues* **2015**, *36*, 97–104.
38. Fielding, K.S.; McDonald, R.; Louis, W.R. Theory of planned behaviour, identity and intentions to engage in environmental activism. *Land Use Policy* **2008**, *28*, 318–326. [CrossRef]
39. Wauters, E.; Bielders, C.; Poesen, J.; Govers, G.; Mathijs, E. Adoption of soil conservation practices in Belgium: An examination of the theory of planned behaviour in the agri-environmental domain. *Land Use Policy* **2010**, *27*, 86–94. [CrossRef]
40. Hair, J.; Howard, M.; Nitzl, C. Assessing measurement model quality in PLS-SEM using confirmatory composite analysis. *J. Bus. Res.* **2020**, *109*, 101–110. [CrossRef]
41. Fang, J.; Wen, Z.; Liang, D.; Li, N. Analysis of regulatory effects based on multiple regression. *Psychol. Sci.* **2015**, *38*, 715–720.
42. Zhou, Y.; Li, X.; Liu, Y. Rural land system reforms in China: History, issues, measures and prospects. *Land Use Policy* **2020**, *91*, 104330. [CrossRef]
43. van Dijk, W.F.A.; Lokhorst, A.M.; Berendse, F.; de Snoo, G.R. Factors underlying farmers intentions to perform unsubsidized agri-environmental measures. *Land Use Policy* **2016**, *59*, 207–216. [CrossRef]
44. Wang, Z.; Zou, W.; Wang, X. Study on the contradiction between the behavior and intention of farmers to withdraw from homestead. *China Land Sci.* **2018**, *32*, 29–35.
45. Willock, J.; Deary, I.J.; McGregor, M.M.; Sutherland, A.; Edwards-Jones, G.; Morgan, O.; Dent, B.; Grieve, R.; Gibson, G.; Austin, E. Farmers' attitudes, objectives, behaviors, and personality traits: The Edinburgh study of decision making on farms. *J. Vocat. Behav.* **1999**, *54*, 5–36. [CrossRef]
46. Adnan, N.; Nordin, S.M.; Ali, M. A solution for the sunset industry: Adoption of Green Fertiliser Technology amongst Malaysian paddy farmers. *Land Use Policy* **2018**, *79*, 575–584. [CrossRef]
47. Cheng, Y.; Zeng, W.; Hu, Y. Analysis on Factors Affecting Farmers' intention to Participate in Underforest Economy Based on the Theory of Planned Behavior. *Rural. Econ.* **2021**, *11*, 62–69.
48. Armitage, C.J.; Conner, M. Efficacy of the theory of planned behavior: A meta–analytic review. *Br. J. Sociol. Psychol.* **2001**, *40*, 471–499. [CrossRef]
49. Bijani, M.; Ghazani, E.; Valizadeh, N.; Haghighi, F. Pro-environmental analysis of farmers' concerns and behaviors towards soil conservation in central district of Sari County, Iran. *Int. Soil Water Conserv. Res.* **2017**, *5*, 43–49. [CrossRef]
50. Wang, C.; Wang, L.; Jiang, F.; Lu, Z. Differentiation of rural households' consciousness in land use activities: A case from Bailin village, Shapingba district of Chongqing municipality, China. *Chin. Geogr. Sci.* **2015**, *25*, 124–136. [CrossRef]
51. Zhao, Q.; Jiang, G.; Ma, W.; Zhou, D.; Qu, Y.; Yang, Y. Social security or profitability? Understanding multifunction of rural housing land from farmers' needs: Spatial differentiation and formation mechanism—Based on a survey of 613 typical farmers in Pinggu district. *Land Use Policy* **2019**, *86*, 91–103.
52. Wang, Y.; Chen, L.; Long, K. Farmers' identity, property rights cognition and perception of rural residential land distributive justice in China: Findings from Nanjing, Jiangsu Province. *Habitat Int.* **2018**, *79*, 99–108. [CrossRef]

sustainability

Article

Does Farmland Transfer Contribute to Reduction of Chemical Fertilizer Use? Evidence from Heilongjiang Province, China

Ningbo Cui *, Xuezhen Ba, Jin Dong and Xiaofan Fan

College of Economics and Management, Northeast Agricultural University, Harbin 150030, China
* Correspondence: struggle_happiness@163.com; Tel.: +86-186-4501-3338

Abstract: Promoting the reduction of chemical fertilizers is an important measure to promote the green and sustainable development of agriculture. Farmland transfer is a new way to minimize the need of chemical fertilizers. However, there is debate over this causality. This paper examines the relationship between farmland transfer and chemical fertilizer reduction. After the theoretical analysis, based on the data of 442 corn farmers in Heilongjiang Province, the study employed the endogenous switching probit model to empirically test the effect of farmland transfer on the reduction of chemical fertilizer. The study finds that in the survey area, the overall actual chemical fertilizer application rate was 12.12 kg/mu higher than the economic optimal application rate, which had more room for chemical fertilizer reduction. Moreover, farmland transfer-in reduced the chemical fertilizer application during corn production. If farmland transfer-in farmers decided not to transfer into the farmland, the chemical fertilizer reduction treatment effect would decrease, while it would increase if farmland non-transfer-in farmers decided to transfer into the farmland. Finally, the chemical fertilizer reduction treatment effect would decrease if farmers who had transferred into farmland concentratedly chose to transfer into farmland dispersedly, while it would increase if farmers who had transferred into farmland dispersedly chose to transfer into farmland concentratedly. These findings can provide experience for achieving more effective farmland transfer and chemical fertilizer reduction.

Keywords: farmland transfer; chemical fertilizer reduction; corn growers; Heilongjiang province

Citation: Cui, N.; Ba, X.; Dong, J.; Fan, X. Does Farmland Transfer Contribute to Reduction of Chemical Fertilizer Use? Evidence from Heilongjiang Province, China. *Sustainability* **2022**, *14*, 11514. https://doi.org/10.3390/su141811514

Academic Editors: Wenwu Tang, Dianfeng Liu and Jianxin Yang

Received: 28 July 2022
Accepted: 4 September 2022
Published: 14 September 2022

1. Introduction

Chemical fertilizers have long played a significant role in China's grain production and agricultural development. However, excessive and inefficient chemical fertilizer application has also had a severely negative impact on environment [1–4]. Moreover, it should be noted that China's chemical fertilizer inputs have entered the stage of diminishing marginal returns [5] and increased chemical fertilizer application can no longer ensure a sustainable increase in grain production. Instead, it may result in soil consolidation, acidification, and water pollution [6–9], endangering food security and the sustainability of agricultural development [10–12]. Therefore, the Chinese government proposed a "zero growth" action plan for chemical fertilizers and has already implemented some measures to promote chemical fertilizer reduction. However, chemical fertilizer reduction is a complex and long-term process. At present, the intensity of chemical fertilizer application in China still exceeds the internationally recognized upper limit of environmental safety in chemical fertilizer application, and the utilization efficiency is much lower than that of developed countries in Europe and the United States [13–16]. How to continue to encourage chemical fertilizer reduction in the future to achieve cleaner output is an essential topic that requires immediate attention for the green and sustainable development of Chinese agriculture.

In fact, the allocation of production elements like chemical fertilizer is an endogenous reflection of changes in important endowment resources like land, meaning that chemical fertilizer inputs are inextricably linked to the scale of farmland [17,18]. In recent years, more and more academics have conducted extensive discussions on the relationship between farmland transfer and chemical fertilizer reduction. Most studies have shown that the expansion of farmland management scale under farmland transfer has a significant negative correlation with chemical fertilizer application, mainly because farmland transfer helps to play the scale management effect and effectively reduces the cost of farmers' acquisition of "new technology and new knowledge", which awakens farmers' ecological consciousness and motivates them to adopt clean production methods, thereby reducing the use of chemical fertilizers [19–23]. However, other studies indicate that the increasing scale of farmland through farmland transfer may push farmers to apply more chemical fertilizer in pursuit of higher yields [24]. At the same time, the time and management costs, as well as the moral hazard and adverse selection of hired workers, may lead to an increase in chemical fertilizer use [25]. That is, large-scale agricultural production may cause environmental pollution and inefficiency [26–28].

Throughout the studies on farmland transfer and chemical fertilizer reduction, we found that the total effect of farmland transfer on chemical fertilizer reduction did not form a consistent conclusion. The reason may be that homogenization treats the connotations of farmers' scale operation and plot scale operation. In actuality, the location of the transferred farmland is different, resulting in different economic scale. In other words, the impact of applying chemical fertilizer will vary depending on whether the farmland rented is adjacent to one another. The land was still fragmented as a result of the dispersive transfer, which makes it challenging to meet the investment threshold for large and medium-sized machinery and advanced technology. Additionally, the corresponding time and financial costs are on the rise, which could encourage farmers to apply excessive chemical fertilizer [29]. Contrarily, if farmers transfer into farmland concentratedly, that will make the transferred farmland adjacent, increasing both the scale of the operation and the size of the plots, which in turn increases the economies of scale impact and reduces the need for chemical fertilizer [30].

Through the above analysis, we found that the spatial difference of farmland transfer has different effects on fertilizer reduction. Unfortunately, previous studies have mainly focused on the impacts of farmland transfer scale on fertilizer reduction, we believe that it is necessary to further explore the internal relationship between farmland transfer and fertilizer reduction from the perspective of spatial differences in farmland transfer, especially to analyze the effect of concentrated transfer of farmland and dispersed transfer of farmland to fertilizer reduction. Secondly, in terms of research methods, since decisions on farmland transfer and fertilizer application are often affected by some unobservable factors at the same time, there may be problems of "simultaneous decision" and "self-selection" that may cause deviations in estimated results. To overcome this defect, we intentionally correct the problem by the endogenous switching probit model to obtain more robust estimation results.

Therefore, in this paper, based on microscopic research, we analyze the impact of farmland transfer on fertilizer reduction in maize production from both theoretical and empirical perspectives, and compare the effects of farmland centralized transfer and farmland decentralized transfer on fertilizer reduction. The results of this study will provide empirical evidence on how the government can promote and support the achievement of more effective farmland transfer and fertilizer reduction, and contribute to green and sustainable agricultural development globally, especially in developing countries.

The paper is organized as follows. Section 2 theoretically analyzes the mechanism of the impact of farmland circulation on fertilizer reduction. Section 3 presents the data sources and econometric methods. Section 4 presents the empirical results and analysis. Section 5 focuses on discussion. Section 6 presents conclusions and implication.

2. Theoretical Analysis and Hypothesis

The allocation of production factors such as chemical fertilizer is an endogenous reflection of changes in key endowment resources, and changes in operation scale and plot size induced by farmland transfer will largely affect farmers' chemical fertilizer input behavior [17,18]. However, the relationship between farmland transfer and chemical fertilizer reduction does not form a consistent conclusion, probably because of the homogeneous treatment of the connotations of farmers' scale operation and plot scale operation. In fact, the different locations of the transferred farmland result in different economies of scale.

Specifically, if the transferred farmland is not adjacent to the farmer's existing farmland, it only indicates that the number of plots managed by farmers has increased, but the size of these plots has not increased [29,30]. Farmland fragmentation akin to that of traditional small farmers may still occur if there are too many plots or if they are too far apart from one another. On the one hand, excessively finely fragmented plots will spatially make it more difficult for large and medium-sized farm machinery to operate, which in turn will reduce the standardization and specialization of mechanical chemical fertilizer application operations and will not only make no contribution to improving chemical fertilizer reduction technology but also incur high time and economic costs. On the other hand, for some farmers to manage non-adjacent farmland, the demand for labor will increase, and the resulting moral hazard and supervision costs may also increase farmers' motivation to increase chemical fertilizer application. It is difficult for farmers to change the spatial distribution of farmland in the short term, and they generally readjust the resource elements they own and reduce high costs such as hired workers, large and medium-sized machinery and new technologies by increasing the amount of chemical fertilizers. Therefore, we believe that if farmland transfer fails to eliminate fragmentation, it will not only exacerbate the loss of production efficiency but also weaken the chemical fertilizer reduction effect of farmland economy of scale.

If the transferred farmland is adjacent to the farmer's existing farmland, or if the transferred farmlands are already adjacent, it can obtain economies of scale on plots by eliminating ridges, dead ends, and compartment ditches. Firstly, the concentrated transfer of farmland achieves the operational space and investment threshold for large and medium-sized machinery and advanced agricultural technology. The deep application method under advanced technology and large machinery can reduce the amount of chemical fertilizer and improve the absorption and conversion rate while improving the application standard and traceability, which is more conducive to the reduction of chemical fertilizer [30–32]. Second, the concentrated transfer of farmland may induce farmers to shift from diversified cropping patterns to more monoculture and specialized cropping patterns. In terms of human capital accumulation effects, horizontal specialization can reduce farmers' time costs and improve their ability to learn specialized planting techniques, especially their ability to learn and apply chemical fertilizer reduction and efficiency technologies. In terms of human capital spillover effects, knowledge, technology, and ability have spillover effects, and the imitation and diffusion of chemical fertilizer reduction technologies among farmers also promotes the diffusion and application of chemical fertilizer reduction technologies to a certain extent [33]. In addition, the concentrated transfer of farmland can reduce the cost of production materials conversion between plots, reduce the apportionment cost per unit area, and encourage farmers to purchase productive service items such as mechanization and chemical fertilizer application. This will further promote the development of the agricultural socialized service market and the deepening of vertical division of labor, thereby promoting the application of specialized and precise chemical fertilizer reduction technology services and enable farmers to obtain the service-scale economy of chemical fertilizer reduction [34–37].

Accordingly, we have the following hypothesis:

Hypothesis 1. *Farmland transfer-in has a positive effect on chemical fertilizer reduction.*

Hypothesis 2. *Compared with farmland transferred dispersedly, the farmland transferred concentratedly is more helpful for the farmers to reduce the application of chemical fertilizers.*

3. Data and Methods

3.1. Data Collection

Data were gathered during a field survey in Heilongjiang province, China, in 2021. The research region was chosen for two major reasons. First, Heilongjiang province, a significant grain-producing region in China, is situated in one of the "three prime maize belts" of the world. In 2020, Heilongjiang Province accounted for 13.28% and 13.99% of China's total maize sown area and production, respectively, significantly contributing to the country's food security. However, the high maize yields rely on high levels of fertilizer inputs, posing a serious threat to the quality of the black land and the ecological environment. After 2015, fertilizer application in Heilongjiang province began to show a downward trend, probably due to the fertilizer reduction initiative implemented by the Chinese government. It is noticed that the development of agricultural land transfer in Heilongjiang Province is rapid, with the proportion of agricultural land transfer and the scale of continuous agricultural land transfer much higher than the Chinese average level. Large-scale operation under agricultural land transfer has facilitated Heilongjiang Province to take the lead in promoting large and medium-sized mechanization, agricultural production services and other agricultural modernization projects. It also provides good conditions for achieving chemical fertilizer reduction. Thus, Heilongjiang province provides a very typical example that can provide China and other developing countries with experience in fertilizer reduction (Figure 1).

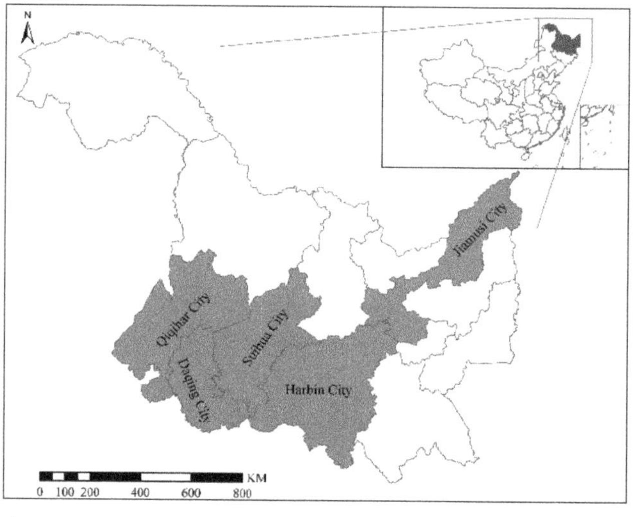

Figure 1. Map of the surveyed area.

To obtain data, we conducted a field survey of maize farmers in Heilongjiang Province from June to September 2021 based on a three-stage random sampling method, and the research process was divided into two phases.

Phase I: Pre-study was launched in June 2021. The research team conducted in-depth interviews with 15 randomly selected farmers in Acheng District, Harbin City, and made adjustments to the interview content and specific questions to improve the shortcomings of the questionnaire on the basis of sorting out the interview content and summarizing experiences.

Phase II: The formal research was launched from July to September 2021. To guarantee the scientific nature of sample selection, a three-stage random sampling was mainly adopted, in which five prefecture-level cities were randomly selected in Heilongjiang province, including Harbin, Qiqihar, Daqing, Jiamusi and Suihua; then, four counties were randomly selected in each prefecture-level city, and two to three villages were randomly selected in each county, and 10 farmers were randomly selected in each village for the questionnaire survey.

Considering the age and educational differences of farmers, the questionnaires were conducted in a one-on-one interview mode. In addition, a reward and punishment mechanism was set up to review and select the quality of questionnaires, to ensure the scientific nature of the questionnaire data. A total of 460 questionnaires were distributed in the survey. After excluding questionnaires with incomplete or missing data, logical contradictions, and irregularities, a total of 442 valid questionnaires were finally obtained, with a sampling efficiency of 96.09%.

3.2. Econometric Methods

3.2.1. Chemical Fertilizer Economic Optimal Application Amount

It is difficult to know whether farmers have reduced the amount of chemical fertilizer application, because the farmer's own answer is subjective and arbitrary. Therefore, in order to more accurately judge whether farmers have reduced fertilizer use, this paper, based on sorting out the views of the rational smallholder school represented by Schultz, tries to estimate the economic optimal chemical fertilizer application by farmers, and to measure whether farmers reduce chemical fertilizer application by comparing the difference between the economic optimal chemical fertilizer application and the actual chemical fertilizer application by farmers. Referring to the past research [38–40], we took corn yield as the dependent variable and chemical fertilizer input, labor input, machinery input and seed input as independent variables, the C-D production function model was constructed to estimate the output elasticity of chemical fertilizers. The C-D production function was set as follows:

$$\ln yield = \alpha_0 + \beta_1 \ln(fertilizer) + \beta_2 \ln(labor) + \beta_3 \ln(machine) + \beta_4 \ln(seed) + \varepsilon \tag{1}$$

In the Equation (1), yield denotes the average corn yield per mu of the farmer, fertilizer denotes the average chemical fertilizer input per mu of the farmer, labor, machine and seed denote the average input per mu of labor, machinery and seed, α and β denote the parameters to be estimated, and ε denotes the random error term.

Based on profit maximization theory, it is known that if farmers want to maximize profit, they should choose a production point where marginal benefit and marginal cost are equal, at which point the marginal benefit of chemical fertilizer on maize yield is equal to the ratio of chemical fertilizer price and maize price. As follows:

$$\frac{\partial yield}{\partial fertilizer} = \frac{p_{fertilizer}}{p_{yield}} \tag{2}$$

Meanwhile, based on the chemical fertilizer output elasticity β_1 measured in Equation (1), the marginal return of chemical fertilizer to maize yield is:

$$\frac{\partial yield}{\partial fertilizer} = \beta_1 \times \frac{yield}{fertilizer} \tag{3}$$

Using the profit maximization Equations (2) and (3), the average economic optimal chemical fertilizer application rate per mu can be obtained (4):

$$fertilizer_{optimal} = \frac{\beta_1 \times yield}{p_{fertilizer}/p_{yield}} \tag{4}$$

3.2.2. Endogenous Switching Probit Model

Since the decision on farmland transfer and the decision on chemical fertilizer application are often affected by some unobservable factors at the same time, there may be problems of "simultaneous decision" and "self-selection", which may lead to biased estimation results. In order to overcome this shortcoming, this paper uses the endogenous switching probit model to correct this problem. The model measures the average processing effect by constructing a counterfactual scenario, and then corrects the selection bias caused by unobservable or observable variables, that is, overcomes the biased estimation problem caused by the "self-selection" of the sample, so as to obtain a more robust estimation result [41–43].

Firstly, the influence model of farmers' decision to transfer into farmland is constructed, and the decision to transfer into farmland is a binary choice variable. It is assumed that the potential benefit to be gained from the transfer of farmland by farmer (i) is D^*_{ia}, and the potential benefits for farmers who have not transferred into farmland is D^*_{in}, and then the condition for farmers to transfer into farmland is $D^*_{ia} - D^*_{in} = D^*_i > 0$, which means that farmers can get a higher return if they transfer into farmland. D^*_i is the latent variable that cannot be observed directly. Then, the decision model for whether a farmer transfers into farmland is:

$$D_\mathrm{i} = \begin{pmatrix} 1, & D^*_\mathrm{i} > 0 \\ 0, & D^*_i \leq 0 \end{pmatrix} \tag{5}$$

In Equation (5), D_i is the farmer's decision whether to transfer into farmland or not. When $D_i = 1$, it means that the farmer transferred into farmland, and when $D_i = 0$, it means that the farmer did not transfer into farmland. Therefore, the following model can be constructed for the effect of transferring into farmland on farmers' chemical fertilizer inputs.

$$Y_i = \alpha_i D_i + \sum_{j=1}^{n} \beta_j X_{ij} + \varepsilon_i \tag{6}$$

In Equation (6), Y_i is the probability of reduced chemical fertilizer application, and X_{ij} is the specific variables such as personal characteristics, household characteristics and production and operation characteristics that affect the chemical fertilizer input. α_i and β_i are both the coefficient to be estimated. ε_i is the random error term. Farmland transfer-in is the result of farmers' self-selection based on expected return analysis and is influenced by other factors which may also affect chemical fertilizer application, and thus generate sample selectivity bias. Therefore, the parameters α_i in model (6) can't accurately reflect the effect of farmland transfer-in on chemical fertilizer reduction.

We chose the endogenous switching probit model (ESP) to scientifically assess the impact of farmland transfer on chemical fertilizer reduction while constructing a counterfactual analytical framework to address issues such as information omissions [44,45].

The ESP model is divided into two stages, the first stage focuses on measuring the probability of farmers transferring into farmland, and the second stage focuses on constructing a decision equation for farmers to reduce chemical fertilizer. The specific equations are as follows:

Select Equation (whether farmers will transfer into farmland):

$$D_i = \gamma Z_i + \delta I_i + \mu_i \tag{7}$$

Result Equation (1) (treatment group, chemical fertilizer reduction equation for farmland transfer-in groups):

$$Y_{ia} = \beta'_a X_{ia} + \varepsilon_{ia} \tag{8}$$

Result Equation (2) (control group, chemical fertilizer reduction equation for farmland non-transfer-in groups):

$$Y_{in} = \beta'_n X_{in} + \varepsilon_{in} \tag{9}$$

In Equation (7), D_i is a binary variable, indicating whether the farmer transfers into farmland, Z_i refers to the influencing factors that affect whether the farmer transfers into farmland, I_i is the identification variable, which is represented by whether the neighbor transfers into farmland; μ_i is the error term, γ and δ indicate the parameter to be estimated; in Equations (8) and (9), Y_{ia} and Y_{in} are the chemical fertilizer reduction behavior of the farmland transfer-in group and the farmland non-transfer-in group, X_{ia} and X_{in} are the factors that affect farmers' chemical fertilizer reduction. ε_{ia} and ε_{in} are the error terms. β'_a and β'_n indicate the parameter to be estimated.

The estimation results of the ESP model show the probability of chemical fertilizer reduction by farmers under the true scenario, and further calculate the expected value of the reduction in chemical fertilizer application by farmers in the counterfactual scenario. Comparing the two, the average treatment effect of farmland transfer-in on the reduction of chemical fertilizer application can be obtained.

$$E[Y_{ia}/D_i = 1] = \beta'_a X_{ia} + \sigma_{\mu a}\lambda_{ia} \tag{10}$$

$$E[Y_{in}/D_i = 0] = \beta'_n X_{in} + \sigma_{\mu n}\lambda_{in} \tag{11}$$

$$E[Y_{in}/D_i = 1] = \beta'_n X_{ia} + \sigma_{\mu n}\lambda_{ia} \tag{12}$$

$$E[Y_{ia}/D_i = 0] = \beta'_a X_{in} + \sigma_{\mu a}\lambda_{in} \tag{13}$$

Equations (10)–(13) are the probability of chemical fertilizer reduction under the four scenarios: farmland transfer-in farmers, farmland non-transfer-in farmers, assumed farmland transfer-in farmers decided not to transfer into farmland, and assumed farmland non-transfer-in farmers decided to transfer into farmland, and thus estimate the average treatment effect of farmland transfer-in on the reduction of chemical fertilizer.

Therefore, the average treatment effect (*ATT*) of farmland transfer-in on the effect of chemical fertilizer reduction, can be expressed as:

$$ATT_i = E[Y_{ia}/T_i = 1] - E[Y_{in}/T_i = 1] = (\beta'_a - \beta'_n)X_{ia} + (\sigma_{\mu a} - \sigma_{\mu n})\lambda_{ia} \tag{14}$$

Correspondingly, the average treatment effect (*ATU*) of farmland non-transfer-in on the effect of chemical fertilizer reduction, can be expressed as:

$$ATU_i = E[Y_{ia}/T_i = 0] - E[Y_{in}/T_i = 0] = (\beta'_a - \beta'_n)X_{in} + (\sigma_{\mu a} - \sigma_{\mu n})\lambda_{in} \tag{15}$$

In summary, this study used the mean of ATT_i and ATU_i to measure the average treatment effect of farmland transfer on chemical fertilizer reduction.

In addition, we applied the endogenous switching probit model, under the counterfactual framework, to further analyze the treatment effects of concentrated and non-concentrated farmland transfer-in on the reduction of chemical fertilizer. Referring to the above studies and models, A_i indicates whether the farmland transfer-in is concentrated, and whether transferring farmland through land transfer intermediary service organizations (T_i) was chosen as the identifying variable.

3.3. Variable Description

Dependent variable. "Chemical fertilizer reduction" was used as the dependent variable, specifically to measure whether farmers reduced chemical fertilizer by comparing the economic optimal chemical fertilizer application rate with the actual chemical fertilizer application rate, and the actual chemical fertilizer application rate below the economic optimal chemical fertilizer application rate was defined as chemical fertilizer reduction. Table 1 shows that the probability of chemical fertilizer reduction for the surveyed farmers was 0.22.

Independent variable. "Farmland transfer-in" and "farmland concentrated transfer-in" were used as independent variables. Specifically, the first independent variable is the behavior of farmers transferring into farmland; the second independent variable is the behavior of farmers transferring into farmland in a concentrated manner, including farmers transferring into farmland adjacent to existing farmland, or the transferred-in farmlands are adjacent to one another. Table 1 shows that the probability of the surveyed farmers transferring into farmland is 0.56, and the probability of farmland concentrated transfer-in is 0.20.

Table 1. Descriptive statistics of variables.

Variables Category	Variables	Definition	Mean	S. D
Input and output variables	Corn yield	kg/mu	674.25	82.97
	Chemical fertilizer input	kg/mu	59.38	13.70
	Labor input	Workdays/mu	22.05	9.29
	Mechanical input	Yuan/mu	92.50	22.28
	Seed input	Yuan/mu	59.96	8.39
Dependent variables	Chemical fertilizer reduction	Yes = 1, No = 0	0.22	0.41
Independent variables	Farmland transfer-in	Yes = 1, No = 0	0.56	0.50
	Farmland concentrated transfer-in	Yes = 1, No = 0	0.20	0.40
Controlled variables	Age	Actual age in 2021	54.08	11.93
	Gender	Male = 1; Female = 0	0.79	0.41
	Education level	Years spent in school	7.43	3.29
	Risk appetite level	Strongly dislike = 1, Strongly prefer = 7	3.46	1.55
	Household income level	Total family income (million yuan)	6.80	3.78
	Main business of household	Agriculture = 1, Other = 0	0.70	0.46
	Chemical fertilizer application technology training	Yes = 1, No = 0	0.34	0.48
	Farmland size	farmland area (mu)	89.13	45.72
	Farmland quality	Very poor = 1, Very good = 7	4.84	1.36
Identifying variables	Neighbors transfer into farmland	Yes = 1, No = 0	0.47	0.50
	Transfer into farmland through intermediaries	Yes = 1, No = 0	0.32	0.47

Controlled variable. The chemical fertilizer input behavior of maize growers will be influenced by other factors besides farmland transfer, and this paper mainly selected control variables from the following perspectives. In terms of personal characteristics of production decision-makers, age, gender, education level, and risk appetite level were selected; in terms of household characteristics, household income level, main business of household and chemical fertilizer application technology training were selected; in terms of production and operation characteristics, farmland size and farmland quality were selected.

Identifying variable. The identifying variables should meet the two conditions of relevance and exogeneity. In this study, whether neighbors transferred into farmland and whether farmers transferred into farmland through intermediaries were selected as identifying variables. Among them, whether neighbors transfer into farmland only affects farmers' behavior of transferring into farmland but not directly affects chemical fertilizer input behavior; whether farmers transfer into farmland through intermediaries only affects whether farmers will transfer into farmland concentratedly but not directly affects chemical fertilizer input behavior; land transfer intermediary organizations, such as village and community organizations and other land transfer service agencies, have the ability to collect information, supervise and coordinate, and are important for matching supply and demand for farmland transfer. They play an important role in matching supply and demand for farmland and in the realization of continuous agricultural land transfer. Table 1 shows that among the surveyed farmers, the average probability of neighbors transferring to farmland is 0.47, and the probability of transferring farmland through intermediaries is 0.32.

4. Empirical Results

4.1. Chemical Fertilizer Economic Optimal Application Amount

The results of estimating Equation (1) using Stata software and ordinary least squares robustness regression are shown in Table 2.

Table 2. C-D production function estimation results.

Variables	Corn Yield	
	Coefficient	Standard Error
Chemical fertilizer input	0.098 **	0.045
Labor input	0.053 *	0.028
Mechanical input	0.023 **	0.011
Seed input	0.036	0.031
Constant	6.532 ***	0.303
Observations	442	
R^2	0.742	

Note: ***, **, and * indicate that it is significant at the 1%, 5%, and 10% levels.

According to the Equation (4), the economic optimal chemical fertilizer application rate of farmers can be calculated, as shown in Table 3. The calculation results show that the average optimal economic chemical fertilizer application rate of 442 households is 47.25 kg/mu, but the actual application rate is 59.38 kg/mu, and the actual chemical fertilizer application rate is 12.12 kg/mu higher than the economic optimal application rate on the whole. According to the analysis of the research results, it can be seen that among 442 farmers, the number of farmers who exceeded the economic optimal chemical fertilizer application amount was as high as 345, accounting for 78.05% of the total sample, and only 97 farmers did not exceed the economic optimal chemical fertilizer application amount, accounting for 21.95% of the total sample number. Therefore, it can be seen that most of the farmers in the survey area have an over-application of chemical fertilizers.

Table 3. Calculation results of economic optimal chemical fertilizer application amount.

Actual Amount of Chemical Fertilizer (kg/mu)	Optimal Amount of Chemical Fertilizer (kg/mu)	Excessive Amount of Chemical Fertilizer (kg/mu)	Percentage of Farmers Who Over-Fertilize (%)
59.38	47.25	12.13	78.05

4.2. Impact of Farmland Transfer on Chemical Fertilizer Reduction

The effects of "farmland transfer-in" and "farmland concentrated transfer-in" on chemical fertilizer reduction separately were studied in this paper.

First, a full-sample regression was performed on 442 farmer households to investigate the effect of farmland transfer on the reduction of chemical fertilizer. From the estimation results of the farmland transfer selection model, age, gender, education level, risk appetite level, main business of household, and farmland size had significant effects on farmers' decision on farmland transfer. Among them, age had a negative effect on farmland transfer-in, while gender, education, risk appetite level, main business of household, and farmland size had a positive effect on farmland transfer-in. From the estimation results of the chemical fertilizer reduction application outcome equation, age, risk appetite level, household income level, and chemical fertilizer application technology training had significant effects on the chemical fertilizer reduction application behavior of farmers. Among them, age and household income level were negatively correlated with chemical fertilizer reduction, and risk appetite level and chemical fertilizer application technology training were positively correlated with chemical fertilizer reduction.

Second, the sample of 248 farmers who transferred into farmland was regressed and the impact of concentrated farmland transfer-in on the reduction of fertilizer application was examined. In terms of the selection equation, age, gender, household income level, main business of household, and farmland size all had significant effects on whether farmers concentrated transferred into farmland. Age had a negative effect on the choice of concentrated transferring into farmland, and gender, household income level, main business of household, and farmland size had a positive effect on farmland transfer-in concentration. In terms of the outcome equation, education level, risk appetite level, and chemical fertilizer application technology training had significant effects on chemical fertilizer reduction, while age had significant effects on chemical fertilizer reduction for farmers who transferred into farmland concentratedly. Specifically, age had a negative effect on the level of chemical fertilizer reduction for farmers who transferred into farmland concentratedly. Education level, risk appetite level and chemical fertilizer application technology training had a greater effect on the level of chemical fertilizer application reduction for farmers who transferred into farmland concentratedly, and a lesser effect for farmers who didn't transfer into farmland concentratedly (Table 4).

To further examine the effect of farmland transfer-in on chemical fertilizer reduction, the endogenous switching probit model was applied to further analyze the level of chemical fertilizer reduction in a counterfactual framework under four scenarios: farmland transfer-in farmers, farmland non-transfer-in farmers, assumed farmland transfer-in farmers who decided not to transfer into farmland, and assumed farmland non-transfer-in farmers who decided to transfer into farmland. The results are shown in Table 5. Farmland transfer has a significant positive treatment effect on chemical fertilizer reduction.

The results of ATT estimation showed that if farmland transfer-in farmers decide not to transfer into farmland, the chemical fertilizer reduction level would be reduced by 18.644%. The ATU estimation showed that if farmland non-transfer-in farmers decide to transfer into farmland, the chemical fertilizer reduction level would be increased by 26.519%. Therefore, farmland transfer-in can significantly increase the level of chemical fertilizer reduction.

Similarly, we applied the endogenous switching probit model, under the counterfactual framework, to further analyze the treatment effects of concentrated and non-concentrated farmland transfer-in on the reduction of chemical fertilizer. Table 6 shows the fertilizer reduction application levels under four scenarios: farmland concentrated transfer-in farmers, farmland non-concentrated transfer-in farmers, assumed farmland concentrated transfer-in farmers who decided not to transfer into farmland concentratedly, and assumed farmland non-concentrated transfer-in farmers who decided to transfer into farmland concentratedly.

We discovered that if farmland concentrated transfer-in farmers decide not to transfer into farmland concentratedly, the chemical fertilizer reduction treatment effect would be reduced by 18.790%. In addition, if farmland non-concentrated transfer-in farmers decide to transfer into farmland concentratedly, the treatment effect would increase by 18.487%. It can be seen that concentrated farmland transfer-in has a significant promoting effect on the reduction of chemical fertilizer.

Table 4. Determinants of farmland transfer and chemical fertilizer reduction.

Variables	Select Equation (Whether to Transfer into Farmland)	Result Equation (Whether to Reduce Chemical Fertilizer)		Select Equation (Whether to Transfer into Farmland Concentratedly)	Result Equation (Whether to Reduce Chemical Fertilizer)	
		Farmland Transfer-in	Farmland Non-Transfer-in		Farmland Concentrated Transfer-in	Farmland Non-Concentrated Transfer-in
Age	−0.021 ** (0.010)	−0.006 ** (0.003)	−0.013 ** (0.005)	−0.062 * (0.032)	−0.023 * (0.014)	−0.045 (0.033)
Gender	0.060 * (0.033)	0.038 (0.027)	0.021 (0.016)	0.038 * (0.021)	0.056 (0.038)	0.050 (0.039)
Education level	0.048 * (0.026)	0.073 (0.055)	0.052 (0.035)	0.145 (0.122)	0.083 ** (0.040)	0.041 * (0.022)
Risk appetite level	0.039 ** (0.017)	0.096 * (0.053)	0.074 * (0.040)	0.072 (0.048)	0.034 ** (0.016)	0.029 ** (0.013)
Household income level	−0.074 (0.055)	−0.019 * (0.011)	−0.023 * (0.013)	0.053 ** (0.023)	−0.036 (0.030)	0.024 (0.017)
Main business of household	0.035 *** (0.013)	−0.162 (0.115)	−0.082 (0.064)	0.068 *** (0.025)	−0.061 (0.050)	−0.032 (0.021)
Chemical fertilizer application technology training	0.008 (0.006)	0.158 ** (0.067)	0.073 *** (0.026)	0.012 (0.009)	0.043 ** (0.019)	0.029 ** (0.013)
Farmland size	0.052 * (0.027)	0.036 (0.024)	0.105 (0.083)	0.051 * (0.027)	0.063 (0.045)	0.103 (0.078)
Farmland quality	0.072 (0.050)	0.372 (0.313)	0.273 (0.226)	0.301 (0.251)	0.141 (0.101)	0.083 (0.068)
Neighbors transfer into farmland/Transfer into farmland through intermediaries	0.593 *** (0.201)	—	—	0.487 *** (0.158)	—	—
Constant	−3.826 *** (1.263)	0.376 *** (0.129)	0.269 *** (0.097)	−4.632 *** (1.544)	0.438 *** (0.164)	0.386 *** (0.133)
ρ_1		−0.736 ** (0.353)			−0.703 * (0.405)	
ρ_0		−0.825 * (0.426)			−0.791 * (0.418)	
Goodness-of-fit test		253.621 ***			238.631 ***	
Log likelihood		−753.621			−642.392	
Observations		442			248	

Note: ***, **, and * indicate significant at the 1%, 5%, and 10% levels.

Table 5. Estimates of the effect of transferring into farmland on chemical fertilizer reduction.

Result Variables	Farmer Type and Treatment Effects	Decision type		Average Treatment Effect	Change (%)
		Farmland Transfer-in	Farmland Non-Transfer-in		
Chemical fertilizer reduction	Farmland transfer-in (ATT)	0.236	0.192	0.044	18.644
	Farmland non-transfer-in (ATU)	0.229	0.181	0.048	26.519

Table 6. Estimates of the effect of concentrated transferring into farmland on chemical fertilizer reduction.

Result Variables	Farmer Type and Treatment Effects	Decision Type		Average Treatment Effect	Change (%)
		Farmland Concentrated Transfer-in	Farmland Non-Concentrated Transfer-in		
Chemical fertilizer reduction	Farmland concentrated transfer-in (ATT)	0.314	0.255	0.059	18.790
	Farmland non-concentrated transfer-in (ATU)	0.282	0.238	0.044	18.487

5. Discussion

Farmland transfer is one of the key ways to promote chemical fertilizer reduction. Based on the theoretical basis of elucidating the intrinsic linkage between farmland transfer and chemical fertilizer reduction, the survey data of 442 corn growers in Heilongjiang Province was used, and the endogenous switching probit model was adopted. Then the effects of transferring into farmland and concentrated transferring into farmland on the reduction of chemical fertilizer were quantitatively analyzed.

Our study clearly shows that farmland transfer-in has a positive effect on chemical fertilizer reduction. This is partly explained by the findings of Ju et al. [19], Zhao et al. [20] and Hu et al. [21] that increasing the size of farmland will greatly reduce the use of chemical fertilizers in agricultural production. One possible explanation is that farmland transfer helps to play the scale management effect and effectively reduces the cost of farmers' acquisition of "new technology and new knowledge", which awakens farmers' ecological consciousness and motivates them to adopt clean production methods, thereby reducing the use of chemical fertilizers.

On the other hand, we also found that compared with farmland transferred dispersedly, the farmland transferred concentratedly is more helpful for the farmers to reduce the application of chemical fertilizers. This finding is the same as the study by Liang et al. [29] and Liang et al. [31]. The possible reason is that the concentrated circulation makes the transferred farmland adjacent, which can not only expand the scale of operation, but also expand the scale of the plot, and then realize the effect of economies of scale through horizontal specialization and deepening vertical division of labor to achieve fertilizer reduction. However, the scattered transfer of agricultural land did not change the size of the plot, and the plot was still in a finely divided state. The increase in time cost and economic cost may lead farmers to apply more chemical fertilizers.

In conclusion, farmland transfer can reduce the amount of chemical fertilizers applied in the process of corn production, but the effect of dispersed transfer to farmland and concentrated transfer to farmland are different. For farmers who transferred into farmland in a decentralized form, they spend more time and money managing the fragmented farmland, which may stimulate them to apply more fertilizers. For the farmers who transferred into the farmland in a centralized way, the application amount of chemical fertilizer was reduced by exerting economies of scale.

The main contributions of this paper are: First, in view of the inconsistent conclusions of existing studies, the intrinsic association between farmland transfer and chemical fertilizer reduction was re-examined, and furthermore compares and analyzes the effects of concentrated transfer of farmland and dispersed transfer of farmland to fertilizer reduction, this provides a new perspective on the fertilizer reduction effect of farmland transfer. Secondly, since the decision on farmland transfer and the decision on chemical fertilizer application may be affected by some unobservable factors at the same time, there may be a problem of "simultaneous decision" and "self-selection", resulting in biased estimation results. We solve this problem perfectly by using an endogenous switching probability model and obtain more robust estimation results.

Finally, it should be pointed out that this research can be further improved upon in the following two aspects in the future: First, this study mainly focuses on corn growers and does not study the behavior of farmers who grow other types of crops in the application of chemical fertilizers in farmland transfer. In view of the differences in the application of chemical fertilizers between commercial crops and food crops, it is necessary to compare and analyze commercial and food crops in the future. Secondly, only taking Heilongjiang Province as the research area, the research results and countermeasures are more applicable to some major grain producing areas with large scale agricultural operations and high levels of farmland transfer. As a result, future research must consider the aforementioned factors in order to conduct more beneficial investigations into issues such as farmland transfer and fertilizer reduction.

6. Conclusions and Implication

This study estimated the effect of farmland transfer on fertilizer reduction. The results showed that farmland transfer could reduce fertilizer application in maize production. In addition, the effects of concentrated transfer to farmland and dispersed transfer to farmland on fertilizer reduction were further analyzed. It was found that farmland concentrated transfer-in farmers were more likely to reduce fertilizer application compared with those who transferred to farmland in a scattered manner.

The results of this study have important implications for more effective realization of farmland transfer and chemical fertilizer reduction. Based on the above analysis, this paper proposes the following suggestions. First, the orderly transfer of farmland management rights should be guided continuously. The government should standardize farmland transfer procedures, increase farmers' willingness and motivation to participate in farmland transfer, dispel farmers' farmland transfer concerns, let farmland transfer drive chemical fertilizer reduction, and promote green and sustainable agricultural development. Second, an information service platform for farmland transfer should be established to strengthen the management, guidance and service of farmland transfer, effectively coordinate and communicate the needs and conflicts of farmland transfer, so as to enhance the effect of farmland transfer on chemical fertilizer reduction. Third, through comprehensive improvement of farmland, a favorable external environment for the continuous circulation of farmland and the reduction of chemical fertilizers should be created by implementing comprehensive measures such as land leveling, soil improvement and agricultural water conservancy construction so as to alleviate the problem of natural fragmentation of farmland. Finally, the scale operation mode should be innovated and improved by means of cooperating with farmers, guiding the appropriate agglomeration of land and providing social services. Simultaneously, convenient conditions for fertilizer reduction should be provided through strengthening the agricultural socialization service system for small farmers, creating a supportive environment, and promoting the scale of agricultural productive services.

Author Contributions: Visualization, X.B. and J.D.; Supervision, N.C.; Writing—Original Draft Preparation, N.C. and X.B.; Writing—Review and Editing, J.D.; Data Curation, Methodology, J.D. and X.F.; Funding Acquisition, N.C. All authors have read and agreed to the published version of the manuscript.

Funding: This research was funded by the National Social Science Foundation of China (Project No. 20BJY149); the Research Planning Project for Heilongjiang Philosophy and Social Science (Project No. 21JYA441).

Informed Consent Statement: Not applicable.

Data Availability Statement: The data presented in this study are available on request from the corresponding author. The data are not publicly available due to restrictions of corporate data open-ness.

Conflicts of Interest: The authors declare that they have no known competing financial interest or personal relationships that could have appeared to influence the work reported in this paper.

References

1. Zhou, S.; Su, S.; Meng, L.; Liu, X.; Zhang, H.; Bi, X. Potentially toxic trace element pollution in long-term fertilized agricultural soils in China: A meta-analysis. *Sci. Total Environ.* **2021**, *50*, 147967. [CrossRef] [PubMed]
2. Sun, Y.; Hu, R.; Zhang, C. Does the adoption of complex fertilizers contribute to fertilizer overuse? Evidence from rice production in China. *J. Clean. Prod.* **2019**, *219*, 677–685. [CrossRef]
3. Wang, X.; Cai, D.; Grant, C.; Hoogmoed, W.B.; Oenema, O. Changes in regional grain yield responses to chemical fertilizer use in China over the last 20 years. *J. Soil Sci. Plant Nutr.* **2018**, *18*, 312–328. [CrossRef]
4. Huang, W.; Jiang, L. Efficiency performance of fertilizer use in arable agricultural production in China. *Chin. Agric. Econ. Rev.* **2019**, *11*, 2–69. [CrossRef]
5. Cong, X.; Shan, J. Research on input reduction of chemical fertilizers and pesticides and soil pollution control of agricultural land. *Jiang-Huai Trib.* **2019**, *62*, 17–23. [CrossRef]
6. Bijay-Singh; Craswell, E. Fertilizers and nitrate pollution of surface and ground water: An increasingly pervasive global problem. *SN Appl. Sci.* **2021**, *3*, 518. [CrossRef]
7. Kliopova, I.; Baranauskaitė-Fedorova, I.; Malinauskienė, M.; Staniškis, J. Possibilities of increasing resource efficiency in nitrogen fertilizer production. *Clean Technol. Environ. Policy* **2016**, *18*, 901–914. [CrossRef]
8. Ghodszad, L.; Reyhanitabar, A.; Maghsoodi, M.R.; Lajayer, B.A.; Chang, S.C. Biochar affects the fate of phosphorus in soil and water: A critical review. *Chemosphere* **2021**, *283*, 131176. [CrossRef] [PubMed]
9. Maghsoodi, M.R.; Ghodszad, L.; Lajayer, B.A. Dilemma of hydroxyapatite nanoparticles as phosphorus fertilizer: Potentials, challenges and effects on plants. *Environ. Technol. Innov.* **2020**, *19*, 100869. [CrossRef]
10. Wu, H.; Ge, Y. Excessive application of fertilizer, agricultural non-point source pollution, and farmers' policy choice. *Sustainablity* **2019**, *11*, 1165. [CrossRef]
11. Nath, R.; Venugopalan, V.K.; Sarath, C.M.A. Smart fertilizers a way ahead for sustainable agriculture. *J. Plant Nutr.* **2022**, *45*, 2068–2076. [CrossRef]
12. Rahman, M.; Haque, K.; Khan, M. A review on application of controlled released fertilizers influencing the sustainable agricultural production: A Cleaner production process. *Environ Technol. Innov.* **2021**, *23*, 101697. [CrossRef]
13. Wu, J.; Wen, X.; Qi, X.; Fang, S.; Xu, C. More land, less pollution? How land transfer affects fertilizer application. *Int. J. Environ. Res. Public Health* **2021**, *18*, 11268. [CrossRef]
14. Pan, D.; Kong, F.; Zhang, N.; Ying, R. Knowledge training and the change of fertilizer use intensity: Evidence from wheat farmers in China. *J. Environ. Manag.* **2017**, *197*, 130–139. [CrossRef] [PubMed]
15. Liu, Q. Spatio-temporal changes of fertilization intensity and environmental safety threshold in China. *Trans. Chin. Soc. Agric. Eng.* **2017**, *33*, 214–221. [CrossRef]
16. Abbas, A.; Zhao, C.; Ullah, W.; Ahmad, R.; Waseem, M.; Zhu, J. Towards sustainable farm production system: A case study of corn farming. *Sustainability* **2021**, *13*, 9243. [CrossRef]
17. Hu, L.; Zhang, X.; Zhou, Y. Farm size and fertilizer sustainable use: An empirical study in Jiangsu, China. *J. Integr. Agric.* **2019**, *18*, 2898–2909. [CrossRef]
18. Chen, Y.; Fu, X.; Liu, Y. Effect of Farmland Scale on Farmers' Application Behavior with Organic Fertilizer. *Int. J. Environ. Res. Public Health* **2022**, *19*, 4967. [CrossRef]
19. Ju, X.; Gu, B.; Wu, Y.; Galloway, J.N. Reducing China's fertilizer use by increasing farm size. *Glob. Environ. Chang.* **2016**, *27*, 26–32. [CrossRef]
20. Zhao, C.; Kong, X.; Qiu, H. Does the expansion of farm size contribute to the reduction of chemical fertilizers?—Empirical analysis based on 1274 family farms in China. *J. Agrotech. Econ.* **2021**, *40*, 110–121. [CrossRef]
21. Hu, X.; Su, K.; Chen, W.; Yao, S.; Zhang, L. Examining the impact of land consolidation titling policy on farmers' fertiliser use: Evidence from a quasi-natural experiment in China. *Land Use Policy* **2021**, *38*, 105645. [CrossRef]
22. Wang, C.; Duan, J.; Ren, C.; Liu, H.; Reis, S.; Xu, J.; Gu, B. Ammonia Emissions from Croplands Decrease with Farm Size in China. *Environ. Technol. Innov.* **2022**, *56*, 9915–9923. [CrossRef] [PubMed]
23. Liu, Y.; Wang, C.; Tang, Z.; Nan, Z. Does farmland rental contribute to reduction of agrochemical use? A case of grain production in Gansu Province, China. *Sustainability* **2019**, *11*, 2402. [CrossRef]

24. Tian, Y.; Zhang, J.; He, K.; Feng, J. Farmers' Behavior of Low-carbon Agricultural Production and Influencing Factors——Taking fertilizer application and pesticide use as examples. *Chin. Rural Surv.* **2015**, *36*, 61–70.
25. Liu, X.; Zhang, D.; Xu, Z. Does grain scale farmers also overuse fertilizer?—Based on the heterogeneity of large-sizeo farmers and small-sized farmers. *J. Agrotech. Econ.* **2020**, *39*, 117–129. [CrossRef]
26. Wiggins, S.; Kirsten, J.; Lambi, L. The future of small farms. *World Dev.* **2010**, *38*, 1341–1349. [CrossRef]
27. Knickel, K.; Redman, M.; Darnhofer, I.; Ashkenazy, A.; Calväo Chebach, T.; Sumane, S.; Tisenkopfs, T.; Zemeckis, R.; Atkociuniene, V.; Rivera, M. Between aspirations and reality: Making farming, food systems and rural areas more resilient, sustainable and equitable. *J. Rural Stud.* **2017**, *59*, 197–210. [CrossRef]
28. Ashkenazy, A.; Chebach, T.C.; Knickel, K.; Peter, S.; Horowitz, B.; Offenbach, R. Operationalising resilience in farms and rural regions–Findings from fourteen case studies. *J. Rural Stud.* **2018**, *59*, 211–221. [CrossRef]
29. Liang, Z.; Zhang, L.; Zhang, J. Land inward transfer, plot scale and chemical fertilizer reduction: An empirical analysis based on main rice-producing areas in Hubei Province. *Chin. Rural Surv.* **2020**, *41*, 73–92.
30. Zhang, L.; Luo, B. Agricultural chemical reduction: The logic and evidence based on farmland operation scale of households. *Chin. Rural Econ.* **2020**, *36*, 81–99.
31. Liang, Z.; Zhang, L.; Zhang, J. Land consolidation and fertilizer reduction: Quasi-natural experimental evidence from China's well-facilitated capital farmland construction. *Chin. Rural Econ.* **2021**, *37*, 123–144.
32. Zhu, C.; Ouyang, Y.; Diao, Y.; Yu, J.; Luo, X.; Zheng, J.; Li, X. Effects of mechanized deep placement of nitrogen fertilizer rate and type on rice yield and nitrogen use efficiency in Chuanxi Plain, China. *J. Integr. Agric.* **2021**, *20*, 581–592. [CrossRef]
33. Adnan, N.; Nordin, S.M.; Rahman, I.; Noor, A. The effects of knowledge transfer on farmers decision making toward sustainable agriculture practices in view of green fertilizer technology. *World J. Sci. Technol.* **2018**, *15*, 98–115. [CrossRef]
34. Liang, Z.; Zhang, L.; Liu, Y.; Zhang, J. Is the agricultural division of labor conducive to the reduction of fertilizer input? Empirical evidence from rice production households in the Jianghan Plain. *Chin. Popul. Resour. Environ.* **2020**, *30*, 150–159.
35. Emmanuel, D.; Owusu-Sekyere, E.; Jordaan, H.; Owusu, V. Impact of agricultural extension service on adoption of chemical fertilizer: Implications for rice productivity and development in Ghana. NJAS-Wagen. *J. Life Sci.* **2016**, *79*, 41–49. [CrossRef]
36. Cai, B.; Shi, F.; Huang, Y.; Abatechanie, M. The impact of agricultural socialized services to promote the farmland scale management behavior of smallholder farmers: Empirical evidence from the rice-growing region of Southern China. *Sustainability* **2022**, *14*, 316. [CrossRef]
37. Rahman, M.; Connor, J. Impact of agricultural extension services on fertilizer use and farmers' welfare: Evidence from Bangladesh. *Sustainability* **2022**, *14*, 9385. [CrossRef]
38. Qiu, H.; Luan, H.; Li, J.; Wang, Y. The impacts of risk aversion on farmer's households' behaviour of overusing chemical fertilizers. *Chin. Rural Econ.* **2014**, *30*, 85–96.
39. Wu, H.; Hao, H.; Lei, H.; Ge, Y.; Shi, H. Farm size, risk aversion and overuse of fertilizer: The heterogeneity of large-scale and small-scale wheat farmers in Northern China. *Land* **2021**, *10*, 111. [CrossRef]
40. Khan, S.; Khan, M.; Jan, I.; Khan, M.; Khan, F.M. Determinants of Sugarcane Yield in District Charsadda, Pakistan. *Sarhad J. Agric.* **2020**, *36*, 1141–1148. [CrossRef]
41. Nkegbe, P.K.; Araar, A.; Abu, B.M.; Alhassan, H.; Ustarz, Y.; Setsoafia, E.D.; Abdul-Wahab, S. Nonfarm activity and market participation by farmers in Ghana. *Agric. Food Econ.* **2022**, *10*, 1–23. [CrossRef]
42. Haile, K.K.; Nillesen, E.; Tirivayi, N. Impact of Formal Climate Risk Transfer Mechanisms on Risk-Aversion: Empirical Evidence from Rural Ethiopia. *World Dev.* **2019**, *130*, 104930. [CrossRef]
43. Issahaku, G.; Abdul-Rahaman, A. Sustainable land management practices, off-farm work participation and vulnerability among farmers in Ghana: Is there a nexus? *Int. Soil Water Conserv. Res.* **2018**, *7*, 18–26. [CrossRef]
44. Gondwe, T.; Tegbaru, A.; Oladeji, A.E.; Khonje, M.; Manda, J.; Gaya, H. Correlates and consequences of women's participation in the cowpea value chain in eastern Zambia. *Arekon* **2017**, *56*, 263–273. [CrossRef]
45. Hao, J.H.; Bijman, J.; Gardebroek, C.; Heerink, N.; Heijman, W.; Huo, X.X. Cooperative membership and farmers' choice of marketing channels–Evidence from apple farmers in Shaanxi and Shandong Provinces, China. *Food Policy* **2018**, *74*, 53–64. [CrossRef]

sustainability

MDPI

Article

Research Progress, Hotspots and Trends of Land Use under the Background of Ecological Civilization in China: Visual Analysis Based on the CNKI Database

Mingyang Nan and Jun Chen *

School of Marxism, China University of Geosciences, Wuhan 430074, China
* Correspondence: chenjun15@cug.edu.cn

Abstract: Land use is a fundamental element of ecological civilization, whose relevant academic results are not only a concentrated expression of the construction of ecological civilization and land use but also an important theoretical basis for guiding land use changes to promote the construction of ecological civilization. Therefore, based on CiteSpace visual software, this paper analyzes the research progress, hotspots, and trends of 558 articles related to land use under the background of ecological civilization in China based on the China National Knowledge Infrastructure (CNKI) database. The results are as follows: (1) The research cycle is characterized by two distinct stages: the nascent stage and the fluctuating growth stage. (2) The number of publications by researchers and institutions is low, the collaborative network is fragmented, and a core of research researchers and institutions has not yet been formed. (3) The journals in which the papers are published indicate that the research is cross-disciplinary in character, while the highly cited journals have a central role, and the research content of the high-frequency cited papers mainly includes three parts: spatiotemporal evolution pattern and measurement, spatial planning, and land reclamation. (4) The research hotspots are grouped into 12 keyword clusters, which can be further grouped into two sections: "ecological civilization construction and land use" and "national spatial planning". (5) The burst of territorial spatial planning has reached 2022 and will continue to be a research hotspot in this field in the future. The results of this study can help relevant scholars clarify the research context and current situation in this field and grasp future research directions.

Keywords: ecological civilization; land use; CiteSpace; visual analysis

Citation: Nan, M.; Chen, J. Research Progress, Hotspots and Trends of Land Use under the Background of Ecological Civilization in China: Visual Analysis Based on the CNKI Database. *Sustainability* **2023**, *15*, 249. https://doi.org/10.3390/su15010249

Academic Editors: Wenwu Tang, Dianfeng Liu and Jianxin Yang

Received: 15 November 2022
Revised: 15 December 2022
Accepted: 20 December 2022
Published: 23 December 2022

1. Introduction

The construction of an ecological civilization is a strategic direction for the well-being of human beings and the long-term development of all countries in the world [1–3]. With the rapid urbanization and industrialization of most countries worldwide, a series of problems, such as environmental degradation and the contradiction between humans and land, have gradually come to the forefront, leading to the frequent introduction of corresponding countermeasures by various countries [4–6]. The Chinese government issued the Opinions of the CPC Central Committee and the State Council on Accelerating Ecological Civilization Construction [7], proposing the establishment of a new development concept of ecological civilization construction and proposing guidance on the layout of productive forces, industrial structure, production methods, lifestyles and institutional systems [8]. By March 2018, China's Ministry of Natural Resources was established, which integrated some of the responsibilities of several departments and highlighted the holistic protection of national land resources and mountains, water, forests, fields, lakes, and grasses to accelerate the construction of ecological civilization [9,10]. Notably, as an important basic element in the construction of ecological civilization, land use plays a pivotal role in promoting the construction of ecological civilization [11–13], and how to promote the coordinated development of land use and ecological civilization construction has become an important challenge that needs to be solved [14–16]. For this reason, the concept of intensive land use has emerged, mainly through increasing the input of land capital, technology, and labor to rationalize and optimize the structure and efficiency of land use [17–19]. At present, various regions in China have successively carried out the preparation of intensive land use and territorial spatial planning and have achieved initial results [20–22].

The research of land use in the context of ecological civilization has been the focus of attention of many researchers, who have carried out extensive research and achieved fruitful results [14,23,24]. Research on land use under the background of ecological civilization in China is divided into the following three areas: the spatial and temporal distribution characteristics of land use, the relationship between the construction of ecological civilization and land use, and the impact of land use in the context of ecological civilization. Specifically, one is in the research of the spatial and temporal distribution characteristics of land use. Based on the grid scale, Yang et al. [25] used the diversity index and other methods to explore the spatiotemporal relationship between land use changes and the value of ecosystem services in Nanchang and found that the growth of land for construction was the largest and the reduction of the amount of arable land was the largest, while land use and the value of ecosystem services were positively correlated and showed a clustering distribution. In addition, Zhang et al. [26] analyzed the spatial and temporal changes and drivers of land use change in India from 2000 to 2020 by using the land use dynamic attitude model and found that the total area of cultivated land in India remained relatively stable, but the rate of conversion of cultivated land to construction land showed a gradual acceleration, while temperature extremes, elevation, population, GDP and policies were the main drivers of the spatiotemporal evolution of land. The second is in the research of the relationship between the construction of ecological civilization and land use. Some scholars argued that in the context of ecological civilization construction, attention should be given not only to the issue of land use efficiency but also to the low-carbon development, green development, and the ability of recycling development in the process of land use to prevent the ecological environment of farmland from being threatened [27–29]. Unlike other researchers, Liang et al. [30] focus on exploring the coupling relationship between intensive land use and healthy urban development in the context of ecological civilization using the entropy value method and coupling coordination degree model and find that the coupling coordination level has obvious regional differences but shows a narrowing trend in time and space, while the proportion of the tertiary industry is the main driving factor. Third, in terms of research on the impact of land use in the context of ecological civilization. Liu et al. [31] believed that the goal of ecological civilization construction will reshape the direction of ecological value-oriented land use transformation, optimize the ecological transformation path in the process of land remediation, and then effectively improve the regional ecological environment and enhance the ecosystem service function; in addition, Wang et al. [32] believed that land resources, as one of the important basic elements in the construction of ecological civilization, by exploring the compilation of a balance sheet in the process of land use is an important guarantee to promote the construction of ecological civilization systems.

Under the background of ecological civilization construction in China, the rational use of land is of great significance. Although the research in this field has been fruitful, there is still a lack of comprehensive research literature, which makes it difficult for some researchers to accurately grasp the research lineage, research hotspots, and future research directions. Therefore, based on bibliometric and content analysis methods, this study uses the China National Knowledge Infrastructure (CNKI) database as the source of literature and uses CiteSpace software to create a knowledge map of researchers, research institutions, and keywords to address the following three questions: (1) What is the lineage of research on land use under the background of the ecological civilization? (2) What is the current status of research on land use under the background of the ecological civilization? (3) What are the future research directions of land use under the background of the ecological civilization? We believe that this study will help readers clarify the research lineage and research hotspots in this field and help them to grasp future research directions.

2. Research Methodology and Data Sources

2.1. Research Methodology

This study explores the current status of land use research under the background of ecological civilization in China by using bibliometric analysis and content analysis [33,34], as follows: (1) Bibliometric analysis method: with the help of Citespace software developed by Prof. Chen's [35] team, by drawing time zone diagrams, cluster diagrams and keyword mutation diagrams, we can sort out the basic situation, important concepts and development lines of land use research under the background of ecological civilization, dig out the hot topics and development patterns in the field, by using methods such as scholar analysis, research institution analysis, keyword co-occurrence analysis and keyword time zone mapping. (2) Content analysis method: Based on the results of the highly cited literature and cluster analysis, the hot topic terms in the field are scientifically discovered, and the contributions made by different researchers in the research hotspots in the field are then elaborated in depth. Based on the above research methodology, this study will provide an in-depth analysis of the literature related to the field of land use research under the background of ecological civilization in China to show the relationship between the various fields of land use research.

2.2. Data Sources

The Chinese government vigorously promotes the construction of ecological civilization [36], and Chinese researchers attach great importance to and make great contributions to the research on land use under the background of ecological civilization construction in China [37]. Therefore, to ensure the comprehensiveness of the literature sample, the source of journal data in this study was selected the CNKI database, which is rich in literature achievements in this field. The search terms were "ecological civilization" and "land use", and the journals were set to "all articles", with no restriction on the year of publication. After the initial search and comparative screening, marginal information such as conference proceedings, newsletters, interviews, and calls for papers were manually excluded, and a total of 558 research articles were retrieved (from 1998 to 19 October 2022). The acquired literature data were stored in the encoding format UTF-8, converted in Refworks format, and imported into CiteSpace software, with the set time slice to one year and the threshold taken as the top 50, and topics such as researchers, research institutions, and keywords were selected to map the knowledge graph for further analysis.

3. Analysis of the Research Trend

To explore the historical span of research themes and the changing trends of research hotspots, this study draws a map of the number of articles published annually (Figure 1) and time zone (Figure 2) on land use research under the background of the ecological civilization of land use research. Additionally, based on the natural breakpoint method [38,39], annual changes in the number of publications, annual changes in keywords, and major historical points, this study divides the research lineage of land use under the background of ecological civilization into two stages: the nascent stage (1998–2011) and the fluctuating growth stage (2012–2022).

Nascent stage (1998–2011): In this stage, a total of 29 articles have been published since the first article was published in 1998, accounting for 5.20% of the total number of articles published in this research field, with an annual average of 1.31 articles, and the main keywords are ecological civilization (7 articles), land ecosystem (4 articles), general land use planning (4 articles), land ecological planning (3 articles), land ecological (3 articles), new rural construction (3 articles), etc. It can be found that, as the budding and starting stage of land use research under the background of ecological civilization, the number of publications is relatively small, the growth of study is slow, and even the number of publications is zero in several years; at the same time, the keywords are scattered, most keywords and clusters have not yet been generated, and the research content has not yet been systematized. Xiang considered land as an important carrier for the construction of ecological civilization and researched how to build ecological civilized cities in the face of the increasingly acute "population–resource–environment" problem [40]. Starting from land eco-ethics, Zhao [41] tries to regulate the behavior of land resource utilization through the binding force of ethics and morality and puts forwards the basic understanding and proposition of land eco-civilization.

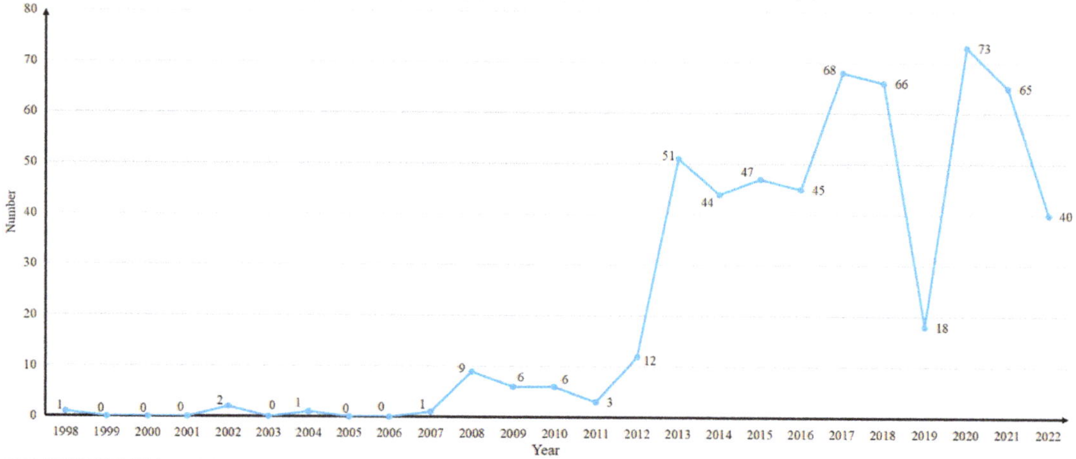

Figure 1. Annual publication volume of land use research under the background of ecological civilization in China.

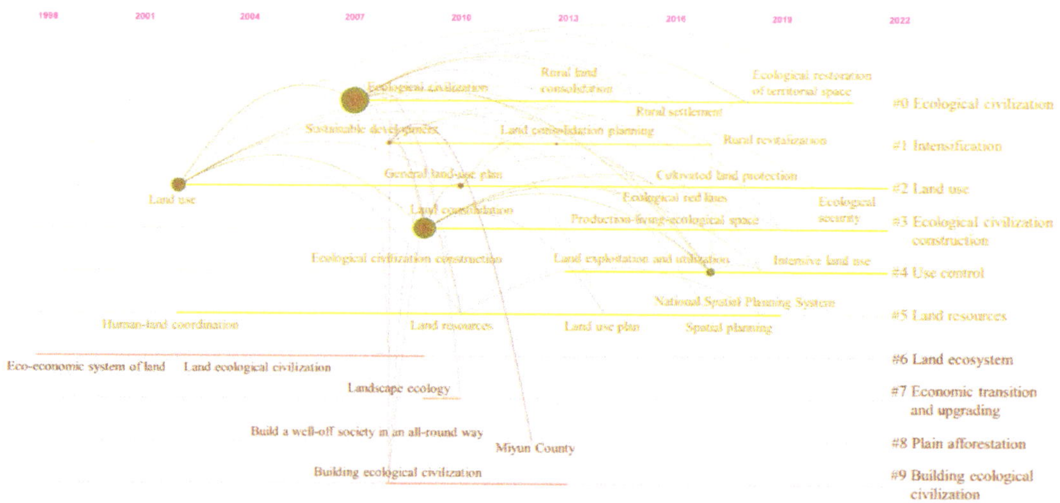

Figure 2. Time zone map for land use research under the background of ecological civilization in China.

Fluctuating growth stage (2012–2022): 529 articles were published in this stage, accounting for 94.80% of the total number of articles, with an average annual volume of 48.09 articles, with the core keywords being ecological civilization (110 articles), ecological civilization construction (86 articles), land use (48 articles), territorial spatial planning (32 articles), land remediation (21 articles) and overall land use planning (19 articles). In

2012, the 18th National Congress of the Communist Party of China made the strategic decision to "vigorously promote the construction of ecological civilization", which led to a high level of attention being paid to the topic of ecological civilization and the number of studies published on land use [42]. Land use, as a fundamental element of ecological civilization, has also seen a rapid increase in the number of articles published around its research [43,44]. This stage of the research focuses on land reclamation, land planning, etc., in the context of ecological civilization. For example, Long et al. [45] explored the coupling of environmental protection planning and land use planning based on the perspective of ecological civilization construction. Li and Gong [46] focused on exploring how to open up various land remediation models in the context of ecological civilization, enhance the ecological service value function of land, and guide the transformation and development of land remediation to ecological governance. Notably, the number of published articles on this topic showed a precipitous decline in 2019, followed by a rapid recovery. The reason for this phenomenon may be that the impact of policy and other factors led to a sudden shift of research focus.

4. Analysis of the Current Status of the Research

4.1. Analysis of Scholar Collaboration Networks

The degree of scholarly collaboration in academic research is an important indicator to judge the research progress of the discipline and can reflect the intensity of collaboration in the field [47]. Therefore, in this study, the word frequency statistics function of CiteSpace is used to summarize the top 10 researchers in terms of the number of papers published (Table 1) and draw the scholar collaboration network map (Figure 3). According to Price law, the minimum number of core scholar publications in a given field is $m = 0.749 * \sqrt{n_{max}} = 1.498$ (n_{max} is the number of papers by the scholar with the highest number of publications). Therefore, researchers with 2 or more publications are positioned as core researchers in the field. The researchers with 3 or more articles are "Lin" (4 articles), "Ye" (3 articles), "Yan" (3 articles), "Zhang" (3 articles), and "Liu" (3 articles), respectively. The centrality of all the researchers is 0, indicating that the degree of cooperation among researchers in this research field is low, and central and scholarly researchers have yet to emerge. In terms of scholarly collaborations, a small number of collaborative relationships exist between researchers, mainly the Lin-centred collaborative network, but the level of collaboration is low, and a close academic community has not yet been formed for inter-team communication. On the whole, the number of papers published in this field is generally low, the core research team has not yet been formed, and the research researchers are scattered and independent.

Table 1. Top 10 researchers of land use research under the background of ecological civilization in China.

Ranker	Number of Papers Published	Centrality	Year	Researchers
1	4	0	2017	Lin, J.
2	3	0	2017	Ye, Z.J.
3	3	0	2013	Yan, J.M.
4	3	0	2013	Zhang, D.
d	3	0	2013	Liu, Y.S.
6	2	0	2016	Liu, W.
7	2	0	2013	Long, H.L.
8	2	0	2018	Lyv, T.G.
9	2	0	2020	Yang, L.S.
10	2	0	2013	Zhang, Y.

Figure 3. Collaborative mapping of land use research researchers under the background of ecological civilization in China.

4.2. Analysis of Research Institution Collaboration Networks

As a platform for researchers to conduct research, issuing institutions can reflect interinstitutional cooperation. The analysis of the issuing institutions led to a table of the top 10 major research institutions (Table 2) and a map of institutional cooperation networks (Figure 4). Comparing the number of papers published by each institution, we found that the highest number of articles (13 articles) was from the Institute of Geographical Sciences and Resources, Chinese Academy of Sciences (IGSR), accounting for 2.33% of the total number of articles published. Other institutions with 7 articles include the "Information Center of Ministry of Natural Resources (9 articles)", the "School of Public Administration and Policy, Renmin University of China (7 articles)" and the "Chinese Academy of Land and Resources Economy (7 articles)". In terms of centrality, "Institute of Geographical Sciences and Natural Resources Research, CAS (0.04)", "College of Urban and Environmental Sciences, Peking University (0.02)" and "Chinese Land Surveying and Planning Institute (0.02)" "have a high degree of centrality and a certain role as a bridge in the field.

Figure 4. Mapping of the cooperative network of land use research institutions under the background of ecological civilization in China.

Table 2. Top 10 research institutions for land use research under the background of ecological civilization in China.

Ranker	Number of Papers Published	Centrality	Year	Research Institutions
1	13	0.04	2008	Institute of Geographical Sciences and Natural Resource Research, Chinese Academy of Sciences
2	9	0.01	2009	Information Center of Ministry of Land and Resources of People's Republic of China
3	7	0.01	2013	School of Public Administration and Policy, Renmin University of China
4	7	0.01	2009	Chinese Academy of Land and Resources Economy
5	6	0.02	2016	College of Urban and Environmental Sciences, Peking University
6	6	0	2013	Information Center of Ministry of Land and Resources of People's Republic of China
7	6	0	2013	School of Geography and Ocean Science, Nanjing University
8	4	0.02	2008	China Land Surveying and Planning Institute
9	4	0	2017	College of Resources and Environment, University of Chinese Academy of Sciences
10	4	0	2014	College of Land Science and Technology, China University of Geosciences (Beijing)

It can be found from the cooperation network of research institutions that there is less cooperation between research institutions but still forms a certain scale of cooperation network. Additionally, among the few collaborating institutions that are mainly concentrated in geographic proximity or different research institutions within universities, the "Institute of Geographical Sciences and Natural Resources Research, CAS", "Information Center of Ministry of Natural Resources", "Chinese Academy of Land and Resources Economy" and other research institutions cooperate more closely, but the volume of publications is insufficient, and the overall academic cooperation network is more dispersed.

4.3. Analysis of Published Papers

CiteSpace software was used to conduct a "journal cocitation" analysis to find the journals that published papers in the field of land use under the background of ecological civilization (Table 3). The journal with the highest number of papers is China Land (20 articles), accounting for 3.58% of the total number of articles, while other journals with 10 articles or more include Natural Resources Information (19 articles), China Land Science (14 papers), Natural Resource Economics of China (11 articles), Land & Resources (11 articles) and Ecological Economy (10 articles). Additionally, an analysis of the top 10 journals in terms of the number of papers published shows that they are mainly related to land science, resource environment, ecological economy, spatial planning, and other related fields, showing a trend of cross-disciplinary research, which will be a hot direction for future research.

Table 3. Top 10 journals for land use research under the background of ecological civilization in China.

Ranker	Journal Name	Number of Articles Published
1	China Land	20
2	Natural Resources Information	19
3	China Land Science	14
4	Natural Resource Economics of China	11
5	Land & Resources	11

Table 3. *Cont.*

Ranker	Journal Name	Number of Articles Published
6	Ecological Economy	10
7	Shanghai Land & Resources	9
8	Zhejiang Land & Resources	8
9	Planners	8
10	Bulletin of Soil and Water Conservation	7

4.4. Analysis of Research Methods

Through the statistical analysis of research methods (Table 4), it is found that qualitative analysis is the main method of land use under the background of ecological civilization in China, and quantitative analysis is the auxiliary method. In qualitative analysis, the research methods used most frequently by scholars were the "policy analysis method" (179 times) and "theoretical analysis method" (49 times). The main research contents were in three aspects: (1) Scholars mainly explored the theoretical sources and rationality of optimizing land use under the background of ecological civilization in China [48,49]. (2) Through the analysis and interpretation of the historical evolution of ecological civilization construction policies, scholars have proposed the guiding significance and content of land use types [50,51]. (3) Under the relevant policies of ecological civilization construction, scholars have formulated land use development plans for different regions [52,53].

Table 4. Top 20 research methods on land use research under the background of ecological civilization in China.

Ranker	Research Methods	Numbers	Ranker	Research Methods	Numbers
1	Policy analysis method	179	11	TOPSIS model	16
2	Theoretical analysis method	49	12	Obstacle degree model	15
3	Entropy method	44	13	Land use dynamics model	15
4	Single case study method	35	14	Kernel density estimation method	12
5	ENVI remote sensing image method	34	15	Cold hot spot analysis	11
6	ArcGIS spatial analysis	27	16	Super-SBM model	11
7	Coupling coordination degree model	25	17	Land transfer matrix method	7
8	Geographic detector	20	18	Land use transfer matrix method	7
9	Multiple regression model	17	19	Factor analysis method	5
10	Bibliometric analysis method	16	20	Analytic hierarchy process method	5

For quantitative analysis, the most frequently used research methods were the "entropy method" (44 times) and the "ENVI remote sensing image method" (34 times). The research contents were divided into two aspects: (1) Scholars evaluated and analyzed the land use status by constructing the index system [54]. (2) With the help of remote sensing, some scholars are devoted to exploring the spatiotemporal distribution characteristics and dynamic trends of land use under the background of ecological civilization in China [55,56].

4.5. Analysis of Highly Cited Literature

The frequency of citations of literature is closely related to the academic influence of the relevant research, and high citations reflect the academic influence and knowledge centrality of the relevant literature in the field of ecological civilization research [57]. Therefore, in this paper, we collate the high-frequency citations of land use research under the background of ecological civilization (Table 5) and analyze the research content of the relevant literature to further understand the knowledge base of this field of research.

As seen from Table 5, among the top 10 high-frequency cited studies, the paper named "Classification Evaluation and Spatial-temporal Analysis of 'Production-Living-Ecological' Spaces in China" published by Liu et al. [58] has the highest number of citations (508 times), which much higher than the number of citations in the subsequent literature. This paper analyzes the dialectical relationship between land use types and land use functions on the basis of the theory of 'production–living–ecological' spaces and establishes a system for classifying and evaluating 'production–living–ecological' spaces, revealing the pattern of 'production–living–ecological' spaces and changing characteristics in China between 1990 and 2010. Other papers with more than 200 citations include Lin et al.'s [52] "Construction of the Spatial Planning System: With Discussions on the Relationship Between Spatial Planning, Territorial Spatial Regulation, and Natural Resources Supervision", Yan et al.'s [59] "Cognition, Direction and Path of Future Spatial Planning based on the Background of Multiple Planning Integration" and "Reshaping and innovation of China land consolidation strategy" by Yun et al. [60].

In general, the research content of the high-frequency cited literature can be divided into three parts: spatiotemporal evolution patterns and measurements, spatial planning, and land reclamation. First, in the area of spatiotemporal evolution patterns and measurements, Liu et al. [58] constructed a classification and evaluation system based on the theory of production-living-ecological space and then analyzed the evolution pattern and change characteristics of production–living–ecological space in China, while Jin et al. [61] focused on the evaluation system of production–living–ecological space for urban clusters to explore the change characteristics of the evolution pattern of production–living–ecological space in urban clusters. Second, in spatial planning, Lin et al. [52] focused on the relationship between the spatial planning system, unified land use control, and improved natural resource supervision system in the context of ecological civilization construction to build a national spatial planning system; additionally, Yan et al. [59] based on the context of "multi-regulation" reform and the perspective of ecological civilization construction analyzed the essential perception of spatial planning, clear reform orientation and proposed reform paths for future spatial planning. In addition, Long et al. [45] used the environmental carrying capacity evaluation method to explore the spatial convergence effect of the "three boundaries and four zones" of land use planning, environmental function zoning, and ecological red lines in environmental protection planning. In contrast to other researchers, Zou [62] focused on the reconstruction of the spatial planning system under the framework of natural resource management and proposes holistic governance ideas, eradicating conflicts among multiple spatial plans and supporting the construction of an ecological civilization. Finally, in terms of land remediation, based on a review of the theoretical foundation and practical exploration of land remediation in China over the past 20 years, Yun et al. [60] proposed reshaping suggestions in terms of innovative concepts and other major problems of land remediation based on the low level of comprehensive land remediation practice and the low level of land ecological remediation practice; additionally, Feng and Yang [63] took China's socioeconomic transformation as the entry point, sorted out the basic direction and strategic focus of comprehensive rural land improvement, and put forwards policy recommendations such as agricultural land improvement as a precursor and integration into ecological civilization construction. In addition, Wang and Zhong [64] focused on analyzing the problems of insufficient ecosystem thinking and imperfect technical systems in land remediation and put forward suggestions for the ecological transformation of land remediation, while Feng [65] focused on analyzing the coordination of human–land relations in the process of land remediation and the sustainable use of land resources.

Table 5. Table of highly cited literature on land use research under the background of ecological civilization in China.

Ranker	Scholar	Paper Title	Journal Name	Cited Volume
1	Liu et al. [58]	Classification evaluation and spatial-temporal analysis of "production-living-ecological" spaces in China	Acta Geographica Sinica	508
2	Lin et al. [52]	Construction of the spatial planning system: With discussions on the relationship between spatial planning, Territorial spatial regulation, and natural resources supervision	City Planning Review	304
3	Yan et al. [59]	Cognition, direction and path of future spatial planning based on the background of multiple planning integration	China Land Science	219
4	Yun et al. [60]	Reshaping and innovation of China land consolidation strategy	Transactions of the Chinese Society of Agricultural Engineering	206
5	Feng and Yang [63]	Key research fields and basic directions of Chinese rural-land comprehensive consolidation in transitional period	Transactions of the Chinese Society of Agricultural Engineering	154
6	Long et al. [45]	Spatial interlinking of land use planning and environmental protection planning From the perspective of ecological civilization construction	Economic Geography	99
7	Jin et al. [61]	Research on the evolution of spatiotemporal patterns of production-living-ecological space in an urban agglomeration in the Fujian Delta region, China	Acta Ecologica Sinica	97
8	Wang and Zhong [64]	Problems and suggestions for developing ecological construction in land management work	Transactions of the Chinese Society of Agricultural Engineering	96
9	Zou [62]	Logic and conception of spatial planning system reconstruction under the framework of natural resources management	Planners	89
10	Feng [65]	Man-nature harmonization theory and regional sustainable land resource use	Journal of Nanjing Agricultural University (Social Sciences Edition)	88

5. Analysis of Research Hotspots

5.1. Keyword Co-Occurrence Analysis

Keywords are the core of research results, and the more frequently the keywords appear, the more frequently the research around the keywords [66]. Therefore, this paper sorts out 20 keywords with high word frequency in land use research under the background of ecological civilization through the word frequency statistics function of CiteSpace, which can reveal the research hotspots in this field to a certain extent (Table 6). In terms of keyword frequency, the keywords with high word frequency are ecological civilization

(117 times), ecological civilization construction (88 times), land use (49 times), national spatial planning (32 times), general land use planning (23 times), land use (22 times), etc. In terms of centrality, ecological civilization (0.42), ecological civilization construction (0.36), general land use planning (0.42), ecological civilization construction (0.36), land use planning (0.17) and land use (0.12) have a strong centrality and play a bridging role among the keywords. The analysis of the keywords and their related literature reveals that the network relationships linked by the keywords span a large period of time and have a complex structure and that the keywords involve ecology, economy, rural areas, agriculture, spatial planning, land improvement, and rural revitalization, which indicates that the research objects and research contents of land use research under the background of ecological civilization show diverse characteristics.

Table 6. Top 20 frequency keywords for land use research under the background of ecological civilization in China.

Ranker	Frequency	Centrality	Year	Keywords	Ranker	Frequency	Centrality	Year	Keywords
1	117	0.42	2007	Ecological civilization	11	9	0.03	2015	Multi-plan integration
2	88	0.36	2009	Ecological civilization construction	12	9	0	2013	Ecological environment
3	49	0.12	2002	Land use	13	8	0.01	2017	Land planning
4	32	0.05	2017	National spatial planning system	14	8	0.01	2018	Rural vitalization
5	23	0.17	2008	General land use planning	15	8	0.02	2016	Use control
6	22	0.07	2010	Land consolidation	16	8	0	2019	Intensive land use
7	15	0.02	2013	Territory resources	17	7	0.01	2013	New urbanization
8	14	0.08	2013	Intensive land use	18	7	0.02	2012	Ministry of land and resources of People's Republic of China
9	13	0.07	2010	Land resources	19	7	0.01	2017	Spatial planning
10	10	0.02	2014	Land use plan	20	6	0	2013	Land consolidation planning

5.2. Keyword Clustering Analysis

The keyword clustering analysis reveals that the keyword clusters of land use research under the background of ecological civilization are mainly divided into "#0 Ecological civilization", "#1 Intensive land use", "#2 Land use", "#3 Ecological civilization construction", "#4 Use control", "#1 Intensive land use", "#2 Land use", "#3 Ecological civilization construction", "#4 Use control", "#5 Land resources", "#8 Land ecosystem", "#9 Economic transformation and upgrading", "#10 Plain afforestation", "#11 Building ecological civilization", "#15 Land use plan" and "#20 System of land and resources of People's Republic of China" (Figure 5). Based on the content of the literature on arable land use under the background of ecological civilization and the keyword nodes shown in Figure 5 and Table 6, the current research themes were grouped into two categories: "ecological civilization construction and land use" and "national spatial planning".

First, in "Ecological civilization construction and land use", specifically "#0 Ecological civilization" in "#0 Ecological civilization", "#1 Intensive land use", "#2 Land use", "#3 Ecological civilization construction", "#11 Building ecological civilization" and five other keyword clusters. As a fundamental element of ecological civilization construction, land use plays an extremely important role in the construction of ecological civilization. Bai et al. [67] used the entropy weight method—the PSR model—to investigate the coordination between ecological civilization construction and intensive land use in the Guanzhong Plain urban agglomeration and found that the level of intensive land use showed a steady upwards trend, while the degree of ecological civilization construction generally showed a fluctuating upwards trend. Moreover, Kong et al. [68] explored the spatial classification system of China's national land based on the perspective of ecological civilization and finally proposed a spatial classification system comprising land use classification, spatial control classification, and the "multiregulation" difference coordination mechanism and provided policy suggestions for the establishment of China's future spatial planning standards. In addition, Yun et al. [69] explored the changes in land use in China and the impacts from an ecological civilization perspective and found that although land use development plays an important role in promoting social and economic development and ensuring food security, overreliance on land resource development has led to a series of ecological problems, making environmental pollution, strict adherence to the ecological red line for arable land, and strengthening territorial spatial planning necessary. What is not used in the same way as other researchers' studies is Zhai [70], who used the construction of ecological civilization as a conceptual guide to deeply analyzed the problems and causes of rural land, and analyzed in depth the problems and causes in the process of rural land use and development, combined with the specific requirements of building an ecological civilization, built a framework for a comprehensive land use system, and put forwards targeted countermeasures and suggestions for the effective use and protection of rural land. In contrast, Guo et al. [71] proposed using and managing land from an ecosystem perspective and shifting from focusing only on quantitative management to placing equal emphasis on quantity, quality and ecology to improve the overall functional characteristics of land.

Figure 5. Cluster analysis of keywords for land use research under the background of ecological civilization.

Second, "National spatial planning" specifically contains 7 keyword clusters: "#4 Use control", "#5 Land resources", "#8 Land ecosystem", "#9 Economic transformation and upgrading", "#10 Plain afforestation", "#15 Land use plan", and "#20 System of land and resources of People's Republic of China". The spatial planning system is not only an important part of the construction of ecological civilization but also the basis for all kinds of development and construction activities. In particular, Zhang [53] proposed revising the current Spatial Planning Law and formulating a Spatial Planning Law that includes the construction of ecological civilization based on the perspective of ecological civilization construction and the Netherlands' Territorial Space Planning Law and focusing on exploring the legislative path of territorial spatial planning law. Feng [72] applied the SWAT model to construct a multiobjective optimization model with hydrology, soil and water conservation, and economic development as the main components and found that the model is applicable in the southern mountainous region of Jinan, helping it to rationalize the use of land use resources in each classification area and protect the ecological environment. As the spatial function of the land is the basis for human–land coupling and territorial spatial planning, in contrast to the concerns of other researchers, Zhou [48] focused on the research of the function of land and space in the context of ecological civilization, and the research suggests that the theoretical construction, weighing mechanism, content expansion, technical innovation and conceptual transformation of the function of land and space should be promoted in the future. Meanwhile, optimizing the spatial development pattern of the country and increasing the natural ecosystem and environmental protection are important tasks in the construction of ecological civilization. Long et al. [45] evaluated the interface between the "three boundaries and four zones" and the environmental function zoning based on the perspective of ecological civilization construction, providing a reference for the preparation of environmental protection planning and land use planning. In addition, Ding [73] analyzed the characteristics of the spatial restructuring of rural areas in southern Jiangsu based on the perspective of ecological civilization, and after an in-depth analysis of the problems of imbalance of interests, spatial chaos, and ecological fragility in the spatial restructuring of rural areas, he proposed countermeasures and suggestions for the spatial restructuring of rural areas in southern Jiangsu from various aspects, such as the construction of an interest coordination mechanism, the intensive use of space, the reshaping of the characteristics of water towns and the construction of ecological patterns.

5.3. Burst Word Analysis

The Burst terms function can focus on the more influential research areas over a period of time [74]. Data analysis through CiteSpace revealed that the only keywords that emerged from this research were "Intensive land use" and "National spatial planning" (Table 7). The analysis found that "national spatial planning" had the highest strength and longest impact period, at 10.09 and 4 years (2019–2022), respectively, while the strength of "intensive land use" was only 4.58, with a duration of 2 years (2013–2014). The emergence of "National spatial planning" lasted from 2019 to 2022, which indicates that under the background of ecological civilization construction, national spatial planning research will become a hot research topic and a frontier direction in the future.

Table 7. Analysis of keyword emergent terms in land use research under the background of ecological civilization.

Keywords	Strength	Begin	End	1999–2022
Intensive land use	4.58	2013	2014	
National spatial planning	10.09	2019	2022	

6. Discussion

6.1. Research Researchers and Institutions Are Scattered, Core Researchers and Institutions Have Been Formed

According to the word frequency statistics function and the function of drawing cooperative network mapping of CiteSpace, it is found that the research researchers and research institutions in the field of land use research under the background of ecological civilization are relatively scattered and have not yet formed research researchers and research institutions with high publication volume, which makes the research in this field lack authoritative and influential research individuals and has not formed mature research teams. The small number of collaborative teams does not change the current research divide in the field, and each researcher and research institution is in a state of independent research, which not only affects the exchange of academic thinking and ideas in the field but also further hinders the expansion of research content in the field. The reason for this phenomenon may be the lack of an effective academic platform for communication in this field, which makes it difficult for scholars to learn from each other. Meanwhile, due to the lack of research depth in this field, leading to the lack of authoritative scholars, academic cooperation is mainly limited to research institutions located close to each other or scholars within the same research institution.

6.2. The Research Fields Show That Interdisciplinary and Highly Cited Literature Has a Central Role

The field of publishing papers is characterized by diversity. The analysis of journal cocitations shows that the research base of land use under the background of ecological civilization shows strong interdisciplinary characteristics, mainly involving land science, resources and environment, ecological economy, spatial planning, and other related fields, and the research methods in a large number of research results are very rich due to the characteristics of each discipline, which also highlights the future development direction of the field. In addition, the highly cited literature highlights, to some extent, the influence and centrality of the paper in the field, which benefits further understanding of the research base in the field. Among them, Liu et al. [58] published the most cited paper, "Classification evaluation and spatiotemporal pattern of 'production–living–ecological space' in China". This paper not only theoretically divides land use types into production, living and ecological spaces but also constructs a relevant evaluation index system, which plays a guiding role in the research of this field. The research content of the top ten highly cited studies can be categorized in detail into three areas: spatiotemporal evolutionary patterns and measurements, spatial planning, and land reclamation, which to some extent indicate the three main lines of current research in the field, around which the rest of the research is mainly focused. Furthermore, this shows that these three research areas are not only the core content of current research in this field but also the hot spots for scholars in this field to carry out research in the future.

6.3. The Research Lineage Has Obvious Stage Characteristics, and the Research Hotspots Show a Diversified Trend

From the characteristics of annual publication volume and keyword distribution, the field of land use research under the background of ecological civilization is characterized by distinct stages. Prior to 2012, there were extremely few publications in the field, keywords were scattered, and most keywords and clusters had not yet been generated. In 2012 and beyond, as the 18th National Congress of the Communist Party of China (CPC) brought "ecological civilization construction" to the forefront of the national strategy, the number of papers published in this field increased dramatically, the number of keywords and categories also increased, and a systematic research system was gradually formed. In this system, the research hotspots are divided into 12 keyword clusters, such as "#0 Ecological civilization" and "#1 Intensive land use", which are further divided into two parts, such as "ecological civilization construction and land use" and "national spatial planning", which shows that although the hotspots of research in this field are scattered, the core content of the research is centred on "ecological civilization construction and land use" and "national spatial planning". This is similar to the analysis results of highly cited literature, indicating that the core research content in this field is relatively stable. The fact that "national spatial planning" is the largest burst intensity until 2022 indicates that ecological civilization construction and national spatial planning will be a hot topic of research in this field for a long time to come.

6.4. Prospect of Land Use under the Background of China's Ecological Civilization Construction

To strengthen the construction of ecological civilization, China's constitutional amendment passed in 2018 has written "ecological civilization" into the Constitution, which marks that the construction of ecological civilization has officially become a national strategy. As the carrier of various terrestrial ecosystems, land use optimization is of great significance for China's ecological civilization construction. Firstly, in view of the soil pollution problem, the Chinese government tries to prevent and control the soil pollution problem by carrying out large-scale land greening actions, adjusting the land use structure and other way, so the pollution in the process of land use should be a hot issue for scholars. Secondly, the academic community should pay attention to the issue of territorial spatial planning and comprehensive land management in different regions. The Chinese government has given full play to the control role of the ecological red line in land development and utilization, organized the preparation of territorial spatial planning for administrative regions at all levels, and accelerated the comprehensive improvement of land by regions to make rational use of land resources. Finally, for regions with serious land ecological environment damage, the government should promote land ecological restoration by issuing relevant policies, and scholars should focus on the theory, technology, and case study of land ecological restoration.

6.5. Research Contributions and Deficiencies

The main contributions of this paper are that bibliometric and content analysis methods are used to clarify the research in this field, analyze the current situation and research hotspots in this field in depth, and indicate future research directions in this field in the absence of comprehensive literature on land use research under the background of ecological civilization. Meanwhile, this paper also has certain shortcomings: (1) this paper only uses the CNKI database as a literature source, which is a relatively single source of data, but it can still adequately summarize the research lineage, current situation, and hotspots in this field. In the future, data sources (such as Web of Science, Scopus, and other databases) can be expanded to carry out the comparison between Chinese and international research. (2) As the literature data are mainly in Chinese papers, the translation into English may have some linguistic deviations, and we have improved the accuracy of the language expression of the paper by various means, such as professional retouching.

7. Conclusions

Based on CiteSpace software, this paper visualizes and analyzes the 558 articles in the field of land use research under the background of China's ecological civilization in the CNKI database, which clarifies the research lineage and current status of the field and promotes the academic community's overall grasp of the research hotspots and future trends in the field. Overall, the findings of this paper are as follows: (1) At present, research in this area is relatively mature, and the scale of research is relatively stable and can be divided into a nascent start-up stage and a fluctuating growth stage. (2) The researchers and research institutions are scattered, the volume of papers published is insufficient, there are only a small number of collaborative networks, and no core researchers and research institutions have been formed. (3) Based on the analysis of the published papers, it is found that research in this field involves many different disciplines, such as land science, resources and environment, ecology and economy, and spatial planning, and shows a cross-disciplinary trend. (4) Highly cited literature has a strong academic influence in the field, and its content as a whole can be categorized into three sections: spatiotemporal evolutionary patterns and measurements, spatial planning, and land reclamation. (5) Although the hotspots of land use research under the background of ecological civilization can be grouped into 12 keyword clusters, they are summarized in detail in two sections: "ecological civilization construction and land use" and "national spatial planning". (6) Intensive land use and territorial spatial planning are the only emergent terms in this field, with territorial spatial planning emerging until 2022 as a future research hotspot and frontier.

In the future, the following three areas will be worthy of continued in-depth research and consideration: (1) Strengthening theoretical research and promoting the construction of a theoretical analytical framework on ecological civilization and land use. (2) Increasing multiscale case research based on differences between different regions and emphasizing cross-disciplinary research. (3) How to ensure that territorial spatial planning is compatible and coordinated with social and economic development and ecological environmental protection under the background of ecological civilization construction.

Author Contributions: Conceptualization, M.N. and J.C.; methodology, M.N.; software, M.N.; validation, M.N.; formal analysis, M.N.; data curation, M.N.; writing—original draft preparation, M.N.; writing—review and editing, M.N. and J.C.; visualization, M.N.; supervision, J.C.; project administration, J.C.; funding acquisition, J.C. All authors have read and agreed to the published version of the manuscript.

Funding: This research was funded by the National Social Science Foundation of China (22BKS158).

Institutional Review Board Statement: Not applicable.

Informed Consent Statement: Not applicable.

Data Availability Statement: Not applicable.

Acknowledgments: The authors express their gratitude to the anonymous reviewers and editors for their helpful comments about this paper.

Conflicts of Interest: The authors declare no conflict of interest.

References

1. Fan, Y.; Fang, C.; Zhang, Q. Coupling coordinated development between social economy and ecological environment in Chinese provincial capital cities-assessment and policy implications. *J. Clean. Prod.* **2019**, *229*, 289–298. [CrossRef]
2. Liu, Y.; Li, J.; Yang, Y. Strategic adjustment of land use policy under the economic transformation. *Land Use Policy* **2018**, *74*, 5–14. [CrossRef]
3. Wang, J.; Zhai, T.; Lin, Y.; Kong, X.; He, T. Spatial imbalance and changes in supply and demand of ecosystem services in China. *Sci. Total Environ.* **2019**, *657*, 781–791. [CrossRef] [PubMed]
4. Li, H.; Liu, D.; He, J. Exploring differentiated conservation priorities of urban green space based on tradeoffs of ecological functions. *Sustainability* **2022**, *14*, 1845. [CrossRef]
5. Pata, U.K. Renewable energy consumption, urbanization, financial development, income and CO_2 emissions in Turkey: Testing EKC hypothesis with structural breaks. *J. Clean. Prod.* **2018**, *187*, 770–779. [CrossRef]
6. Li, J.; Gong, J.; Guldmann, J.-M.; Yang, J.; Zhang, Z. Simulation of land-use spatiotemporal changes under ecological quality constraints: The case of the Wuhan Urban Agglomeration Area, China, over 2020–2030. *Int. J. Environ. Res. Public Health* **2022**, *19*, 6095. [CrossRef] [PubMed]
7. Zhang, Y.; Fu, B. Impact of China's establishment of ecological civilization pilot zones on carbon dioxide emissions. *J. Environ. Manag.* **2023**, *325*, 116652. [CrossRef] [PubMed]
8. Gu, Y.; Wu, Y.; Liu, J.; Xu, M.; Zuo, T. Ecological civilization and government administrative system reform in China. *Resour. Conserv. Recycl.* **2020**, *155*, 104654. [CrossRef]
9. Gao, M.; Hu, Y.; Bai, Y. Construction of ecological security pattern in national land space from the perspective of the community of life in mountain, water, forest, field, lake and grass: A case study in Guangxi Hechi, China. *Ecol. Indic.* **2022**, *139*, 108867. [CrossRef]
10. Wang, Z. Land spatial development based on carrying capacity, land development potential, and efficiency of urban agglomerations in China. *Sustainability* **2018**, *10*, 4701. [CrossRef]
11. Du, Y.; Qin, W.; Sun, J.; Wang, X.; Gu, H. Spatial pattern and influencing factors of regional ecological civilisation construction in China. *Chin. Geogr. Sci.* **2020**, *30*, 776–790. [CrossRef]
12. Zhou, Z.; Liu, D.; Sun, Y.; He, J. Predicting joint effects of multiple land consolidation strategies on ecosystem service interactions. *Environ. Sci. Pollut. Res.* **2022**, *29*, 37234–37247. [CrossRef] [PubMed]
13. Liu, D.; Li, H.; Qiu, M.; Liu, Y. Understanding coupled coordination relationships between social and ecological functions of urban green spaces. *Geo-Spat. Inf. Sci.* **2022**. [CrossRef]
14. Feng, W.; Li, Y. Measuring the ecological safety effects of land use transitions promoted by land consolidation projects: The case of Yan'an City on the Loess Plateau of China. *Land* **2021**, *10*, 783. [CrossRef]
15. Li, C.; Wu, J. Land use transformation and eco-environmental effects based on production-living-ecological spatial synergy: Evidence from Shaanxi Province, China. *Environ. Sci. Pollut. Res.* **2022**, *29*, 41492–41504. [CrossRef]
16. Tang, W.; Zheng, M.; Zhao, X.; Shi, J.; Yang, J.; Trettin, C.C. Big geospatial data analytics for global mangrove biomass and carbon estimation. *Sustainability* **2018**, *10*, 472. [CrossRef]
17. Wang, Y.; Feng, Y.; Zuo, J.; Rameezdeen, R. From "Traditional" to "Low carbon" urban land use: Evaluation and obstacle analysis. *Sustain. Cities Soc.* **2019**, *51*, 101722. [CrossRef]
18. Qian, Q.; Liu, H.; Zheng, X. A regional sustainable intensive land use evaluation based on ecological constraints: A case study in Jinan City. *Sustainability* **2019**, *11*, 1434. [CrossRef]
19. Zhang, Y.; Liu, Y.; Wang, Y.; Liu, D.; Xia, C.; Wang, Z.; Wang, H.; Liu, Y. Urban expansion simulation towards low-carbon development: A case study of Wuhan, China. *Sustain. Cities Soc.* **2020**, *63*, 102455. [CrossRef]
20. Zhang, Y.; Jiang, P.; Cui, L.; Yang, Y.; Ma, Z.; Wang, Y.; Miao, D. Study on the spatial variation of China's territorial ecological space based on the standard deviation ellipse. *Front. Environ. Sci.* **2022**, *10*, 982734. [CrossRef]

21. Liu, Y.; Zhou, Y. Territory spatial planning and national governance system in China. *Land Use Policy* **2021**, *102*, 105288. [CrossRef]
22. Cheng, P.; Qin, Y.; Zhu, S.; Kong, X. Ecological disturbance of rural settlement expansion: Evidence from Nantong, Eastern China. *Land* **2022**, *11*, 1741. [CrossRef]
23. Zhao, L.; Liu, G.; Xian, C.; Nie, J.; Xiao, Y.; Zhou, Z.; Li, X.; Wang, H. Simulation of land use pattern based on land ecological security: A case study of Guangzhou, China. *Int. J. Environ. Res. Public Health* **2022**, *19*, 9281. [CrossRef] [PubMed]
24. Kong, X.; Fu, M.; Zhao, X.; Wang, J.; Jiang, P. Ecological effects of land-use change on two sides of the Hu Huanyong Line in China. *Land Use Policy* **2022**, *113*, 105895. [CrossRef]
25. Yang, G.; Lyu, K.; Li, F. Spatial and temporal correlation analysis of land use change and ecosystem service value in Nanchang City based on grid scale. *China Land Sci.* **2022**, *36*, 121–130. (In Chinese) [CrossRef]
26. Zhang, H.; Deng, W.; Wang, Y. Analyses on the spatio-temporal evolution and driving factors of land use in India during 2000–2020. *World Reg. Stud.* **2022**, *31*, 786–799. (In Chinese) [CrossRef]
27. Zhang, J.; Zhang, G. Spatiotemporal evolution of urban land intensive utilization in Guangdong Province aiming at ecological civilization construction. *Bull. Soil Water Conserv.* **2022**, *42*, 233–238. (In Chinese)
28. Liu, J.; Feng, H.; Wang, K. The low-carbon city pilot policy and urban land use efficiency: A policy assessment from China. *Land* **2022**, *11*, 604. [CrossRef]
29. Xiao, Y.; Chen, J.; Wang, X.; Lu, X. Regional green development level and its spatial spillover effects: Empirical evidence from Hubei Province, China. *Ecol. Indic.* **2022**, *143*, 109312. [CrossRef]
30. Liang, L.; Liang, Y.; Huang, Z.; Shi, S. Research on relationship of coupling coordination between the development of healthy city and the intensive use of land under the background of ecological civilization: A case from Beijing-Tianjin-Hebei Region. *Ecol. Econ.* **2020**, *36*, 88–94. (In Chinese)
31. Liu, Y.; Dai, L.; Long, H.; Feng, X. Land consolidation mode and ecological oriented transformation under the background of rural revitalization: A case research of Zhejiang Province. *China Land Sci.* **2021**, *35*, 71–79. (In Chinese) [CrossRef]
32. Wang, J.; Shi, W. Research on the compilation of land resources balance sheet under the background of ecological civilization. *Stat. Decis.* **2021**, *37*, 32–36. (In Chinese) [CrossRef]
33. Cheng, P.; Tang, H.; Dong, Y.; Liu, K.; Jiang, P.; Liu, Y. Knowledge mapping of research on land use change and food security: A visual analysis using CiteSpace and VOSviewer. *Int. J. Environ. Res. Public Health* **2021**, *18*, 13065. [CrossRef] [PubMed]
34. Jiang, Y.; Li, M.; Dennis, A.; Liao, X.; Ampaw, E.M. The hotspots and trends in the literature on cleaner production: A visualized analysis based on citespace. *Sustainability* **2022**, *14*, 9002. [CrossRef]
35. Chen, C. CiteSpace II: Detecting and visualizing emerging trends and transient patterns in scientific literature. *J. Am. Soc. Inf. Sci. Technol.* **2006**, *57*, 359–377. [CrossRef]
36. Yan, L.; Zhang, X.; Pan, H.; Wu, J.; Lin, L.; Zhang, Y.; Xu, C.; Xu, M.; Luo, H. Progress of Chinese ecological civilization construction and obstacles during 2003-2020: Implications from one set of emergy-based indicator system. *Ecol. Indic.* **2021**, *130*, 108112. [CrossRef]
37. Zuo, Z.; Guo, H.; Cheng, J.; Li, Y. How to achieve new progress in ecological civilization construction? Based on cloud model and coupling coordination degree model. *Ecol. Indic.* **2021**, *127*, 107789. [CrossRef]
38. Cheng, P.; Tang, H.; Lin, F.; Kong, X. Bibliometrics of the nexus between food security and carbon emissions: Hotspots and trends. *Environ. Sci. Pollut. Res.* **2022**. [CrossRef]
39. Sun, W.; Mao, L. Evolution of research on Beijing-Tianjin-Hebei cooperative development based on CiteSpace method. *Acta Geogr. Sin.* **2018**, *73*, 2378–2391. (In Chinese) [CrossRef]
40. Xiang, E. Research on the construction of "ecological civilization city". *China Land* **1998**, 25–26. (In Chinese)
41. Zhao, Z.; Wu, C.; Sheng, Y. On land ecological ethics and ecological civilization. *Sci. Technol. Manag. Land Resour.* **2004**, *1*, 50–54. (In Chinese)
42. Gu, S.; Hu, Y.; Zhou, H. Ecological civilization construction: Scientific connotation and basic paths. *Resour. Sci.* **2013**, *35*, 2–13. (In Chinese)
43. Meng, F.; Guo, J.; Guo, Z.; Lee, J.C.K.; Liu, G.; Wang, N. Urban ecological transition: The practice of ecological civilization construction in China. *Sci. Total Environ.* **2021**, *755*, 142633. [CrossRef] [PubMed]
44. Huang, P.; Westman, L. China's imaginary of ecological civilization: A resonance between the state-led discourse and sociocultural dynamics. *Energy Res. Soc. Sci.* **2021**, *81*, 102253. [CrossRef]
45. Long, H.; Liu, Y.; Li, T.; Wan, J. Spatial interlinking of land use planning and environmental protection planning from the perspective of ecological civilization construction. *Econ. Geogr.* **2014**, *34*, 1–8. (In Chinese) [CrossRef]
46. Li, H.; Gong, J. Thoughts on land regulation under the background of ecological civilization. *China Land* **2018**, *11*, 45–46. (In Chinese)
47. Chen, C.; Hu, Z.; Liu, S.; Tseng, H. Emerging trends in regenerative medicine: A scientometric analysis in CiteSpace. *Expert Opin. Biol. Ther.* **2012**, *12*, 593–608. [CrossRef]
48. Zou, L.; Zhang, L.; Liu, Y. Territorial space function research in the context of ecological civilization: From the past into the future. *Prog. Geogr.* **2022**, *41*, 1312–1324. (In Chinese) [CrossRef]
49. Li, R.; Li, Y.; Hu, H. Support of ecosystem services for spatial planning theories and practices. *Acta Geogr. Sin.* **2020**, *75*, 2417–2430. (In Chinese)

50. Wang, Q.; Lang, H.; Zhong, J.; Chen, M.; Zhang, B. Progress review on land science research in 2019 and prospects for 2020. *China Land Sci.* **2020**, *34*, 79–91. (In Chinese) [CrossRef]
51. Li, Y.; Qi, S. Ecological civilization small watershed of construction of soil and water conservation. *China Popul. Environ.* **2018**, *28*, 14–17. (In Chinese) [CrossRef] [PubMed]
52. Lin, J.; Wu, Y.; Wu, J.; Liu, S. Construction of the spatial planning system: With discussions on the relationship between spatial planning, territorial spatial regulation, and natural resources supervision. *City Plan. Rev.* **2018**, *42*, 9–17. (In Chinese)
53. Zhang, Z. Research on ways of spatial planning legislation under the background of the development of an ecological civilization. *Hebei Law Sci.* **2018**, *36*, 45–58. (In Chinese) [CrossRef]
54. Liu, M.; Zhang, A.; Xiong, Y. Spatial differences of urban land use ecological efficiency in the Yangtze River Economic Belt and its interactive spillover effects with industrial structure upgrading. *China Popul. Environ.* **2022**, *32*, 125–139. (In Chinese)
55. Luo, W. Spatial and temporal differentiation of urban park land growth and its driving factors: Empirical analysis based on urban panel data of Hunan Province. *Econ. Geogr.* **2021**, *41*, 74–83. (In Chinese)
56. Zhang, K.; Wei, W.; Zhou, J.; Yin, L.; Xia, J. Spatial-temporal evolution characteristics and mechanism of "Three-Function Space" in the Three-Rivers Headwaters' Region from 1992 to 2020. *J. Geo-Inf. Sci.* **2022**, *24*, 1755–1770. (In Chinese) [CrossRef]
57. Han, G.; Luo, R.; Sa, K.; Zhuang, M.; Li, H. Bibliometric analysis of highly cited papers on resources and the environment in China. *Sustainability* **2022**, *14*, 604. [CrossRef]
58. Liu, J.; Liu, Y.; Li, Y. Classification evaluation and spatial-temporal analysis of "production-living-ecological" spaces in China. *Acta Geogr. Sin.* **2017**, *72*, 1290–1304. (In Chinese) [CrossRef]
59. Yan, J.; Chen, H.; Xia, F. Cognition, direction and path of future spatial planning based on the background of multiple planning integration. *China Land Sci.* **2017**, *31*, 21–27+87. (In Chinese) [CrossRef]
60. Yun, W.; Zhu, D.; Tang, H. Reshaping and innovation of China land consolidation strategy. *Trans. Chin. Soc. Agric. Eng.* **2016**, *32*, 1–8. (In Chinese) [CrossRef]
61. Jin, X.; Lu, Y.; Lin, J.; Qi, X.; Hu, G.; Li, X. Research on the evolution of spatiotemporal patterns of production-living-ecological space in an urban agglomeration in the Fujian Delta region, China. *Acta Ecol. Sin.* **2018**, *38*, 4286–4295. (In Chinese) [CrossRef]
62. Zou, B. Logic and conception of spatial planning system reconstruction under the framework of natural resources management. *Planners* **2018**, *34*, 5–10. (In Chinese)
63. Feng, Y.; Yang, Q. Key research fields and basic directions of Chinese rural-land comprehensive consolidation in transitional period. *Trans. Chin. Soc. Agric. Eng.* **2014**, *30*, 175–182. (In Chinese)
64. Wang, J.; Zhong, L. Problems and suggestion for developing ecological construction in land management work. *Trans. Chin. Soc. Agric. Eng.* **2017**, *33*, 308–314. (In Chinese) [CrossRef]
65. Feng, N. Man-nature harmonization theory and regional sustainable land resource use. *J. Nanjing Agric. Univ. Soc. Sci. Ed.* **2002**, *2*, 29–34. (In Chinese)
66. Lu, X.; Peng, W.; Huang, X.; Fu, Q.; Zhang, Q. Homestead management in China from the "separation of two rights" to the "separation of three rights": Visualization and analysis of hot topics and trends by mapping knowledge domains of academic papers in China National Knowledge Infrastructure (CNKI). *Land Use Policy* **2020**, *97*, 104670. [CrossRef]
67. Bai, Y.; Hong, Z.; Xue, X.; Shi, W. Research on coupling of intensive land use and ecological civilization construction of urban agglomeration in Guanzhong Plain. *Res. Soil Water Conserv.* **2021**, *28*, 272–280. (In Chinese) [CrossRef]
68. Kong, J.; Zeng, J.; Gao, M. Categorization of national space from the perspective of ecological civilization. *Planners* **2019**, *35*, 60–68. (In Chinese)
69. Yun, W.; Gao, L.; Zhang, C.; Lyv, Y. The change and influence of land use in China from the perspective of ecological civilization. *Environ. Prot.* **2018**, *46*, 31–35. (In Chinese) [CrossRef]
70. Zhai, X. A probe into the comprehensive rural land use from the perspective of ecological progress. *Frontiers* **2017**, *19*, 88–91. (In Chinese) [CrossRef]
71. Guo, W.; Zhang, D.; Chen, J. Put ecological civilization into the main tone of land management. *China Land* **2017**, *7*, 4–7. (In Chinese) [CrossRef]
72. Feng, B.; Liang, X.; Zeng, Z. Optimizing land usage in southern mountain areas of Jinan based on the SWAT model. *J. Irrig. Drain.* **2018**, *37*, 121–128. (In Chinese) [CrossRef]
73. Ding, J. Regulation strategy analysis of rural space reorganization in SUNAN Country from the perspective of ecological civilization. *Ecol. Econ.* **2015**, *31*, 177–180. (In Chinese)
74. Long, H.; Zhang, Y.; Ma, L.; Tu, S. Land use transitions: Progress, challenges and prospects. *Land* **2021**, *10*, 903. [CrossRef]

sustainability

MDPI

Article

Spatio-Temporal Evolution Characteristics, Development Patterns, and Ecological Effects of "Production-Living-Ecological Space" at the City Level in China

Yukun Cao [1,2], Xianqiao Huang [1,2,*], Xiangyue Liu [1,2] and Bo Cao [1,2]

1 College of Economics and Management, Northeast Forestry University, Harbin 150040, China
2 Center for Forestry Economics and Management, Northeast Forestry University, Harbin 150000, China
* Correspondence: hxq@nefu.edu.cn; Tel.: +86-451-8219-2205

Abstract: Effective production, living, and ecological space allocation is essential for advancing territorial policy optimization and improving the sustainability of land resource use. Based on the theory of the "production-living-ecological" space (PLES), the present study uses the spatial transfer matrix model, the coupling degree model, and ecosystem service value measurement to analyze the changes in the number and structural characteristics of the PLES and the evolution pattern of development in 336 cities in China from 2000 to 2020 and to evaluate the resulting ecological effects. The results are as follows: the living space is growing; the agricultural production space is decreasing; and the ecological space has been decreasing and then increasing. The evolution of the city space structure has five distinct patterns of development. Cities in the southeast with high urbanization rates have shifted from the pure economic expansion development pattern to the coordinated diversified development pattern. In contrast, the cities in the northeast and northwest, where ecological space accounts for an absolute proportion, still prefer the economic expansion development pattern. There is still a struggle between the "impulse of local development" and the "objective of central coordination". The development patterns of ecological protection and the coordinated diversified development patterns have higher ecological effects among the five development approaches, confirming the effectiveness of the territorial spatial planning policy under the coordinated development objective. Meanwhile, the optimization of future spatial planning policies should consider not only the rational allocation of space but also the quality development of space.

Keywords: land use allocation; the "production-living-ecological" space (PLES); development pattern; ecological effects; spatial optimization

Citation: Cao, Y.; Huang, X.; Liu, X.; Cao, B. Spatio-Temporal Evolution Characteristics, Development Patterns, and Ecological Effects of "Production-Living-Ecological Space" at the City Level in China. *Sustainability* **2023**, *15*, 1672. https://doi.org/10.3390/su15021672

Academic Editors: Wenwu Tang, Dianfeng Liu and Jianxin Yang

Received: 29 November 2022
Revised: 5 January 2023
Accepted: 13 January 2023
Published: 15 January 2023

1. Introduction

China's land use has a long history of spatial evolution; its development priority has shifted from economic development to urban and rural construction to ecological conservation, and the "production-living-ecological" space (PLES) structure has undergone significant change [1,2]. With the development of urbanization, the contradiction of an imbalance in the structural configuration of land for production space (PS), living space (LS), and ecological space (ES) has become prominent, resulting in the continuous deterioration of the ecological environment [3,4]. Meanwhile, the hierarchical structure of China's spatial planning system and the conflict between the central government's land development allocation model and local development needs make the system vulnerable to spatial mismatch and regional development disorder. Incomplete numbers indicate that there are 83 different types of plans and that there is a severe issue with "many conflicting plans" that hinders the execution of spatial policies [5,6]. The Chinese government has proposed the objective of "building the production space that is intensive and efficient, the living space that is livable and appropriate, and an ecological space that is beautiful and unspoiled" through the thorough construction of an ecological civilization in an effort to address these issues [7]. The Chinese government has also established a unified national territory spatial planning (NTSP) system that incorporates multiple regulations in the hope of achieving efficient and reasonable allocation of national resources [8].

Effective PLES identification is a need for achieving a fair allocation of the PLES. Clarifying the space's definition and classifying it are the first two steps in this process. Chinese scholars developed the PLES concept, which has been used to examine territorial spatial planning. The idea was inspired by the multifunctionality of land notion [9]. The theory of land multifunctionality first appeared in agricultural research, where an interdisciplinary research project was designed to verify the "sustainability" and "multifunctionality" of agriculture [10]. It was later extended to land with multiple social, economic, and environmental functions, and the EU Sixth Framework Project "Sustainability Impact Assessment: Tools for Environmental Social and Economic Effects of Multifunctional Land Use in Europe Regions" (SENSOR) formally proposed a conceptual framework for land multifunctionality [11,12]. There are two main approaches to classifying spaces based on the multiple functions of land. On the one hand, from the perspective of the dominant function of land, the land classification system of the PLES is constructed by refining the land classification criteria [13]. Different study scales classify land use differently in PLES, with the classification of living space and producing space showing the most variation, but they all stress how crucial it is to define ecological space precisely [14,15]. On the other hand, a multi-source data index system was constructed to identify the functions of the PLES by selecting indicators such as total labor productivity, urbanization rate of the resident population, and per capita water ownership [16,17]. Currently, it is common practice to identify the spatial functions of a country's territory using a multi-source data index system; however, due to the difficulty of accessing data sources, this method cannot evaluate the functional evolution of long time series [18], and the evaluation index system of regional main function judgment, although it takes into account natural resource elements and demographic and socio-economic elements, mostly relies on the statistical results of indicators within administrative districts, ignoring the importance of the influence of spatial structure on regional functions [19,20].

The evolutionary process of land space is a spatial mapping of the coupled human–land relationship in the study region. Additionally, the quantitative and spatial redistribution of land resources among functions is a dynamic game involving regional economic, social, and environmental resources [21] that, in the end, reflects linkages such as trade-offs and synergies [22,23]. The convergence of land use types in the PLES has directly resulted in this situation. Policy, in turn, has a significant impact on how land resources are allocated based on function. Government intervention and regulation of resource allocation are typical occurrences in China's institutional system, particularly in the allocation of land resources [5]. For instance, the Chinese government has improved PS and LS expansion in the central and western regions by putting Western Development Strategy into practice; however, this necessitates an increase in the ecological land area through protection projects such as the Grain-for-Green Project and the Natural Forest Protection Project. The compromise and cooperation between the PLES's functions ultimately lead to the development direction of the city.

The current research on the function of the PLES and its trade-offs and synergy in spatial and temporal dimensions is primarily focused on the coupling and coordination relationship between the "space" function within the research unit [24] and the difference in the coupling and coordination levels of different research units [20]. The coupling characteristics, distribution characteristics, spatial and temporal characteristics, and influencing factors among the PLES functions within the study area were investigated by building an evaluation index system for spatial functions using the coupling coordination degree model and the gray correlation degree model [25,26]. However, due to the differences in function definition standards, research scales, and the inconsistency of evaluation index systems, the currently constructed PLES function evaluation index systems are not uniform [17], and the logic behind regional function judgments ignores the current situation of territorial spatial development and protection [27]. At the research scale, provinces [28], cities (urban clusters) [29], and counties [30] are involved, but the larger the administrative unit at the macroscopic scale, the more pronounced the functional differences within the region. The spatial function determination with a single administrative unit of a county often results in functional bias, and it is also difficult to provide support for promoting regional economic development and guiding the efficient development of population clusters [27]. Therefore, there is a lack of different regions at different levels of the research scale and a lack of systematic research on the differential characteristics of spatial-temporal evolution at the local and municipal levels as well as attention to the overall spatial integration and regional land use conflicts.

The ecological environment effect has received more attention as territorial space function has developed. Exploring the impact of urban land use change on ecosystem service value has important practical significance for identifying urban regional ecological environment status and optimizing territorial spatial patterns [28,31]. Current research assesses ecological effects by measuring indicators such as the ecological quality index [32], ecological contribution ratio [33], and ecosystem service value [34]. The first two respond to environmental quality better, while the latter can more accurately reflect the worth of spatial functions. Ecosystem service value assessment is considered a powerful tool to facilitate spatial management and land use optimization [35]. There are many ways to measure the value of ecosystem services, and ecosystem functions and services in China were estimated by using Costanza et al.'s classification and economic parameters [36] based on Gaodi Xie's equivalence factor method [37]. Fewer studies have used the value of ecosystem services as an assessment tool to determine the scope of the ecological effects of various spatial development patterns, despite the fact that there is an expanding body of research on measuring and analyzing the factors that influence the value of ecosystem services. Therefore, this paper analyzes the spatial and temporal evolution characteristics of the PLES at the scale of 336 prefecture-level cities from 2000 to 2020 according to the classification identification of the PLES functions. We first obtain the development patterns of the PLES based on the coupling degree model and then calculate the value of ecosystem services under different urban space development patterns to provide decision-making suggestions for future territorial spatial planning policies in spatial layout and function optimization.

2. Materials and Methods

2.1. Data Source and Processing

2.1.1. Data Source and Study Area

The study area was the major cities in mainland China, with prefecture-level cities as the study unit, including 336 prefecture-level cities, regions, autonomous regions, and municipalities directly under the central government (Figure 1). Due to data acquisition reasons, Taiwan, Hong Kong, Macao, and Sansha City in Hainan Province were not included in the study.

The data for the natural resources category included city boundary data and land use data of mainland China. City boundary data were downloaded from the National Geographic Information Center of China (http://ngcc.sbsm.gov.cn, accessed on 18 January 2022). Land use/cover data were obtained from the European Space Agency's (ESA) CCI-LC project (http://maps.elie.ucl.ac.be/CCI/viewer/index.php, accessed on 10 June 2021) [38] that uses the CCI-LC dataset containing two versions due to the long-term continuous time-series nature of the sample. The land use data atlas (2000–2015) uses version 2.0.7, and the land use data atlas (2016–2020) uses version 2.1.1. Both versions were produced with the same processing chain, with a spatial resolution of 300 m and a temporal resolution of 1 year.

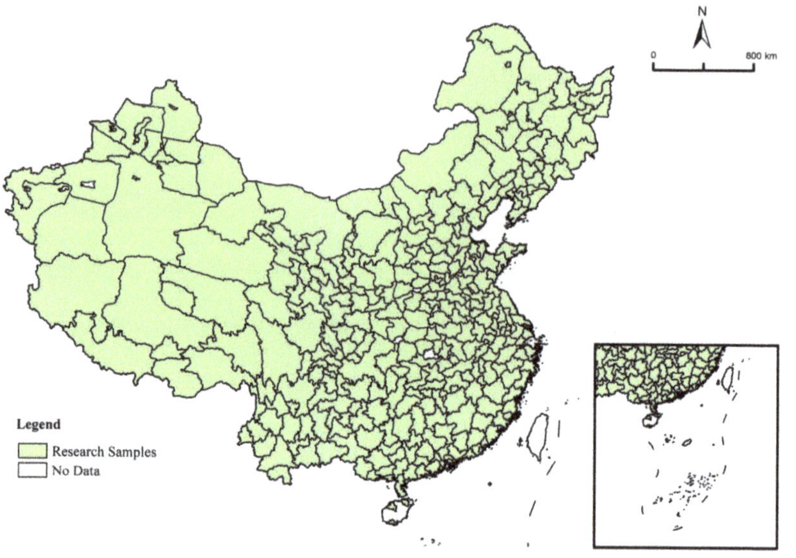

Figure 1. Distribution of the sample cities.

The economic and social data were obtained from the China Statistical Yearbook, the National compilation of agricultural cost-benefit information, provincial statistical yearbooks, and the official website of the National Bureau of Statistics (https://data.stats.gov.cn, accessed on 18 January 2022). The main sown areas and unit area production of rice, wheat, and corn for each year nationwide and by the province as well as the consumer price index for residents were obtained from the China Statistical Yearbook (2001–2021), and the average prices of rice, wheat, and corn for each year in each region nationwide were obtained from the National Compilation of Costs and Benefits of Agricultural Products (2001–2021).

2.1.2. Data Processing

Reclassification of land use types was performed under the definition of the PLES. Seven land cover types, including arable land, urban land, and forest land, were reclassified in terms of their functional dominance. The ES refers to the national land space with natural attributes and the main function of providing ecological services or ecological products, including various ecological elements such as forests, grasslands, and wetlands. The LS is the space used for people's daily life activities, and the PS is the specific functional area where people carry out production activities [15]. In the further classification criteria of secondary space, we refer to the study of Ziyan Ling et al. [15,29]. We further divided the PLES index system into agricultural PS, urban LS, and ES. It was improved according to the "three zones and three lines" (ecological red line, urban development boundary, and basic farmland protection red line) proposed by the Chinese government, and the division of agricultural space, urban space, and ecological space in this system can fit well with ESA's land classification standards. The spatially specific land use types are based on the classification of land cover data provided by ESA [39] (Table 1). In addition, although the CAS statistics cover industrial and mining construction land and rural living land, it is difficult to analyze the spatial evolution pattern under long time series due to the limited time-series data. In the work of spatial classification corresponding to land cover data, urban space and rural settlements were included in the LS according to ESA's land classification standards [39,40].

Table 1. The production–living–ecological space classification of Chinese cities corresponding to the land cover categories.

Classification of PLES	Space Secondary Classification	The Number of PLES	Land Cover Type	Specific Categories	Land Data Code
Production space (PS)	Agricultural production space	1	Cropland	Rain-fed cropland (dry land)	10
				Irrigated or post-flooding cropland (paddy field)	20
				Mosaic cropland and natural vegetation	30, 40
Living Space (LS)	Living Space	2	Built-up land	Urban	190
Ecological Space (ES)	Forestry Ecological Space	3	Forest land	Coniferous Forest	70, 80
				Mixed Coniferous Forest	90
				Broadleaf Forest	50, 60
				Shrubland	120
				Mixed forest land	100
				Mangrove Forest	160, 170
	Grassland Ecological Space	4	Grassland	Grassland	130
				Mosaic herbaceous cover (>50%) Land	110
				Lichens and mosses	140
	Wetland Ecological Space	5	Wetland	Wetland	180
	Other Ecological Spaces	6	Desert	Sparse vegetation	150
				Bare areas	200
	Water ecological space	7	Water area	Water bodies	210
				Permanent snow and ice	220

2.2. Methods

2.2.1. PLES Transfer Matrix Model

The PLES transfer matrix model refers to the land use transfer matrix model based on the functions of the PS, LS, and ES, and the model portrays the quantitative relationship between the three spatial types of conversion. The net conversion area is the difference between the area of two different types of spaces that are interconverted, and the positive or negative value of the difference reflects the final direction of transformation between two of the three types of PS, LS, and ES within the study area. For the PLES, this means that the direction of transformation can be unidirectional or multidirectional. The PLES transfer matrix and the net conversion area can reveal the characteristics of the changing spatial pattern of each city in a given period [28,41]. The logical diagram of the PLES transfer matrix model is shown in Figure 2.

The PLES transfer matrix is given by

$$F_{ij} = \begin{bmatrix} f_{11} & f_{12} & \dots & f_{1n} \\ f_{21} & f_{22} & \dots & f_{2n} \\ \dots & \dots & \dots & \dots \\ f_{n1} & f_{n2} & \dots & f_{nn} \end{bmatrix} \quad (1)$$

where F_{ij} is the area of the space type i at the beginning of the period that is transformed into the space type j at the end of the period; n is the number of space types, and according to the number of space types in PLES, the value of n is taken as 3.

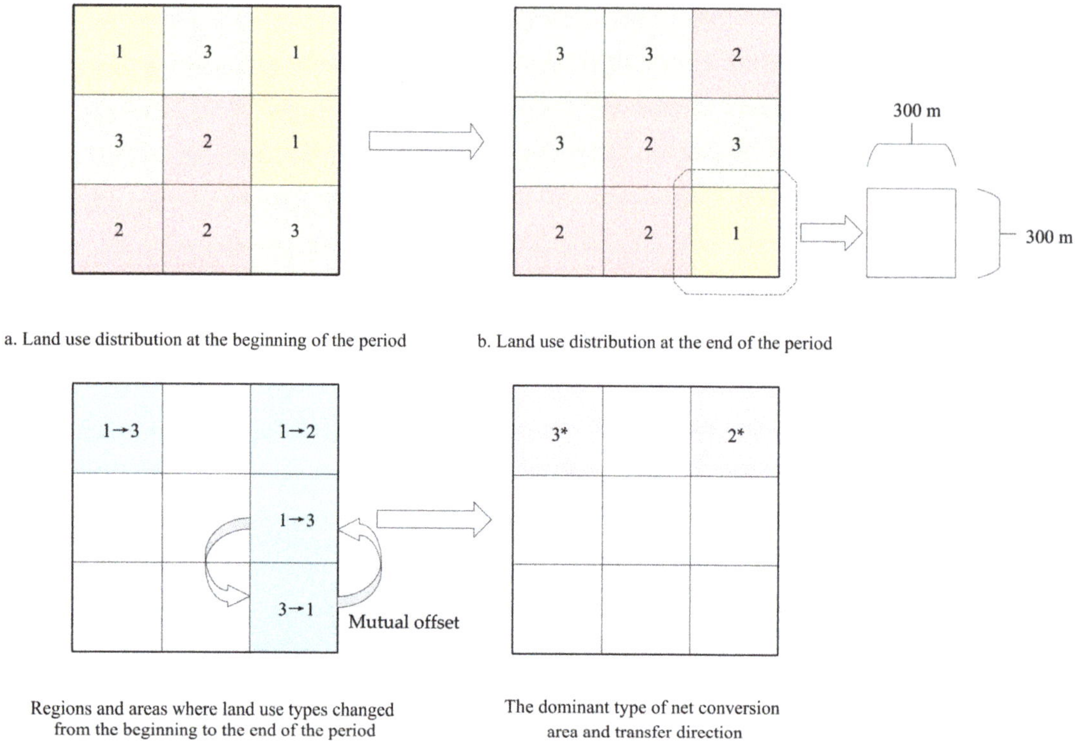

Figure 2. Logical diagram of the PLES transfer matrix model. Note: 1, 2, and 3 denote three different types of land. The"→" shows that the land's land type has changed. The"1→2" denotes that the land type changed from 1 to 2 at the end of the period. The "*" denotes the conversion area, and the number denotes the final conversion direction of the land type.

The equations of the net conversion area of the PLES are as follows:

$$F_{i.j} = f_{ji} - f_{ij} \tag{2}$$

$$F_i = \begin{cases} 0, F_{i.j} < 0 \\ F_{i.j}, F_{i.j} \geq 0 \end{cases} \tag{3}$$

$$F_j = \begin{cases} |\ F_{i.j}\ |, F_{i.j} < 0 \\ 0, F_{i.j} \geq 0 \end{cases} \tag{4}$$

where f_{ji} is the area transferred from spatial type j to spatial type i; f_{ij} is the area transferred from spatial type i to spatial type j; and $F_{i.j}$ is the difference of the converted area between spatial type i and j, which is the net converted area of spatial type i and j. When $F_{i.j} > 0$, this means that the conversion result between spatial type i and j in the study area is a larger area converted from j to i. The direction of spatial structure development in this area is dominated by type i. At this time, the net conversion area between spatial type i and j is F_i, and F_i refers to the net conversion area converted to type i. When $F_{i.j} < 0$, the conversion between spatial types i and j is larger than for conversion to j. The direction of spatial structure development in the region is dominated by type j, and the net spatial conversion area is F_j.

2.2.2. PLES Coupling Degree Model

The PLES coupling degree model is based on the concept of coupling degree in physics and is chosen to quantitatively analyze the coupling interaction between the functions of the PLES in the city. The process of spatial redistribution of the functions of the PLES (in terms of the net conversion area between spaces in spatial structure) and the strength of the interconversion relationship can further reflect the trade-offs and decision-making results of cities in terms of economic development and ecological conservation-oriented development and thus summarize the similarities and differences of urban spatial development approaches [24]. Since this study focuses more on analyzing the interaction between economic functions dominated by the PS and the LS ($F_1 + F_2$) and ecological functions dominated by the ES (F_3) in regional development, the following coupled coordination model was constructed based on the above principles, and different development patterns and characteristics are classified according to the value of the coupling coordination degree S.

The PLES coupling degree model can be expressed as follows:

$$S = 2 \times \sqrt[2]{(F_1+F_2) \times F_3} \,/\, [(F_1+F_2)+F_3] \tag{5}$$

where F_1 represents the net transfer area of the PS in the PLES of the city during the study period; F_2 represents the net transfer area of the LS; and F_3 represents the net transfer area of the ES. S is the coupling degree, a value between 0 and 1. When S = 1, the coupling degree is maximal; the system reaches a high level of coordinated development between the elements within the system, and the system tends to develop in a coordinated manner. In contrast, when S = 0, the system will tend to develop in an uncoordinated manner [42].

The urban development patterns characterized by different coupling degree values are shown in Table 2.

Table 2. Characteristics of development patterns under different coupling degree values.

The PLES Coupling Degree	Assessment	Patterns of Development	Characteristics
S = 0	$F_1 + F_2 \neq 0, F_3 = 0$	Pure economic expansion (S1)	The city is in the development mode of economic and urban construction with the reduction of ecological space.
	$F_1 + F_2 = 0, F_3 \neq 0$	Pure ecological protection (S2)	The city has a spatial structure aimed at slowing down the economy and has taken a development approach to ecological restoration and protection with expanded ecological space.
S ∈ (0,0.5]	$(F_1 + F_2) > F_3$	Unbalanced, biased economic expansion (S3)	A diversified approach has been used for the spatial development of the city, where the choice of spatial function is in a dynamic game state, with the economic development function dominating and the ecological function playing a smaller role.
	$(F_1 + F_2) < F_3$	Unbalanced, biased ecological protection (S4)	A diversified approach has been used for the spatial development of the city, where the choice of spatial function is in a dynamic game state, with a greater preference for ecological protection functions and a smaller role for production and living functions.
S ∈ (0.5,1]	-	Coordinated and diversified development pattern (S5)	The three spatial structures of the city are transformed into each other in a more orderly manner and are in a coordinated development pattern that can meet the needs of different subjects.

2.2.3. Ecosystem Service Value (ESV)

The ecological and environmental effects of the PLES can be characterized as the differences in ecosystem services under different coupled and coordinated development approaches, and the rationality of the layout of the PLES can be judged by comparing the differences in ESV growth and growth rates under different patterns for the reference of spatial layout optimization and sustainable use of resources under the construction of an ecological civilization [43,44]. Reviewing the results of Costanza's study [45], Chinese ecologist Gaodi Xie et al. [46,47] developed a table of equivalent weighting factors to measure the value of ecosystem services in China. Compared with the InVEST model and ARIES model, which measure the quality of things [48], the equivalent factor method is more applicable to assessing the value of ecosystem services at regional and global scales [49]. This study used these equivalence factor measures to calculate the value of ecosystem services for 336 cities in China from 2000 to 2020. A standard unit ESV equivalent factor was defined as the ecological service value of 1 hm^2 of farmland food production with an equivalent value of 1. The equivalent factors of other ecosystem types of ecological service values were the magnitude of their contribution relative to the service function of farmland food production [50]. Also, for the value of the standard unit ESV equivalent factor, the magnitude of the factor depends on its location in space, and the value of the unit varies from region to region [37]. The impact of the correction and determination of the equivalent factor on the value of ecosystem services is significant [51]. Therefore, this study revised the ESV coefficients per unit area in different regions by counting the annual grain prices in each province.

The equation for the value of one standard unit of ESV equivalent factor is as follows:

$$C = 1/7 \times P \times Q \tag{6}$$

where C is the value of one standard unit of ESV equivalent factor (CNY/hm^2); P is the average price of grain in each province (CNY/kg); and Q is the yield per unit area of grain in each province (kg/hm^2). The average prices and unit area yields of the three-grain crops of wheat, corn, and rice from 2000 to 2020 can be obtained based on the National Compilation of Costs and Benefits of Agricultural Products (2000–2020). Wheat, corn, and rice are the three major grains in China, and their cultivation areas are distributed in all provinces [52]. Therefore, these three grain crops were chosen as representatives to calculate the value of one standard unit of ESV equivalent factor. To eliminate the impact of price fluctuations on value changes, this study introduced the consumer price index to adjust the average grain price data of each year to the price level of 2000, and the value of one standard unit of ESV equivalent factor in each province of the country in each year was calculated.

The equation of the ESV coefficient per unit area is as follows:

$$C_i = EC_i \times C, i = 1, 2, \ldots, 18 \tag{7}$$

where C_i is the ESV of land use type i per unit area (CNY/hm^2); C is the value of one standard unit of ESV equivalent factor (CNY/hm^2); EC_i is the equivalent value of ecosystem services per unit area, which can be obtained from the equivalence factor table in the research results of Gaodi Xie et al. [46]; and i is the type of land cover, including seven indicators, namely, cropland, urban, forest land, grassland, wetland, desert, and water area. The secondary index contains 18 types of rain-fed cropland, irrigated or post-flooding mosaic cropland, urban area, coniferous forest land, mixed coniferous forest land, broadleaf forest land, shrubland, mixed forest land, mangrove forest land, grassland, mosaic herbaceous cover (>50%) land, lichens and mosses land, wetland, sparse vegetation land, bare areas, water bodies, and permanent snow and ice.

The ESV is given by

$$ESV = \sum_{i=1}^{18} A_i \times Ci,\ i = 1, 2, \ldots, 18 \quad (8)$$

where ESV is the total ecosystem service value (CNY); A_i is the area of land cover type I; C_i is the ESV per unit area of land cover type I (CNY/hm^2); and I is the land cover type.

3. Results

3.1. Evolutionary Characteristics of the PLES

3.1.1. Spatio-Temporal Variation Characteristics of the PLES

After reclassifying the land use types according to the criteria for defining the PLES, the distribution of the PLES in Chinese cities from 2000 to 2020 was assessed (Figure 3) using ArcGIS 10.7, and the ratio of the area of each type of space from 2000 to 2020 was calculated by extracting the data (Table 3).

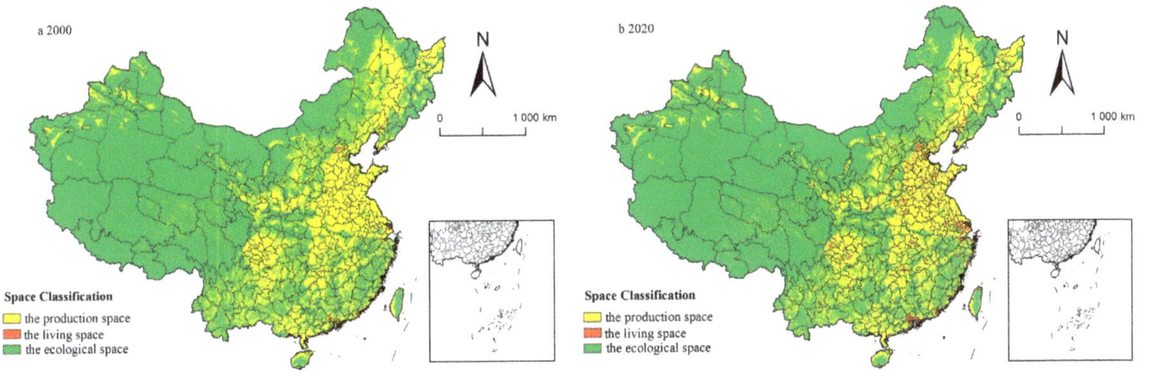

Figure 3. Spatial distribution of PLES from 2000 to 2020.

Table 3. The ratio and shift in each sort of space area in China from 2000 to 2020.

Year	Share of the PS Area (%)	Share of the LS Area (%)	Share of the ES Area (%)
2000	29.031	0.464	70.504
2005	28.905	0.744	70.351
2010	28.879	1.023	70.098
2015	28.697	1.314	69.989
2020	28.298	1.561	70.141
2000–2005	−0.126	0.279	−0.153
2005–2010	−0.026	0.284	−0.253
2010–2015	−0.181	0.287	−0.109
2015–2020	−0.399	0.290	0.152
2000–2020	−0.733	1.097	−0.364

According to time characteristics, from 2000 to 2020, the ES made up the majority of the PLES in China from 2000 to 2020, with the LS making up the smallest portion. The LS was growing as the PS was shrinking. Initial declines in the ES were followed by increases. Additionally, the ES distribution pattern of "high in the west and low in the east" continues to exist. According to the process of national urbanization and industrialization, the LS, particularly the urban LS, is growing; from 0.464% in 2000 to 1.561% in 2020, the area proportion has increased. Between 2000 and 2015, the ES fell, with a 0.515% drop in the proportion. Under the policies of a complete prohibition on commercial cutting of natural forests in 2014 and the creation of China's ecological redline (ER) policy of the NTSP, the proportion increased by 0.152% between 2015 and 2020, and the area of ES was effectively increased. In contrast, the LS has been gradually annexing the PS as a result of urbanization and environmental conservation initiatives such as the Grain-for-Green Project. As a result, the PS's agricultural PS has decreased from 29.031% in 2000 to 28.298% in 2020, a drop of 0.733%.

The PLES's spatial characteristics (Figure 3) demonstrate that there was little change in the PLES's distribution pattern between 2000 and 2020. The principal line separating China's east and west in terms of population density and natural resources is the Hu Line (HL, also known as the Heihe-Tengchong Line). The PS and the LS are mainly situated to the east of the line. The majority of the ES are located west of the line. The LS moves from the focal point outward to the surrounding area over time.

The particular characteristics are as follows: first, the PS (agricultural PS) is consistent with the "seven regions and 23 belts" as the primary agricultural strategy in the functional area planning policy, which is distributed primarily in the cities where the main agricultural production areas are located, such as the Northeast Plain, the Yellow Huaihai Plain, the Yangtze River Basin, the Fenwei Plain, the Hetao Irrigation Area, South China, and Xinjiang. Second, the LS is primarily distributed in the Bohai Rim, the Yangtze River Delta, and the Pearl River Delta regions centered in Beijing, Shanghai, and Guangzhou. The proportion of urban LS is rising in these regions, and it is followed by a dotted distribution in the densely populated areas and provincial capitals. Third, the ES is largely distributed in the major cities in the Qinghai–Tibet Plateau, the Loess Plateau, Northeast China, Southwest China, and the southern hilly region.

3.1.2. Structural Transformation Characteristics of the PLES

The structural transformation of the PLES is reflected in the transformation between the types of PLES, including the quantitative transfer and the final transfer direction. According to Equations (1)–(4) of the transfer matrix model of the PLES in the research methodology, the area of inter-transformation (i.e., the transformation of production space to ecological space and vice versa) and the net transformed area between the three spaces of the national territory in different periods were obtained (Table 4), and a graph was drawn using ArcGIS 10.7 that depicts the transformation between the three functional spaces of Chinese cities from 2000 to 2020 (Figure 4).

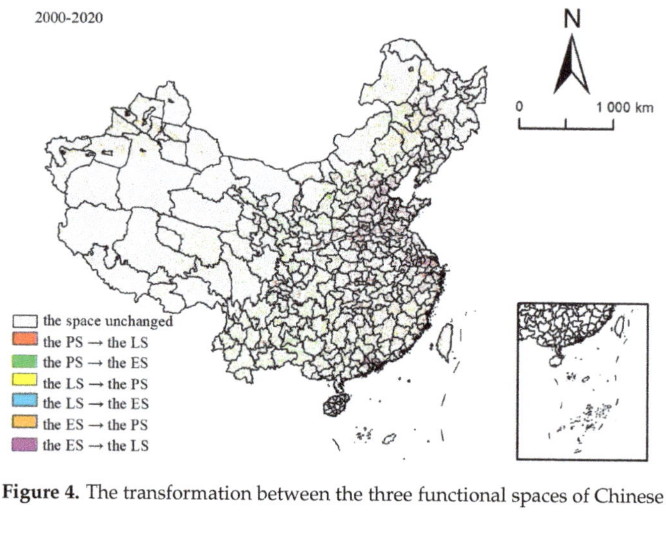

Figure 4. The transformation between the three functional spaces of Chinese cities from 2000 to 2020.

Table 4. Table of the interconversion of the area of the PLES in China.

Period	Type	PS–LS (km^2)	PS–ES (km^2)	LS–ES (km^2)
2000–2020	The former → the latter conversion area (f_{ij})	86,753	206,347	417
	The latter → the former conversion area (f_{ji})	2022	213,164	19,899
	The net conversion area ($F_{i.j}$)	84,731	−6817	−19,482
	Dominant Type	The PS → the LS	The ES → the PS	The ES → the LS
2000–2005	The former → the latter conversion area (f_{ij})	20,588	25,149	0
	The latter → the former conversion area (f_{ji})	0	33,709	6033
	The net conversion area ($F_{i.j}$)	20,588	−8560	−6033
	Dominant Type	The PS → the LS	The ES → the PS	The ES → the LS
2005–2010	The former → the latter conversion area (f_{ij})	22,478	10,444	0
	The latter → the former conversion area (f_{ji})	0	30,407	4187
	The net conversion area ($F_{i.j}$)	22,478	−19,963	−4187
	Dominant Type	The PS → the LS	The ES → the PS	The ES → the LS
2010–2015	The former → the latter conversion area (f_{ij})	24,322	7165	0
	The latter → the former conversion area (f_{ji})	0	14,199	3331
	The net conversion area ($F_{i.j}$)	24,322	−7034	−3331
	Dominant Type	The PS → the LS	The ES → the PS	The ES → the LS
2015–2020	The former → the latter conversion area (f_{ij})	31,480	174,268	1568
	The latter → the former conversion area (f_{ji})	14,023	145,618	7524
	The net conversion area ($F_{i.j}$)	17,457	28,650	−5956
	Dominant Type	The PS → the LS	The PS → the ES	The ES → the LS

In terms of temporal characteristics, the PLES can be quantitatively transformed from one type to another at macroscopic study scales, and the end direction of the spatial structure was from the PS and the ES to the LS. Between the ES and the PS, the ES was more frequently transformed into the PS. Specifically, from 2000 to 2015, the PS and ES were continuously transferred to the LS whereas the LS was not transferred out. The PS occupied a larger portion of the ES. From 2015 to 2020, the PLES types were transferred among each other, but finally, the PS was shrunk and transformed into the ES and LS.

In terms of spatial characteristics, the transition from the PS and the ES to the LS most frequently occurred in the eastern region, while the PS and the ES were interchanged in the central and northwestern regions. First, there has been an increase in the area of urban LS in important cities and provincial capitals such as Beijing, Shanghai, and Guangzhou as well as in the more prosperous provinces of Hebei, Shandong, Jiangsu, and Zhejiang. The top five cities with the highest area of PS converted to LS are Baoding, Hebei (1396 km^2), Shanghai (1330 km^2), Suzhou, Jiangsu (1320 km^2), Linyi, Shandong (1282 km^2), and Cangzhou, Hebei (1216 km^2). The cities with the largest area of ES converted to LS are Tianjin (1014 km^2), Tangshan, Hebei (562 km^2), Baotou, Inner Mongolia (391 km^2), Suzhou, Jiangsu (361 km^2), and Beijing (342 km^2). Second, there is spatial exchange between PS and ES in the same area, meaning that the region that was production space at the beginning of the period transforms into ecological space at its conclusion, while the area that was ecological space at the beginning of the period becomes production space. This spatial shift is focused in less developed and populous regions such as Inner Mongolia, Qinghai Province, Tibet, and Xinjiang Province. The top five cities in the conversion of PS to ES are Hulunbeier City, Inner Mongolia (4712 km^2), Yushu City, Qinghai (4636 km^2), Chifeng, Inner Mongolia (3919 km^2), Naqu City, Tibet (3632 km^2), and Xi'an League, Inner Mongolia (3613 km^2). The top five cities in terms of area converted from ES to PS in descending order are Hulunbeier City, Inner Mongolia (6155 km^2), Chifeng City, Inner Mongolia (4781 km^2), Tongliao City, Inner Mongolia (4727 km^2), Yushu City, Qinghai (4362 km^2), and Xilinguole City, Inner Mongolia (3890 km^2). These cities have a rich resource base, large administrative areas, and most of them are located in ecological function areas. Third, the majority of the cities that have switched from PS to ES are located within the Grain-for-Green Project's application region, including significant cities in the Guangxi, Yunnan, and Shaanxi Provinces.

3.2. Transformation of the City-Level Development Pattern of the PLES

To judge the choice between economic development and ecological protection made by cities under the policy guidance of the NTSP, the spatial distribution and transformation characteristics of the spatial development patterns of cities in different periods were determined according to the coupling degree model. According to Equation (5), the number of cities under different development patterns of the PLES from 2000 to 2020 (Table 5) and the spatial distribution of different development patterns in the four periods (Figure 5) were obtained.

Table 5. Number of cities under different development patterns in the PLES from 2000 to 2020.

Pattern	2000–2005	2005–2010	2010–2015	2015–2020
The pure economic expansion development pattern (S1)	204	217	201	87
The pure ecological protection development pattern (S2)	2	0	1	1
The unbalanced, biased economic expansion development pattern (S3)	29	29	38	35
The unbalanced, biased ecological protection development pattern (S4)	20	4	8	31
The coordinated and diversified development pattern (S5)	81	86	88	182
Total number of cities	336			

In terms of temporal characteristics, the PLES's development pattern changed from the pure economic expansion to the coordinated and diversified pattern between 2000 and 2020. The number of cities with the pure ecological protection development pattern was the lowest, with a maximum of two cities every period. There were only a small number of cities with two imbalanced development patterns—a minimum of 4 and a maximum of 38—and there was little variation in the total number. Prior to 2010, there were 2.5 times as many cities with the pure economic expansion development pattern as there were cities with the coordinated and diversified development pattern. More cities began to follow the coordinated and diversified development pattern after 2010, and the number of cities following this pattern quickly expanded from 88 in the years 2010–2015 to 182 in the years 2015–2020. This marked a substantial shift in the development pattern. This partially reflects the efficacy of the State Council's 2010 policy known as "the Main Function Area Plan".

In terms of spatial characteristics, from 2000 to 2020, the coordinated and diversified development pattern spread and relocated from major cities in the center to the more economically developed regions in the east, while the pure economic expansion development pattern shifted from the east to the central and northwestern regions. The pure ecological protection development pattern and the unbalanced, biased ecological protection development pattern were located in the western regions with larger ES.

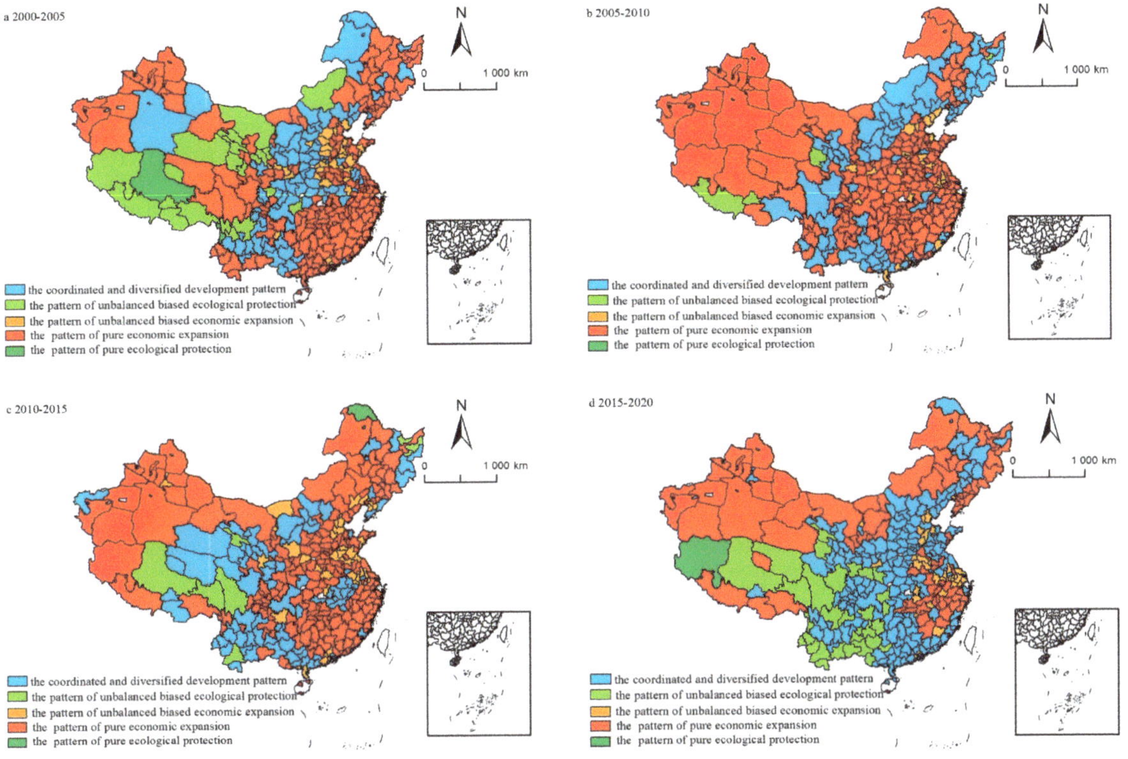

Figure 5. Development patterns of the PLES in different cities from 2000 to 2020.

Specifically, from 2000 to 2005, cities with the pure economic expansion development pattern were more scattered, being concentrated in economically developed cities along the east coast, major cities south of the Yangtze River basin, provincial capitals in the northeast and northwest China, and major cities with high agricultural production. The major cities with the unbalanced, biased economic expansion development pattern were concentrated in four provinces, namely, Hebei, Henan, Shandong, and Xi'an, with the representative cities being Shijiazhuang, Jinan, Zhengzhou, and Xi'an. Most of the cities with the coordinated and diversified development pattern overlap with the "Hu Line". Most of the cities with ecological protection development pattern are concentrated in Inner Mongolia, Gansu, Qinghai, Tibet, and Guangxi Province. The representative cities are Alashan, Zhangye, and Linzhi.

From 2005 to 2010, cities with the pure ecological protection development pattern and those with the unbalanced, biased ecological protection development pattern have been replaced, and the number of cities with the pure economic expansion development pattern has been rising, with a clear trend of increase in the northwest region. The representative cities are Xilinguole, Alashan, Zhangye, and Naqu City. The majority of the cities with the coordinated, diversified development pattern may be found in the provinces of Shanxi, Heilongjiang, Sichuan, Yunnan, and Guangxi. Baise City, Yulin, Xinanmeng, Yichun, and Yuxi are the representative cities. These cities are located in the construction zones for ecological protection projects, and the policy's effect over their development has compelled them to diversify throughout this time.

From 2010 to 2015, cities with the coordinated diversified development pattern began to expand to the southeast, and the coordinated diversified development pattern began to emerge in Jiangxi, Hubei, Anhui, Jiangsu, and Chongqing, represented by cities such as Nanchang, Wuhan, Hefei, and Zhenjiang. Most cities in Yunnan and Guangxi provinces also belong to this development pattern. Several cities in Northeast China where state-owned forest areas are located have shifted to the ecological protection development pattern, and some cities in Yunnan, Sichuan, and Tibet have reverted to the ecological protection development pattern.

From 2015–2020, most of the more economically developed cities in the eastern region began to shift to a coordinated and diversified development pattern, except for some cities in the southeast; most of the northwest and northeast regions remain in the pure economic expansion development pattern, with economic development as the main goal. The number of cities with the pure ecological protection development pattern and the unbalanced, biased economic ecological protection development pattern has increased, becoming concentrated in several provinces such as Guizhou, Guangxi, Yunnan, Tibet, Sichuan, and Qinghai, with representative cities being Zunyi, Baise, Pu'er, Naqu, and the Ali region.

3.3. Evolution Law of Ecological Effects under the Development Pattern of the PLES

Based on the ecological effect of ecosystem service value measurement, the ecosystem service value of each city in each year was calculated using Equations (6)–(8). Based on the different development patterns of the PLES, the average annual growth of ecosystem service value and the average annual growth rate were calculated for all cities under each pattern, and the average annual growth and average annual growth rate of each development pattern in different periods were obtained (Table 6).

Table 6. Mean values of ESV growth and growth rates under different development patterns.

Pattern	Average Annual Growth of ESV (Billion, CNY)				The Average Annual Growth Rate of ESV (%)			
	2000–2005	2005–2010	2010–2015	2015–2020	2000–2005	2005–2010	2010–2015	2015–2020
National average value	13.14	11.15	2.61	−2.78	7.38	7.31	0.89	−0.94
(S1)	12.07	19.38	2.21	−3.21	8.29	7.70	0.80	−0.93
(S2)	109.02	–	12.20	−71.40	8.48	–	0.74	−2.20
(S3)	4.16	7.38	0.60	−3.18	8.03	7.63	0.28	−1.57
(S4)	36.40	41.08	−3.09	−4.74	10.23	8.33	0.62	0.39
(S5)	14.90	21.60	4.78	−1.78	9.32	8.00	1.59	−0.56

To assess their ecological effects, the average annual growth rates for each development pattern and the average annual ESV growth were rated from highest to lowest. Over multiple periods, it was discovered that the unbalanced, biased ecological protection development pattern had the maximum ecological value effect. The coordinated, varied development pattern came next. The lowest positive impact on ecological value was caused by the uneven, biased economic expansion development model. The specific characteristics are as follows.

First, the ecologically constrained development approach and the coordinated diversified development pattern have higher ecological effects among the five development patterns. A comprehensive ranking of the values of ecosystem service value growth and growth rate shows that the coordinated diversified development pattern ranked steadily in the top three for several periods. Among the uncoordinated development patterns, the unbalanced, biased ecological protection development pattern was more often ranked first in ecological effects, and the unbalanced, biased economic expansion development pattern was ranked lowest in several periods. Combined with the evolution pattern of the PLES development patterns of cities from 2000 to 2020 (Table 5), the number of cities under the coordinated diversified development pattern has been increasing to the highest proportion, and the ecological effects provided have had the greatest comprehensive value, reflecting the rationality of the evolutionary direction of the PLES under the NTSP.

Second, there are temporal fluctuations in the ecological effects of different development patterns. Before 2010, the mean ESV growth and growth rates under the two development patterns with ecological protection were ranked in the top two, with the pure economic expansion development pattern ranked next. After 2010, however, the pure economic expansion development pattern ranked second for the period 2010–2015, and the pure ecological protection development pattern ranked fifth for the period 2015–2020, contrary to expectation. The reason for this was that the composition of cities belonging to different development approaches has changed over time. Combining the spatial evolution characteristics of the development patterns of the PLES from 2000 to 2020 (Table 5), before 2010, cities with the pure economic expansion development pattern were mostly in economically developed regions such as Beijing, Shanghai, and Guangzhou. After 2010, the cities that belonged to the pure economic expansion development pattern changed to places such as Yichun City in Heilongjiang Province, Hulunbeier City in Inner Mongolia, and Hami City in Xinjiang that were rich in natural resources but less economically developed.

Third, there are breakpoints in the temporal evolution of ecological effects. Before 2015, the average value of ESV growth volume and growth rate under the five development patterns was positive. In contrast, the average values of ESV growth volume and growth rate were negative for all patterns except for the unbalanced, biased ecological protection development pattern during the period 2015–2020. The national average ESV growth volume decreased from CNY 13.14 billion (2000–2005) to CNY –2.78 million (2015–2020) and the average annual growth rate decreased from 7.38% (2000–2005) to −0.94% (2015–2020). In addition, ES is internally subdivided into several types, including woodland ES, grassland ES, and wetland ES. The amount of ecological effect value that can be assigned to different land types also varies, and the conversion between ESs can lead to fluctuations in ecological effects; for example, during 2015–2020, the city with a single purely ecologically bound development approach was the Ali region of Tibet, where the total ES area increased, but within the ES, grassland ES decreased (by an area of 62,563 km^2), which was converted to other ESs such as a desert (where the area increased by 61,268 km^2), resulting in the loss of ecological effects.

4. Discussion

4.1. Scientific Optimization of Spatial Structure

This study analyzed the spatio-temporal evolutionary characteristics of the layout of the PLES in China under the improved classification system of the PLES. It concluded that LS in China has been expanding from 2000 to 2020, and these findings are consistent with the findings of the existing study by Dongyan Kong et al. [28]. Ecological space shows a trend of first decreasing and then increasing; specifically, after 2015, ES continued to expand, and the ecological attributes of national space were given more attention. Moreover, this study found an overlooked problem, which is the continuous decrease in agricultural PS. China has been in the process of rapid urbanization for the past 20 years, and the living space has been expanding as a result of China's past economic development. But ensuring urbanization through the continuous expansion of urban living space is an unsustainable approach [53,54]. Environmental protection and food supply are equally important. Arable land resources are related to a country's food security, especially for developing countries with large populations. In general, China's agricultural PS faces shrinkage due to the demand for land for urban development and construction and the impact of the environmental protection policy of returning farmland to the forest [55], a result that was effectively verified in this study. Economical and concentrated utilization of land resources is the fundamental strategy for realizing the common development of the social economy and ecosystem service value [53]. While the future territorial spatial planning policy optimizes the goal of promoting the expansion of ES, it should more strictly control the red line of basic farmland protection while paying more attention to the intensive and efficient agricultural land production model and being aware of the country's food security problems.

4.2. Spatial Differences in Development Patterns and Policy Orientation

Exploring development patterns of the city that can promote sustainable development has been a topic of interest in the field of spatial planning, and the last two decades have been a period of development transition and large changes in spatial patterns in China, where the endogenous dynamics of economic development and the external constraints of ecological protection have jointly influenced the spatial development of a city [56,57]. There are clear characteristics of regional differences in urban development patterns, and our results are consistent with the above theoretical expectations. Before the promulgation of China's National Functional Area Plan policy in 2010, most cities in China chose a single development pattern of pure economic expansion, and the transfer of space basically followed the principle of maximum benefit or minimum cost in economics; regardless of the presence or absence of environmental policy intervention, agricultural PS and ES would be transferred to LS with higher economic value, while the probability of transferring LS to other spaces was less [58]. Only under strong environmental protection policy constraints will cities choose to convert agricultural PS to ES; for example, there are more coordinated and diversified development patterns of ecological protection in the cities belonging to the Grain-for-Green Project and the Natural Forest Protection Project areas. Since 2010, the Chinese government has proposed a development strategy of zoning the main functions, delineated a stricter spatial control of the PLES, and put forward the development goals of coordinated spatial development and regional balance. Under this policy guidance, most cities have shifted from the pure/unbalanced biased economic expansion development pattern to the diversified and coordinated development pattern, and the urbanization process in the more developed regions of China has begun to slow down, taking the lead in completing this paradigm shift. However, the game between the "goal of central co-ordination" and the "impulse of local development" still exists [59]. Under the national macro planning, each region's development and protection are prioritized according to its primary role and resource endowment, while the cities continue to take into account the practical requirements of urban growth while making decisions, a circumstance in which the more economically developed eastern regions begin to raise the ES more as a result of an uptick in public demand for the environment. In the contrast, cities in the ecologically abundant but economically underdeveloped ecological function regions of the northeast and west continue to adopt a shoddy economic expansion model that depends on growing LS and PS because there is insufficient endogenous development momentum in these regions. Therefore, to fundamentally improve the ecological environment, the best path is to complete the upgrade of the development stage as fast as possible to achieve the transformation from resource-driven growth to environmentally friendly quality- and innovation-driven growth [13] and actively explore the asset value of ecological resources and the diversified models of payment for ecosystem services [60]. Promoting the realization of the value of ecological products is an effective path to truly realize a coordinated and diversified development approach.

4.3. Better Increase of Ecological Effects

The issue of concern in the field of spatial planning revolves around the ecological and environmental effects arising from the urban land use process, and it has become an academic consensus to use ESV as a measure of ecological and environmental effects [61,62]. Compared to the existing literature on measuring the value of ecosystem services within cities and between regions and the influencing factors, this paper focuses more on the assessment of the advantages and disadvantages of the ecological effects measured by ecosystem values under different development patterns. Given that the magnitude of ESV is primarily driven by land use cover change and the value per unit area coefficient, the resulting ecological effects reveal some beneficial findings. According to our comparison of the mean values of ESV growth volume and growth rate under the five development patterns, first, the ecological effects of the city under the coordinated and diversified development pattern were able to stabilize at a high level. Moreover, the continuous rapid urban expansion negatively affected the value of ecosystem services [61], and the average annual growth volume and growth rate of ESV under the unbalanced biased economic expansion development pattern in each period were at a low level, as confirmed by our study. Second, we also found that some cities have maintained a high ESV growth and growth rate, despite the purely economic expansion pattern of development over a certain period, i.e., a constant expansion of production and living areas, due to the rich resource base of the city that can guarantee a stable growth of ESVs over a short period, masking to some extent the drawbacks of this pattern. Finally, the pure/unbalanced biased ecological protection development pattern is at a high level most of the time, consistent with the academic consensus that the expansion of ES can have an effective ecological improvement effect. However, our study also found that most of the cities that chose this development pattern had a dominant role in the spatial allocation of ES and the value of ecosystem services does not increase infinitely with the expansion of ES. The marginal ecological benefits generated by the expansion of ES show a law of diminishing returns; this is also related to the scarcity value of ES [63,64]. The process of improving territorial spatial planning policy should not only stay at the level of "restriction and control" but should also play the role of "leading and guiding" via formulating refined development plans according to the differences of regional development, considering the threshold value of ecological effects of regional development, and allowing the reasonable transformation of the internal structure of the PLES within a scientific scope. Furthermore, it is necessary to pay more attention to the improvement of ES quality and enhancement of the ecological effect by increasing the value of ecosystem services per unit area and enhancing the synergy with LS. Furthermore, in contrast to the conclusion that the value of ecosystem services in China is decreasing, as concluded by an existing study [38], our study through the revision of the equivalent factor coefficients concluded that the value of ecosystem services as a whole has shown an increase, despite the decreasing growth rate of ESV.

4.4. Limitations and Further Work

Our study also has some limitations. This is reflected in the following two aspects. First, this study focuses more on the configuration of the primary classification of land functions in the PLES, while the more detailed secondary classifications, such as ES, are also transformed into forest ES and grassland ES, and the layout and transformation between these types also have an impact on the value of ecosystem services. However, for the sake of brevity, this study only provides some descriptive statistical analysis of the interconversion of land functions of secondary classifications in some chapters, and the analysis of the spatial transformation mechanisms between secondary indicators, especially forest ES and other ESs within the ES, and their influencing factors need to be studied more thoroughly in the future. Second, the value of ecosystem services is closely related to biodiversity and ecosystem functions, and it is not easy to quantify and assess the value of ecosystem services accurately. Although this study also revised the equivalence factor coefficients, there are various ways to revise the equivalence factor coefficients in China, and no unified standard has been formed, suggesting that the accuracy of the results needs to be further improved in the future.

5. Conclusions

Spatial development patterns at the city level in China have evolved, with great variance, driven by both the development goal of maximizing benefits and ecological protection policies. The PLESs of cities are progressing toward sustainable development under the direction of territorial spatial planning strategies, but there are still some areas that need to be optimized. The following are the study's main findings.

(1) The area occupied by ES is the greatest and the area occupied by LS is the smallest percentage in terms of the spatial composition and distribution of the PLES. While the ES is concentrated in the western hilly regions, the LS and PS are clustered in the eastern plains. The LS has been growing, the agricultural PS has been declining, and the ES has demonstrated a trend of first dropping and then increasing over the evolution of spatial layout during the previous 20 years. The LS spreads out and encroaches on the PS and ES in the eastern region. Between the PS and the ES, they are switched around spatially in the central and northwest regions.

(2) The regional distribution of the PLES's structural change over time varies greatly. The pure economic growth development pattern is giving way to the coordinated, diversified development pattern in China, and an increasing number of cities are adopting the ecological protection development pattern. The more economically developed eastern cities, particularly the coastal cities in the southeast, take the lead in changing from the pattern of economic expansion to the coordinated diversified development pattern. The economically underdeveloped but environmentally rich cities in the northeast and northwest still support the economic expansion development pattern, despite the fact that they are recognized by national policy guidelines as ecological functioning zones.

(3) There are differences in the ecological effects of different development patterns, with the biased ecological protection development pattern and the coordinated diversified development pattern exerting higher ecological effects among the five development patterns. There are temporal fluctuations in the ecological effects of different development approaches. The mean values of ESV growth and growth rate under all development modes showed decreasing trends, and the ecological effects are influenced not only by the ecological area but also by the quality of ES.

Thus, the protection policies of the red line of basic farmland protection and the red line of ecological protection established by the Chinese government in 2019 should be strictly implemented. Particular attention should be given to the question of decreasing arable land in the future, particularly the further reduction of arable land caused by the expansion of urban boundaries, and concentrating on high-quality urban construction. Cities should focus on the guiding principles of spatial planning policies, support research into methods for recognizing the value of ecological products in resource-rich regions, and advocate for the inherent strength of sustainable development. Policies should not only optimize the allocation among PLESs but also pay more attention to the intensive management of land resources to bring about positive outcomes.

Author Contributions: Conceptualization, Y.C. and X.H.; methodology, Y.C.; software, X.H. and B.C.; validation, Y.C., X.H. and X.L.; formal analysis, Y.C.; investigation, X.H.; resources, Y.C. and X.H.; data curation, X.L. and B.C.; writing—original draft preparation, Y.C. and X.H.; writing—review and editing, Y.C, X.H., X.L. and B.C.; visualization, B.C.; supervision, Y.C. and X.L.; project administration, Y.C.; funding acquisition, Y.C. and X.H. All authors have read and agreed to the published version of the manuscript.

Funding: This research was funded by Heilongjiang Province Philosophy and Social Science Key Fund Project, grant number 21GLA449. This research was supported by "the Fundamental Research Funds for the Central Universities", grant number 2572021AW33.

Institutional Review Board Statement: Not applicable.

Informed Consent Statement: Not applicable.

Data Availability Statement: Publicly available datasets were analyzed in this study. The city boundary data were downloaded from the National Geographic Information Center of China (http://ngcc.sbsm.gov.cn, accessed on 18 January 2022). The land use/cover data (2000–2020) were obtained from the European Space Agency's (ESA) CCI-LC project (http://maps.elie.ucl.ac.be/CCI/viewer/index.php, accessed on 10 June 2021).

Conflicts of Interest: The authors declare no conflict of interest.

References

1. Yue, J.; Fan, P. Value Transmission and Goal Evolution of Territorial Spatial Governance in the New Era: Response of the Territorial Spatial Planning Index System at the City and County Scale. *Trop. Geogr.* **2021**, *41*, 676–684.
2. Li, L.H.; Zhao, R.F.; Ming, S.Y.; Han, Z.Y.; Jia, Z.B. Evolvement characteristics and coupling coordination analysis of "production-living-ecological space" function in the upper reaches of the Yellow River. *J. Earth Environ.* **2022**, *13*, 449–464.
3. Zhu, J.M.; Guo, X.; Guo, Y. Compact city restructured by territorial planning: Differentiation between villages and spatial justice. *City Plan. Rev.* **2022**, *46*, 48–56.
4. Fan, J. Perspective of China's Spatial Governance System after 19th CPC National Congress. *Bull. Chin. Acad. Sci.* **2017**, *32*, 396–404.
5. Peng, S.; Wang, J.; Sun, H.; Guo, Z. How Does the Spatial Misallocation of Land Resources Affect Urban Industrial Transformation and Upgrading? Evidence from China. *Land* **2022**, *11*, 1630. [CrossRef]
6. Liu, Y.; Zhang, Y.; Zhang, Y.; Liu, Y.; Wang, H.; Liu, Y. Conflicts between three land management red lines in Wuhan City: Spatial patterns and driving factors. *Prog. Geogr.* **2018**, *37*, 1672–1681.
7. Dou, Y.; Ye, W.; Li, B.; Liu, P. Tourism Adaptability of Traditional Village Based on Living-Production-Ecological Spaces: A Case Study of Zhangguying Village. *Econ. Geogr.* **2022**, *42*, 215–224. [CrossRef]
8. Han, Z.; Shi, J.; Wu, J.; Wang, J. Recognition Method of "The Production, Living and Ecological Space" based on POI Data and Quad-tree Idea. *J. Geo-Inf. Sci.* **2022**, *24*, 1107–1119.
9. Zhang, Z.; Li, J. Spatial Suitability and Multi-Scenarios for Land Use: Simulation and Policy Insights from the Production-Living-Ecological Perspective. *Land Use Policy* **2022**, *119*, 106219. [CrossRef]
10. Aubry, C.; Ramamonjisoa, J.; Dabat, M.-H.; Rakotoarisoa, J.; Rakotondraibe, J.; Rabeharisoa, L. 'Urban Agriculture and Land Use in Cities: An Approach with the Multi-Functionality and Sustainability Concepts in the Case of Antananarivo (Madagascar)'. *Land Use Policy* **2012**, *2*, 429–439. [CrossRef]
11. Paracchini, M.L.; Pacini, C.; Jones, M.L.M. An aggregation framework to link indicators associated with multifunctional land use to the stakeholder evaluation of policy options. *Ecol. Indic.* **2011**, *11*, 71–80. [CrossRef]
12. Meng, J. Spatial-Temporal Trade-Offs of Land Multi-Functionality and Function Zoning at Finer Township Scale in the Middle Reaches of the Heihe River. *Land Use Policy* **2022**, *2*, 1–12. [CrossRef]
13. Huang, J.; Lin, H.; Qi, X. A literature review on optimization of spatial development pattern based on ecological-production-living space. *Prog. Geogr.* **2017**, *36*, 378–391. [CrossRef]
14. Zhang, H.; Xu, E.; Zhu, H. An ecological-living-industrial land classification system and its spatial distribution in China. *Resour. Sci.* **2015**, *37*, 1332–1338.
15. Wang, Y.; Liu, X.; Wei, X.; Yu, H. The Method and Practice of Regional Spatial Planning from "Three Basic Spaces" to "Three-zones and Three-lines". *Urban Plan. Forum.* **2018**, *4*, 65–74. [CrossRef]
16. Lu, C.; Ji, W.; Liu, Z.; Mao, J.; Li, J.; Xue, B. Spatial-temporal Pattern and Influencing Factors of the "Production-Living-Ecological" Functional Space of the Yellow River Basin at County Level in Gansu, China. *Sci. Geogr. Sin.* **2022**, *42*, 579–588. [CrossRef]
17. Shan, Y.; Wei, S.; Yuan, W.; Miao, Y. Spatial-temporal differentiation and influencing factors of coupling coordination of "production-living-ecological" functions in Yangtze River Delta urban agglomeration. *Acta Ecol. Sin.* **2022**, *42*, 6644–6655.
18. Feng, Y.; Zhen, J. Study on the adaptation of the land's functions and its synergy/trade-off relationship: A case study of Hohhot-Baotou-Ordos region, Inner Mongolia. *Arid Zone Res.* **2022**, *in press*. Available online: https://kns.cnki.net/kcms/detail/65.1095.X.20220831.13.19.007.html. (accessed on 20 December 2022).
19. Fan, J. Spatial organization pathway for territorial function-structure: Discussion on implementation of major function zoning strategy in territorial spatial planning. *Geogr. Res.* **2019**, *38*, 2373–2387. [CrossRef]
20. Wu, T.; Yue, W.; Xia, H.; Xiong, J. Optimization of Major Function Zoning Strategy from the Perspective of Territorial Spatial Planning. *Econ. Geogr.* **2022**, *42*, 11–17, 73. [CrossRef]
21. Li, B.; Zeng, C.; Dou, Y.; Liu, P.; Chen, C. Change of human settlement environment and driving mechanism in traditional villages based on living-production-ecological space: A case study of Lanxi Village, Jiangyong County, Hunan Province. *Prog. Geogr.* **2018**, *37*, 677–687.
22. Li, D.; Fan, K.; Lu, J.; Wu, S.; Xie, X. Research on Spatio-Temporal Pattern Evolution and the Coupling Coordination Relationship of Land-Use Benefit from a Low-Carbon Perspective: A Case Study of Fujian Province. *Land* **2022**, *11*, 1498. [CrossRef]
23. Allan, E.; Manning, P.; Alt, F. Land use intensification alters ecosystem multifunctionality via loss of biodiversity and changes to functional composition. *Ecol. Lett.* **2015**, *18*, 834–843. [CrossRef]
24. Wang, C.; Tang, N. Spatio-temporal characteristics and evolution of rural production-living-ecological space function coupling coordination in Chongqing Municipality. *Geogr. Res.* **2018**, *37*, 1100–1114.
25. Wang, F.; Dong, Y. Dynamic evaluation of land use functions based on gray relation projection method and diagnosis of its obstacle indicators: A case study of Guangzhou City. *J. Nat. Resour.* **2015**, *30*, 1698–1713.
26. Zhao, S.; Li, S.; Xu, X.; Li, Y.; Yang, W.; Qiu, L. Production-Life-Ecological Space Pattern Optimization of the Urban Agglomeration in Central Yunnan Based on Coupled MOP and FLUS Model. *Res. Soil Water Conserv.* **2022**, *29*, 322–328. [CrossRef]
27. Fan, J.; Zhao, Y. China's Regional Development Pattern Oriented Toward Modernization: The Scientific Connotation and Strategic Priorities. *Econ. Geogr.* **2021**, *41*, 1–9. [CrossRef]

28. Kong, D.; Chen, H.; Wu, K. The evolution of "Production-Living-Ecological" space, eco-environmental effects and its influencing factors in China. *J. Nat. Resour.* **2021**, *36*, 1116–1135. [CrossRef]
29. Ling, Z.; Li, S.; Jiang, W.; Liao, C.; Ling, Y. Dynamic Change Characteristics of "Production-living-ecological Spaces" of Urban Agglomeration Interlaced with Mountains, Rivers and Sea: A Case Study of the Beibu Gulf Urban Agglomeration in Guangxi. *Econ. Geogr.* **2022**, *42*, 18–24. [CrossRef]
30. Chen, H.; Yang, Q.; Peng, L.; Su, K.; Zhang, H.; Liu, X. Spatiotemporal evolution characteristics and scenario simulation of production-living-ecological space at county level in Three Gorges reservoir areas. *Trans. Chin. Soc. Agric. Eng.* **2022**, *38*, 285–294. [CrossRef]
31. Zhang, X.Y.; Zhang, X.; Li, D.H.; Lu, L.; Yu, H. Multi-Scenario Simulation of the Impact of Urban Land Use Change on Ecosystem Service Value in Shenzhen. *Acta Ecol. Sin.* **2022**, *42*, 2086–2097.
32. Gong, Y.; Han, S.; Shi, X.; Tang, X. Temporal and Spatial Evolution and Associated Eco-environment Effects of the Land Use Transformation of Ecological-Production-Living Spaces in Guangdong Province. *Res. Soil Water Conserv.* **2020**, *27*, 203–209.
33. Gao, X.; Liu, Z.X.; Li, C.X.; Cha, L.S.; Song, Z.Y.; Zhang, X.R. Land use function transformation in the Xiong′an New Area based on ecological-production-living spaces and associated eco-environment effects. *Acta Ecol. Sin.* **2020**, *40*, 7113–7122.
34. Tan, D.M.; Ding, S.Y.; Han, B.L.; Pan, X.L.; Liu, H.X. Impact of different development patterns on the value of regulating services of urban ecosystems. *Acta Ecol. Sin.* **2022**, *42*, 6554–6564.
35. Zhou, Y.; Yin, D.; Huang, Q.; Zhang, L.; Bai, Y. Spatial optimization based on ecosystem services and the participatory mapping: A case study in Baiyangdian watershed. *J. Nat. Resour.* **2022**, *37*, 1988–2003. [CrossRef]
36. Costanza, R.; D'Arge, R.; De, G.R.; Farber, S.; Grasso, M.; Hannon, B.; Limburg, K.; Naeem, S.; O'Neill, R.V.; Paruelo, J.; et al. The value of the world's ecosystem services and natural capital. *Ecol. Econ.* **1998**, *25*, 3–15. [CrossRef]
37. Xie, G.D.; Lu, C.; Leng, Y.; Zheng, D.; Li, S. Ecological assets valuation of the Tibetan Plateau. *J. Nat. Resour.* **2003**, *18*, 189–196.
38. Yang, R.; Ren, F.; Xu, W.; Ma, X.; Zhang, H.; He, W. China's Ecosystem Service Value in 1992–2018: Pattern and Anthropogenic Driving Factors Detection Using Bayesian Spatiotemporal Hierarchy Model. *J. Environ. Manag.* **2022**, *302*, 114089. [CrossRef] [PubMed]
39. Song, H.L.; Zhang, X.N. Precision validation of multi-sources land cover products derived from remote sensing. *Remote Sens. Land Resour.* **2018**, *30*, 26–32.
40. Zhou, Y.; Li, X.H.; Liu, Y.S. Land use change and driving factors in rural China during the period 1995–2015. *Land Use Policy* **2020**, *99*, 105048. [CrossRef]
41. Song, Y.; Xue, D.; Xia, S.; Mi, W. Change characteristics and formation mechanism of the territorial spatial pattern in the Yellow River Basin from 1980 to 2018, China. *Geogr. Res.* **2021**, *40*, 1445–1463.
42. Wang, Q.; Tang, F. The Evaluation of Coulpling Coordination Development of Ecology-Economy-Society System in Dongting Lake Area. *Econ. Geogr.* **2015**, *35*, 161–167, 202.
43. Zhu, Y.; Yao, S.B. The coordinated development of environment and economy based on the change of ecosystem service value in Shaanxi Province. *Acta Ecol. Sin.* **2021**, *41*, 3331–3342.
44. Qiao, W.; Hu, Y.; Jia, K.; He, T.; Wang, Y. Dynamic Modes and Ecological Effects of Salt Field Utilization in the Weifang Coastal Area, China_ Implications for Territorial Spatial Planning. *Land Use Policy* **2020**, *99*, 104952. [CrossRef]
45. Costanza, R.; D'Arge, R.; De, G.R.; Farber, S.; Grasso, M.; Hannon, B.; Limburg, K.; Naeem, S.; O'Neill, R.V.; Paruelo, J.; et al. The value of the world's ecosystem services and natural capital. *Nature* **1997**, *387*, 253–260. [CrossRef]
46. Xie, G.D.; Zhang, C.X.; Zhang, L.M.; Chen, W.H.; Li, S.M. Improvement of the Evaluation Method for Ecosystem Service Value Based on Per Unit Area. *J. Nat. Resour.* **2015**, *30*, 1243–1254. [CrossRef]
47. Wang, H.; Qin, F.; Zhu, Y.; Zhang, C.C. The effects of land use structure and landscape pattern change on ecosystem service values. *Acta Ecol. Sin.* **2017**, *37*, 1286–1296.
48. Wu, Y.; Tao, Y.; Yang, G.; Ou, W.; Steven, P.; Sun, X.; Chen, G.; Tao, Q. Impact of Land Use Change on Multiple Ecosystem Services in the Rapidly Urbanizing Kunshan City of China: Past Trajectories and Future Projections. *Land Use Policy* **2019**, *85*, 419–427. [CrossRef]
49. Wang, A.; Liao, X.; Tong, Z.; Du, W.; Zhang, J.; Liu, X.; Liu, M. Spatial-Temporal Dynamic Evaluation of the Ecosystem Service Value from the Perspective of "Production-Living-Ecological" Spaces: A Case Study in Dongliao River Basin, China. *J. Clean. Prod.* **2022**, *333*, 130218. [CrossRef]
50. Chen, Y.; Li, J.; Xu, J. The Impact of Socio-economic Factors on Ecological Service Value in Hubei Province: A Geographically Weighted Regression Approach. *China Land Sci.* **2015**, *29*, 89–96.
51. Wei, J.; Wen, Y.; Gong, Z.; Wang, X.; Cai, Y. Land use changes and ecosystem service value in the buffer zone of Poyang Lake in recent 30 years. *Acta Ecol. Sin.* **2022**, 1–13. [CrossRef]
52. Huang, H.C.; Lei, M.; Kong, X.B.; Wen, L.Y. Changing spatial patterns of cropland and its ecosystem service value response in China. *Res. Soil Water Conserv.* **2022**, *29*, 339–348. [CrossRef]
53. Cao, Y.N.; Kong, L.Q.; Zhang, L.F.; Ouyang, Z.Y. The balance between economic development and ecosystem service value in the process of land urbanization: A case study of China's land urbanization from 2000 to 2015. *Land Use Policy* **2021**, *108*, 105536. [CrossRef]

54. Zhu, P.H.; Liu, X.X. A New Type of Urbanization with People at the Core: Development Goals and Practical Strategies for 2035. *Reform* **2022**, 1–15. Available online: https://kns.cnki.net/kcms/detail//50.1012.F.20221219.0957.002.html (accessed on 20 December 2022).
55. Qian, F.; Xiang, Z.; Wang, H.; Sun, F. Research on cultivated land zoning protection around the main urban area based on the demarcation of territorial spatial three control lines. *J. Agric. Resour. Environ.* **2022**, in press. [CrossRef]
56. Ma, S. Towards Balanced Development Stage: Regulating the Spatial Pattern of Agglomeration with Collaborative Optimal Allocation of Urban Land. *Cities* **2022**, *126*, 13. [CrossRef]
57. Xie, T.; Wang, Y. Environmental Quality and Chinese Urban Development: An Interpretation from the Perspective of Human Capital. *J. World Econ.* **2022**, *45*, 133–157. [CrossRef]
58. Ding, Z.M.; Yao, S.B. Effects of cultivated land transition on ecosystem service value in Shaanxi Province of China. *Resour. Sci.* **2019**, *41*, 1070–1081. [CrossRef]
59. Zhang, J.X.; Xia, T.C. The change and reconstruction of spatial planning system under the goal of modern national governance. *J. Nat. Resour.* **2019**, *34*, 2040–2050. [CrossRef]
60. Feng, L.; Guo, J.X.; Wang, L.E. Marketization path of tourism ecological compensation and theoretical analysis. *Resour. Sci.* **2020**, *42*, 1816–1826. [CrossRef]
61. Diogo, G.V.; Ricardo, C.D.; Catarina, P.T.; Claudia, O.F.; Walter, L.F.; Nelson, B.; Rui, L.M. Clustering Public Urban Green Spaces through Ecosystem Services Potential: A Typology Proposal for Place-Based Interventions. *Environ. Sci. Policy* **2022**, *132*, 262–272. [CrossRef]
62. Rachael, D.; Adrian, E. Deliberative-Analytic Approaches to Ecosystem Services as a Way Forward for the Land Sparing/Sharing Debate. *Land Use Policy* **2022**, *116*, 12. [CrossRef]
63. Chen, H.J. Land use trade-offs associated with protected areas in China: Current state, existing evaluation methods, and future application of ecosystem service valuation. *Sci. Total Environ.* **2020**, *711*, 134688. [CrossRef]
64. Ding, Z.M.; Yao, S.B. Theoretical analysis and model design of search for ecological conservation redlines based on the sensitivity assessment of ecosystem services. *Land Use Policy* **2020**, *97*, 104745. [CrossRef]

sustainability

MDPI

Article

Spatiotemporal Changes in Ecosystem Services Value and Its Driving Factors in the Karst Region of China

Liu Yang [1,*] and Hongzan Jiao [2,3]

1 College of Architecture and Urban Planning, Guizhou University, Guiyang 550025, China
2 Department of Urban Planning, School of Urban Design, Wuhan University, Wuhan 430072, China; jiaohongzan@whu.edu.cn
3 Engineering Research Center of Human Settlements and Environment of Hubei Province, Wuhan 430072, China
* Correspondence: lyang3@gzu.edu.cn

Abstract: Over the last few decades, most regional ecosystem services (ESs) have significantly deteriorated, primarily driven by an increase in human dominance over the natural environment. Creating an assessment framework of ESs and identifying its driving factors at the regional scale is challenging for researchers, administrators, and policy-makers. In this study, we attempt to quantify the economic value of ESs (ESV) in Guizhou Province from 2000 to 2018, one of the most prominent areas of karst landforms in China. We identified the major factors affecting ESs using the geographical detector (GD) model. Then, we conducted a multiscale geographically weighted regression (MGWR) analysis to examine the spatial differentiation of the causal effects of both natural and anthropogenic factors on ESs. Our results demonstrate the following: (1) the total ESV of Guizhou Province was approximately USD 81,764.32 million in 2000, USD 82,411.06 million in 2010, and USD 82,065.31 million in 2018, and the increase of USD 300.99 million from 2000 to 2018 was the result of the remarkable conversion from cultivated land to forestland; (2) significantly considerable differentiation existed in the spatial distribution of ESV at the county level, with a higher value in the eastern region and a lower value in the western region; (3) among the driving factors, population density had a more significant effect on the spatial differentiation of ESV than did natural factors; and (4) agricultural output value was the dominant factor influencing the ESV during the study period, with a significantly positive correlation, whereas per capita GDP and population density had significantly negative impacts on ESV, according to the effective performance of the MGWR model that evaluated the spatial heterogeneity in geospatial relationships between the driving factors of ESV. Our findings can provide notable guidance to land administrators and policy-makers for effective land resource conservation and management plans, thereby improving regional sustainability.

Keywords: ecosystem services value; driving factors; geographical detector model; multiscale geographically weighted regression; karst areas

Citation: Yang, L.; Jiao, H. Spatiotemporal Changes in Ecosystem Services Value and Its Driving Factors in the Karst Region of China. *Sustainability* **2022**, *14*, 6695. https://doi.org/10.3390/su14116695

Academic Editor: Åsa Gren

Received: 16 March 2022
Accepted: 24 May 2022
Published: 30 May 2022

1. Introduction

ESs refer to life-supporting products and services obtained directly or indirectly through the structures, processes, and functions of an ecosystem [1,2]. Since the 1990s, severe environmental issues have created impediments for future food security and national development strategies, and thus ESs have attracted the attention of researchers and governments. The scientific community has revealed that intensive human activity has both a direct and indirect impact on numerous environmental factors (e.g., climate, landscapes, socioeconomic factors) and is responsible for altering the structures, processes, and functions of ESs [3,4]. A multiscale assessment framework at the local, regional, and global levels is crucial to more deeply understanding the benefits or damages that result from the alteration of Ess [5]. Since Costanza et al. (1997) first mapped the global values of ESs and put forth the concept of ESV, defined as a range of goods and services important for human well-being from a monetary-unit perspective [6], ESV has been applied in numerous studies that assess the changes to ecological services occurring across regions [7–10]. Subsequently, various classifications have been developed and adjusted for scientific assessment of ESV over the last decades, e.g., the Millennium Ecosystem Assessment (MA) [11,12], the Economics of Ecosystems and Biodiversity (TEEB) [13,14], the Intergovernmental Science-Policy Platform on Biodiversity and Ecosystem Services (IPBES) [15], the Common International Classification of Ecosystem Services (CICES) [16,17], and the National Ecosystem Services Classification System (NESCS) [18]. It is concluded that there are differences in the details among these classification systems, that is, Costanza's method includes seventeen services, while MA includes twenty-three and the TEEB includes twenty-two while the CICES was developed to provide a hierarchically consistent and science-based classification to be used for natural capital accounting purposes [19].

Since then, a growing number of studies on ecosystem service values (ESV) and the impact of land use land cover (LULC) on ESV have been performed at different levels all over the world [20,21]. However, the methods of valuation have been challenged due to their limited use and regional characteristics, such as, regional area, changes of ecological protection, observation and survey of the environment, which have affected the data that were required in many classification systems (e.g., MA, or TEEB) and have made data acquisition extraordinarily difficult [5,22]. Therefore, Costanza's method is considered a relatively simple approach to quantifying the spatial distribution of ESV and rapidly obtaining sustained momentum as a framework to communicate values and benefits to scientists, stakeholders, policy-makers, and the public [23–27]. In China, a series of coefficients for Chinese ESV, namely China's ESV system, at the national scale were developed from Costanza's research according to China's characteristics by Xie et al. [28,29]. Owing to more flexibility and less requirement for basic data, many scholars have widely utilized the China's ESV system to quantitatively estimate ecological functions and the benefits that land use transition may provide [10,30–34].

The relationships between ESs and their driving factors, particularly human activity, have gained increased attention over the last several years, and several studies have found that a multitude of factors influence ESs and ESV [35–38]. For instance, Cai et al. demonstrated that the rapid expansion of urban areas has resulted in a dramatic decline in ecosystem services and that the ESV of cultivated lands and wetlands has had a significant negative correlation with total GDP [39]. Zhu et al. explored global and local factors impacting ESV in the Beijing–Tianjin–Hebei region and found that the primary industry-related factors were socioeconomic factors [40]. The extensive land use transition from rural to urban areas, accompanied by rapid industrialization and the intensification of human activities, is the dominant factor leading to changes in ESs. Pilogallo et al. revealed that the greatest loss in ESV in the Basilicata region occurred within wooded areas and agricultural mosaics, whereas bare and arable lands increased in ESV [41]. Berihun et al. evaluated the impact of human-driven LULC changes on ESV and concluded that the population growth leading to the expansion of cultivated land had a negative impact on ESV in the Upper Blue Nile basin area of Ethiopia [42]. Given that different sets of environmental characteristics will generate different ESs, it is necessary to understand how these climatic and natural and human-induced socioeconomic factors ultimately decide which ESs will be sustainable.

The relationship between ES and its driving factors is not linear but rather has significant spatial heterogeneity [34,43]. To understand the spatial variation in ES and ESV, due to the spatial heterogeneity of driving factors such as topography, soil, vegetation, climate, and landscape structures, an increasing number of studies have concentrated on spatial autoregressive and multivariate regression methods [44–46]. Among them, the ordinary least square (OLS) model has been utilized to identify the interactions between ESV and its driving factors and ESV [47–49] but does not reflect the essential autocorrelation or homogeneity in space [50]. Compared with this model, the geographically weighted regression (GWR) model, developed from a linear regression with the weighted least squares (WLS) method, is a simple yet beneficial approach for identifying the spatial characteristics of relationships by measuring spatial variations in spatial association for each unit in the study area [51–56]. Therefore, we implemented the GWR model in this study to produce varying local attributes throughout the feature space by establishing local regression equations for ESV and its driving factors.

Over the last few decades, the karst area in southwestern China has suffered from a sequence of anthropogenic and natural adversities, including rocky desertification and soil erosion, which have resulted in a rapid decrease in ESs. Guizhou Province, located in Southwest China and known as one of the most prominent karst landform areas, has confronted notable pressure to balance its ecological protection and economic growth. Its vulnerable ecosystem provides a distinctive landscape and necessary habitat for rare plants and animals, which can contribute to the ecotourism industry, one of the province's predominant economic services [57–59]. Despite its profound ecological significance, researches on ESs changes and their driving mechanisms in karst areas have been explored in only a few studies, among which, the assessments and development trends of ESV based on MA, TEEB, IPBES, were so rarely involved that it was difficult to gain approximate parameters from the literature. Therefore, we attempted to investigate the spatial and temporal variability of the ESV and identify the primary driving factors for ESV in Guizhou Province via the following: (1) assessing the spatial and temporal variability of the ESV by using China's ESV system; (2) identifying potential driving factors by adopting GD; and (3) exploring the spatially heterogeneous relationship between the ESV and its driving factors based on MGWR. The outcomes of this study may not only enrich the current existing research on ESV in ecologically fragile areas and easily compare with other regions of China, but they may also provide suggestions for safeguarding both ecology and development.

2. Materials and Methods

2.1. Study Area

Guizhou, which lies at the eastern end of the Yungui Plateau in Southwest China (103°36′–109°35′ E, 24°37′–29°13′ N), is a crucial ecologically protected area in the upper reaches of the Yangtze and Pearl rivers and is an important part of the Yangtze River Economic Belt. Covering approximately 176,000 km^2 (Figure 1), Guizhou's orography is high in the west and low in the east. There are four major mountains in the province—Wumeng, Dalou, Miaoling, and Wuling—altogether accounting for 92.5% of the total area. Known as one of the most prominent karst areas in the world, 61.9% of the total area comprises karst landscapes and holds abundant resources, including water resources and coal mines. Guizhou Province currently governs six county-level cities—Guiyang, Zunyi, Liupanshui, Anshun, Bijie, and Tongren—and three autonomous prefectures—Qiandongnan, Qiannan, and Qianxinan)—encompassing 88 counties in total.

With the implementation of various national strategies (e.g., the development of the western region in China, the National Big Data Strategy, and the rise of the Yangtze River Economic Zone), Guizhou has undergone rapid economic development, with an average annual 8.32% increase in GDP from 2000 to 2019. However, as one of the nation's ecological civilization pilot zones, as well as the first of these zones in western China, Guizhou faces significant disturbances to its fragile ecosystems that occur as a result of rapid urbanization, economic growth, and drastic increases in land use. An assessment of the spatiotemporal status of ESs in this region can provide a scientific reference for ecological management; furthermore, it has important guiding implications for territorial spatial planning.

2.2. Data Sources and Descriptions

In this study, we utilized three types of data: land-use remote image data, ESV coefficient data, and social and economic development data for Guizhou Province. These included the following: (1) land-use maps (Shapefile data) from 2000, 2010, and 2018 (Figure A1, Appendix A) provided by the Resource and Environment Science and Data Central (http://www.resdc.cn, accessed on 19 September 2020); using the ESV coefficient established by the Chinese Academy of Sciences, we converted original land-use maps to TIFF images with 30 m × 30 m spatial resolution, extracted and generalized seven land classification types (cultivated land, forestland, grassland, water bodies, construction land, and unused land) in Guizhou Province; (2) a digital elevation model (DEM) (30 m × 30 m spatial resolution), the average annual temperature (30 m × 30 m spatial resolution), and the average annual precipitation (30 m × 30 m spatial resolution) obtained from the Resource and Environment Science and Data Central (http://www.resdc.cn, accessed on 19 September 2020); and (3) statistical data, including grain yield, grain price, GDP data primarily collected from the Statistical Yearbooks of Guizhou Province (http://data.cnki.net, accessed on 29 September 2020), and the statistical bulletin of national economic and social development of Guizhou Province (http://stjj.guizhou.gov.cn, accessed on 29 September 2020).

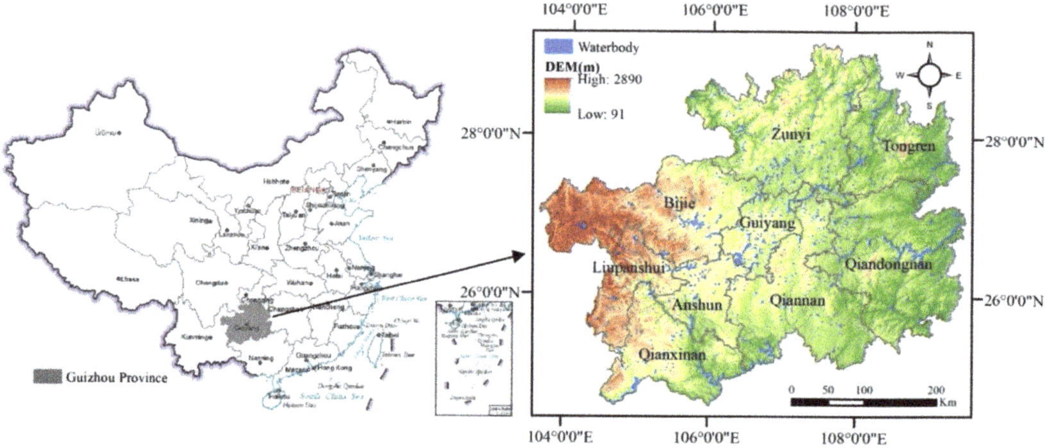

Figure 1. Location of the study area.

2.3. *Assessment of Ecosystem Services Value*

We adopted the China's ESV system which was based on the equivalent value factor per unit ecosystem area originated from Costanza et al. [6] and developed by Xie et al. [28,29] to quantify the ESV in China. Integrating Costanza's research and China's characteristics, Xie et al. divided the country's ESs into nine functions and adjusted the ESV coefficient; here, the function of food production from farmland represented the net profit of grain production per unit area of farmland and defined it as the standard ESV coefficient of China, with its equivalent value deemed as 1; meanwhile, the other function coefficients were all equivalent values based on the standard value of 1 (Table 1) [16]. Moreover, it should be noted that we evaluated the ESV coefficient framework at the national level and provincial or local ESV coefficients should be therefore revised to comply with local characteristic factors. Accordingly, Xie et al. proposed various biomass factors for different provinces in China to revise the national ESV coefficients; the biomass factor for Guizhou Province was 0.63 (Table 1) [17]. The economic value of the standard ESV coefficient is the average natural food production of farmland per unit area per year, which was assumed to be one seventh of the actual food production without any labor input. In Guizhou Province, the average actual food production of farmland was 3704.91 kg/ha between 2000 and 2018, and the average market price for grain in 2018 was USD 0.72/kg (CNY4.77/kg) (Note: the average exchange rate between USD and CNY in 2018 was 6.6174 (http://www.gov.cn, accessed on 28 April 2022). Hence, the economic value of the standard ESV coefficient is USD 381.53 /ha (CNY 2524.76/ha). We calculated the ESV for each land-use type per hectare using Equation (1). Table 2 (or CNY see Table A1) displays the results.

$$VC_{kf} = EC_{kf} \times 381.53 \quad (1)$$

where VC_{kf} is the ESV per hectare for land-use type k and service function f, and EC_{kf} is the equivalent ESV coefficient for land-use type k and service function f in Table 1.

Table 1. Equivalent value per unit area of ecosystem services in China and Guizhou Province.

Ecosystem Service and Functions	Cultivated Land		Forestland		Grassland		Water Body		Barren Land	
	China	Guizhou	China	Guizhou	China	Guizhou	China	Guizhou	China	Guizhou
Food production	1	0.63	0.33	0.21	0.43	0.27	0.53	0.33	0.02	0.01
Raw material	0.39	0.25	2.98	1.88	0.36	0.23	0.35	0.22	0.04	0.03
Gas regulation	0.72	0.45	4.32	2.72	1.5	0.95	0.51	0.32	0.06	0.04
Climate regulation	0.97	0.61	4.07	2.56	1.56	0.98	2.06	1.30	0.13	0.08
Water supply	0.77	0.49	4.09	2.58	1.52	0.96	18.77	11.83	0.07	0.04
Waste treatment	1.39	0.88	1.72	1.08	1.32	0.83	14.85	9.36	0.26	0.16
Soil formation and retention	1.47	0.93	4.02	2.53	2.24	1.41	0.41	0.26	0.17	0.11
Biodiversity protection	1.02	0.64	4.51	2.84	1.87	1.18	3.43	2.16	0.4	0.25
Recreation and culture	0.17	0.11	2.08	1.31	0.87	0.55	4.44	2.80	0.24	0.15
Total	7.9	4.98	28.12	17.72	11.67	7.35	45.35	28.57	1.39	0.88

Table 2. The annual ESV for each land use type per hectare in Guizhou Province (USD/ha yr).

Ecosystem Service and Functions	Cultivated Land	Forestland	Grassland	Water Body	Construction Land	Unused Land
Food production	240.37	80.12	103.01	125.91	0	3.82
Raw material	95.38	717.28	87.75	83.94	0	11.45
Gas regulation	171.69	1037.77	362.46	122.09	0	15.26
Climate regulation	232.73	976.73	373.90	495.99	0	30.52
Water supply	186.95	984.36	366.27	4513.54	0	15.26
Waste treatment	335.75	412.06	316.67	3571.15	0	61.05
Soil formation and retention	354.83	965.28	537.96	99.20	0	41.97
Biodiversity protection	244.18	1083.56	450.21	824.11	0	95.38
Recreation and culture	41.97	499.81	209.84	1068.29	0	57.23
Total	1903.85	6756.96	2808.09	10,904.23	0	331.93

Table 2 exhibits the ESV of one unit area of each land use type in Guizhou Province assigned based on the nearest equivalent ecosystems. For instance, cultivated land falls under the category of "farmland," forestland falls under "forest," and unused land falls under "barren land." We suppose that the ESV for construction land is 0 as a result of the transformation to construction land. The service value for each land use type and service function are provided in Equation (2):

$$ESV = \sum_{k}\sum_{f} A_k \times VC_{kf} \quad (2)$$

where ESV refers to the total ecosystem service value. A_K is the area for land-use type k, and VC_{kf} is the ESV per hectare for land-use type k and service function f in Table 2.

We analyzed the spatial changes in ESV in each county by using the ESV per unit area, which can be calculated as follows:

$$ESVA_i = ESV_i / Area_i \quad (3)$$

where $ESVA_i$ is the ESV per unit area of county i. ESV_i is the total ESV of county i, and Areai is the total area of county i.

2.4. Potential Driving Factor System of ESV

The potential driving factor system of ESV reveals the degree to which natural, economic, and social factors have potential impacts on ecosystems. Because no factor system is universal and the relevant criteria of ESV vary, depending on local conditions, we established a potential driving factor system of ESV for Guizhou Province, with four natural factor variables, five economic factor variables, and three social factor variables (Table 3).

Each factor, based on counties or administrative districts, has unique attributes, which allows for more area-specific results.

Table 3. Factors and their data sources in the primary driving factor system of ESV.

Factors	Variables	Data Resources	Variable Number
Natural factors	Elevation	DEM	X_1
	Terrain slope	DEM	X_2
	Average annual temperature	Meteorological map	X_3
	Average annual precipitation	Meteorological map	X_4
Economic factors	Gross domestic product (GDP)	Statistical annual	X_5
	Per capita GDP	Statistical annual	X_6
	Per capita disposable income of rural residents	Statistical annual	X_7
	Agricultural output value	Statistical annual	X_8
	Forestry output value	Statistical annual	X_9
Social factors	Resident population	Statistical annual	X_{10}
	Population density	Statistical annual	X_{11}
	Rural employment	Statistical annual	X_{12}

2.5. Exploring the Driving Factors of ESV Using GD and GWR

2.5.1. Geographical Detector Model

The geographical detector (GD) model is a relatively novel statistical technique for detecting spatial heterogeneity and revealing the driving force behind it. Its core hypothesis is that if independent variable X has an important impact on dependent variable Y, then a similar spatial distribution exists between them [60,61]. The GD model has unique advantages for dealing with both numerical and quantitative data and has been gradually used over the last several years in various research fields, such as environmental [62], social, and health sciences. The GD model includes four detectors: the factor detector, interaction detector, risk detector, and ecological detector. In this study, we used the factor detector to quantify the degree of impact of each explanatory variable X (the potential driving factors in Table 3) on dependent variable Y (ESV) for each year by calculating the q-statistic. In a range from 0 to 1, the higher its value, the greater the explanatory variable contributes to the dependent variable. The q-statistic is calculated by Equation (4):

$$q = 1 - \frac{\sum_{h=1}^{L} N_h \sigma_h^2}{N\sigma^2} \tag{4}$$

where h = 1, 2, ..., L is a certain stratum of each explanatory variable X (potential driving factor) and, of the dependent variable Y (ESV), L is the number of strata, N_h and N are the number of samples in stratum h and the entire study area, respectively, and σ^2 is the variance of dependent variable Y in stratum h and the entire study area. A *p* value, as the significance indicator of each explanatory variable, is also calculated through the noncentral F-distribution.

2.5.2. MGWR

Multiscale geographically weighted regression (MGWR) has become a popular approach for local spatial statistical analysis since it was first proposed by Brunsdon et al. [51]. This method can obtain spatially nonstationary relationships between dependent and independent variables by incorporating geographical information. Based on the Tobler Law [63], which states that "everything is related to everything else, but near things are more related than distant things", the GWR model predicts different weights for each location. The model is mathematically expressed as follows:

$$y_i = \beta_0(u_i, v_i) + \sum_m \beta_m(u_i, v_i) x_{im} + \varepsilon_i \tag{5}$$

where v_i is the dependent variable at location i (ESV), x_{im} is the m-th potential driving factor at location i, (u_i, v_i) are the geographical coordinates at location i, $\beta_0(u_i, v_i)$ is the intercept coefficient at location i, $\beta_m(u_i, v_i)$ is the m-th local regression coefficient for x_{im} and ε_i represents the random error term associated with location i.

MGWR is an improved version of GWR that considers spatial multiscale effects and heterogeneity and reflects those differences in ESV [64]. The MGWR model expression is Equation (6) as follows:

$$y_i = \beta_0(u_i, v_i) + \sum_m \beta_{bwm}(u_i, v_i)x_{im} + \varepsilon_i \tag{6}$$

where bwm in β_{bwm} indicates the bandwidth used for the calibration of the m-th conditional relationship. MGWR allows for the estimation of local regression coefficients of dependent and independent variables on different spatial scales [64–66].

In this study, both the MGWR and GWR models used a fixed Gaussian kernel function and were calibrated using a golden section search bandwidth selection routine [66]. All model calibrations were undertaken using MGWR 2.2 software [64].

3. Results

3.1. Historical Changes in Land Use and ESV

To explore the substantial magnitude of land use transitions that significantly affected the total ESV in Guizhou Province, we produced statistics of land use changes by using "Analysis Tools" in ArcGIS 10.2 software (Figure A2) and ESV changes between 2000 and 2018 (Table 4), which combine the data in Table 2 with Equation (2). Cultivated land and forestland comprised the largest portions of the total area (over 80%). The area of cultivated land was 493.73 × 10^4 ha in 2000 and 484.46 × 10^4 ha in 2018, decreasing by 1.88% to an average annual decrease of 5.15 × 10^3 ha. In contrast, forestland dramatically increased to 1.49 × 10^5 ha before 2010, owing to the policy mandating the conversion of farmland to forests, but it slightly dropped to 2.87 × 10^4 ha from 2010 to 2018; however, the increased area was higher than the decreased area during the period from 2000 to 2018. Grassland experienced the most conspicuous change from 2000 to 2018: with a continuous decline, the area decreased by 5.06% to 16.70 × 10^4 ha. The water area increased marginally, experiencing a continuous rise from 2000 to 2010 of 0.76 × 10^4 ha and 0.48 × 10^4 ha from 2010 to 2018. Construction land saw a dramatic ascension as a result of urban development—from 8.80 × 10^4 ha in 2000 to 21.61 × 10^4 ha in 2018, increasing by 145.71%. From 2000 to 2010, unused land decreased by 25.45%, nearly 0.10 × 10^4 ha, whereas from 2010 to 2018, the area increased by 2.61%, equal to 0.01 × 10^4 ha.

The total ESV of Guizhou Province was approximately USD 81,764.32 million in 2000, USD 82,411.06 million in 2010, and USD 82,065.31 million in 2018 (Table 4, or CNY see Table A2). Because of its larger equivalent ESV coefficient value and larger area, forestland ESV was the highest, representing approximately 77% of the total value. Although the equivalent ESV coefficient value of the water body areas was the highest among the six land use types, these areas were small and thus generated low ESV. Therefore, forestland played the most important role in Guizhou Province ESs. From 2000 to 2010, the ESV increment due to the increase in forestland was offset by a value decline in grassland and cultivated land. As a result, the total ESV increased by USD 646.74 million in last decade; however, the total ESV of Guizhou Province from 2010 to 2018 shrank by USD 345.76 million, primarily as a result of the decrease in forestland and the continuous decline in grassland and cultivated land. Overall, the net growth of the province's ESV was approximately USD 300.99 million from 2000 to 2018, primarily because of the significant conversion from cultivated land to forestland over the past 18 years.

Our value calculation results of different ES functions in Guizhou Province (Table 5, CNY see Table A3) revealed that the ESV of food production exhibited a downwards trend during the study period, whereas changes in other functions were overall consistent, increasing from 2000 to 2010 and declining from 2010 to 2018. The total values of each ES function from 2000 to 2018 from largest to smallest were as follows: biodiversity protection, soil formation and retention, water supply, gas regulation, climate regulation, raw material, waste treatment, recreation and culture, and food production. Due to the large area of forestland in Guizhou Province, changes in biodiversity protection and soil formation and retention were similar to those of forestland. Meanwhile, widely distributed water bodies throughout the area—namely, the Wujiang River, Nanpan River, and Beipan River—play crucial roles in microclimate improvements and ecosystem regulation services; however, the influence of the forestland to built-up land conversion on gas regulation, climate regulation, and water supply was heightened. These results reflect that, although human socioeconomic activities negatively affected the ESV of the Guizhou Province, the changes and structural distributions of the ESV for single functions, which came with changes in land use patterns, were relatively stable. Obviously, biodiversity protection has been the dominant ES function of the Guizhou Province over the last few decades, which has been closely related to the protection of forestland and the ecological civilization policy.

Table 4. Land use and ESV changes in Guizhou Province in 2000, 2010, 2018, 2000–2010, and 2010–2018.

Land Use Types		Cultivated Land	Forestland	Grassland	Water Body	Construction Land	Unused Land	Total
2000	Land area (10^4 ha)	493.73	918.99	329.99	9.18	8.8	0.4	1761.09
	ESV (Million USD)	9399.94	62,095.74	9266.50	1000.82	0.00	1.33	81,764.32
2010	Land area (10^4 ha)	491.82	933.88	315.55	9.94	9.6	0.3	1761.09
	ESV (Million USD)	9363.57	63,102.07	8861.00	1083.44	0.00	0.99	82,411.06
2018	Land area (10^4 ha)	484.46	931.01	313.29	10.41	21.61	0.31	1761.09
	ESV (Million USD)	9223.35	62,908.15	8797.45	1135.34	0.00	1.01	82,065.31
2000–2010	Land area (10^4 ha)	−1.91	14.89	−14.44	0.76	0.8	−0.1	0
	ESV (Million USD)	−36.37	1006.33	−405.50	82.62	0.00	−0.34	646.74
2010–2018	Land area (10^4 ha)	−7.36	−2.87	−2.26	0.48	12.01	0.01	0
	ESV (Million USD)	−140.22	−193.92	−63.55	51.90	0.00	0.03	−345.76
2000–2018	Land area (10^4 ha)	−9.28	12.02	−16.7	1.23	12.82	−0.09	0
	ESV (Million USD)	−176.58	812.41	−469.05	134.52	0.00	−0.31	300.99

Table 5. Value changes of different ecosystem service functions in Guizhou Province in 2000, 2010, and 2018.

Ecosystem Service Functions	ESV (Million USD)			Changes of ESV (Million USD)		
	2000	2010	2018	2000–2010	2010–2018	2000–2018
Food production	2274.59	2268.01	2246.27	−6.58	−21.73	−28.32
Raw material	7360.02	7452.98	7423.78	92.96	−29.20	63.76
Gas regulation	11,592.05	11,691.90	11,641.85	99.85	−50.05	49.80
Climate regulation	11,404.60	11,495.35	11,444.08	90.75	−51.27	39.48
Water supply	11,592.17	11,716.49	11,687.67	124.32	−28.82	95.50
Waste treatment	6817.47	6853.69	6826.98	36.22	−26.72	9.51
Soil formation and retention	12,407.22	12,467.23	12,401.70	60.01	−65.53	−5.53
Biodiversity protection	12,725.04	12,822.89	12,767.55	97.85	−55.34	42.51
Recreation and culture	5591.15	5642.52	5625.43	51.37	−17.09	34.27
Total ESV	81,764.32	82,411.07	82,065.31	646.74	−345.76	300.99

3.2. Historical Transitions of Land Use and ESV

As shown in Table 6, we found that between 2000 and 2018, 2.49% of the total land area had been transformed, whereas the ESV increased by USD 300.99 million. The most notable land use change in Guizhou Province was the transition from grassland to forestland, during which forestland increased by approximately 136,522.14 ha over 18 years, leading to an ESV increase of USD 539.11 million. The conversion of cultivated land (74,856.53 ha) to built-up land caused an ESV loss of USD 142.52 million. Meanwhile, the water body areas displayed a significant increase in both area and ESV, mostly due to the conversion of cultivated land (4880.96 ha), which was primarily the result of the implementation of an ecological protection policy that mandated the return of farmland to lakes and wetlands. However, both the cultivated land and grassland showed declining trends—specifically, the grassland area decreased as a result of forestland and cultivated land encroachment. The transformation of 48,589.04 ha from grassland to cultivated land resulted in an ESV decrease of USD 43.94 million; however, the transition from grassland to forestland subsequently increased the ESV to USD 539.11 million. The area of cultivated land shrank due to the increase in built-up land (74,856.53 ha), representing an ESV decline of USD 142.52 million. The transition from cultivated land to forestland, equal to 43,461.11 ha, gave rise to an increase in ESV of USD 210.92 million.

Table 6. Transition of land use and ecosystem service value from 2000 to 2018 in Guizhou Province.

2000–2018	Cultivated Land	Forestland	Grassland	Water Body	Construction Land	Unused Land	Total
			Land use transition (ha)				
Cultivated land	4,776,695.05	43,461.11	37,411.82	4880.96	74,856.53	19.30	4,937,324.77
Forestland	18,926.50	9,129,075.19	10,148.60	3243.34	28,389.13	111.67	9,189,894.43
Grassland	48,589.04	136,522.14	3,084,531.42	4188.45	26,047.77	55.95	3,299,934.76
Water body	55.25	2.19	27.57	91,654.94	42.59	0.00	91,782.53
Construction land	117.75	289.83	736.85	135.07	86,675.00	0.00	87,954.50
Unused land	190.16	777.30	42.39	16.64	99.39	2868.79	3994.67
Total	4,844,573.74	9,310,127.74	3,132,898.65	104,119.40	216,110.42	3055.70	17,610,885.66
			ESV transition (Million USD)				
Cultivated land	0.00	210.92	33.83	43.93	−142.52	−0.03	146.13
Forestland	−91.85	0.00	−40.08	13.45	−191.82	−0.72	−311.02
Grassland	−43.94	539.11	0.00	33.91	−73.15	−0.14	455.80
Water body	−0.50	−0.01	−0.22	0.00	−0.46	0.00	−1.19
Construction land	0.22	1.96	2.07	1.47	0.00	0.00	5.72
Unused land	0.30	4.99	0.10	0.18	−0.03	0.00	5.54
Total	−135.76	756.97	−4.30	92.94	−407.98	−0.89	300.99

3.3. Spatial Distribution of Land Use Changes and ESV Changes at the County Level

As shown in Figure 2, calculated with Equation (3), from 2000 to 2018, the maximum value of ESVA at the county level increased by 5.57%, totaling USD 327.89/ha, whereas the minimum value was raised by 9.73% (USD 69.71/ha). Lower ESV was primarily located in the western regions (Bijie and Liupanshui) and the south and east of Zunyi where vegetation coverage was low. The capital of Guizhou Province, Guiyang city, had the lowest ESVA owing to the rapid urban expansion in the area. In contrast, the eastern and southern areas with higher vegetation mountains, including Tongren, Qiandongnan, and Qiannan, had higher ESV due to the extensive forest areas within them.

In terms of spatial variation, the ESVA in most counties experienced a significant increase from 2000 to 2010 as a result of the conversion of grassland to forestland. Moreover, the growth of ESVA (with an increase of less than 10%) distinctly emerged in seven counties (Chishui, Shuicheng, Guanshanhu, Longli, Fuquan, and Jianhe) while the growth of more than 10% was only located in Xingyi. For instance, the increased ESVA in southwestern Qianxinan, was the result of an increase in grassland transitioning to waterbodies and forestland. In contrast, areas with a marked decline in ESVA were largely found in the western and central regions as a result of the rapid increase in built-up land and notable losses of forestland and grassland; however, from 2010 to 2018, ESVA throughout the study region dropped to less than 10% due to urban expansion and cultivated land conservation, particularly in Tongren and Qiannan. Of note, the ESVA in southern Guiyang and southern Zunyi (Bozhou) was reduced to more than 10% because of the urbanized regions in Guizhou Province. Nevertheless, the areas with specific increases in ESVA were mainly scattered in 18 counties along rivers (e.g., the Nanpan River and Beipan River in Qianxinan and the Sancha River in Liupanshui) as a result of lowered human activity and the implementation of protection policies for water resources.

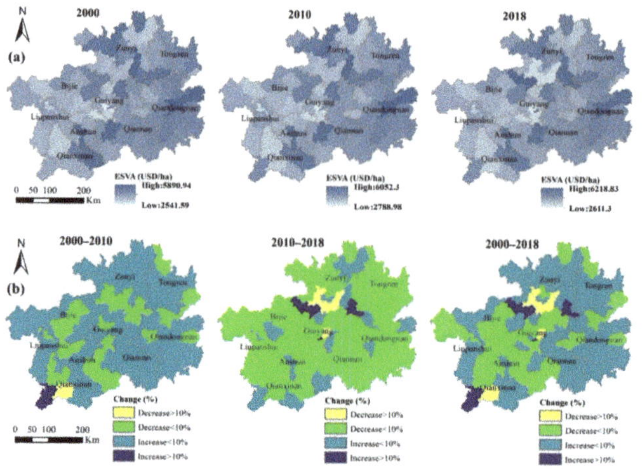

Figure 2. Spatial distribution of ESV changes at the county level from 2000 to 2018. Notes: (**a**) ESV per unit area in each county (ESVA); (**b**) ESVA changes.

3.4. Driving Factors of ESV

Using the factor detector (Table 7), we identified the impact of driving factors on ESV. In 2000, seven possible influencing factors (i.e., GDP, per capita GDP, per capita disposable income of rural residents, agricultural output value, forestry output value, population density, and rural employment) were selected and found to be statistically significant at the 99% significance level ($p < 0.01$). The q-statistic values were distributed from largest to smallest as follows: population density (0.651), per capita GDP (0.592), per capita disposable income of rural residents (0.497), agricultural output value (0.493), forestry output value (0.481), rural employment (0.427), and GDP (0.184). The results indicate that the highest q-statistic value was derived from population density, followed by per capita GDP and the per capita disposable income of rural residents. The remaining examined factors were found to be statistically insignificant at the 95% significance level. In 2010, six possible factors affected the ESV (i.e., per capita GDP, per capita disposable income of rural residents, agricultural output value, forestry output value, population density, and rural employment) with statistical significance at the 99% level. The q-statistic values were distributed from largest to smallest as follows: population density (0.600), forestry output value (0.574), per capita GDP (0.517), rural employment (0.450), per capita disposable income of rural residents (0.427), and agricultural output value (0.383). Population density still had the highest q-statistic value, followed by forestry output value and per capita GDP.

Accordingly, in 2018, six possible factors were detected with statistical significance at the 99% level, which were sequenced in q-statistic order: population density (0.572), forestry output value (0.547), agricultural output value (0.484), rural employment (0.449), per capita GDP (0.417), and per capita disposable income of rural residents (0.320). The highest q-statistic value was found in population density, followed by forestry output value and agricultural output value. From 2000 to 2018, population density as a social factor played a decisive role in ESV, and the contributions of per capita GDP and per capita disposable income of rural residents became more trivial. In contrast, the effects of forestry output value and rural employment on ESV grew more significant. The p value of GDP indicated that GDP was crucial to ESV in 2000 but insignificant in 2010 and 2018.

3.5. Spatial Variability of Driving Factors on ESV

We noticed heterogeneity in the geospatial relationships between driving factors and ESV. The results of spatial autocorrelation analysis showed that all the Moran's I values of ESV in this study were greater than 0, and the p values were all less than 0.01, indicating significant positive spatial autocorrelations in the ESV of Guizhou Province from 2000 to 2018 (Table 8).

Table 7. The q-statistic values and p values for the driving factors of ESV from the factor detector between 2000 and 2018 in Guizhou Province.

Factor Number	2000		2010		2018	
	p Value	q-Statistic Value	p Value	q-Statistic Value	p Value	q-Statistic Value
X_1	0.807	0.040	0.816	0.040	0.788	0.043
X_2	0.014	0.379	0.015	0.373	0.020	0.377
X_3	0.956	0.038	0.864	0.044	0.791	0.057
X_4	0.400	0.131	0.340	0.111	0.040	0.128
X_5	0.007	0.184	0.024	0.130	0.026	0.137
X_6	0.000	0.592	0.000	0.517	0.000	0.417
X_7	0.000	0.497	0.000	0.427	0.000	0.320
X_8	0.000	0.493	0.000	0.383	0.000	0.484
X_9	0.000	0.481	0.000	0.574	0.000	0.547
X_{10}	0.500	0.115	0.785	0.415	0.932	0.024
X_{11}	0.000	0.651	0.000	0.600	0.000	0.572
X_{12}	0.000	0.427	0.000	0.450	0.000	0.449

Elevation (X_1); terrain slope (X_2); average annual temperature (X_3); average annual precipitation (X_4); GDP (X_5); per capita GDP (X_6); per capita disposable income of rural residents (X_7); agricultural output value (X_8); forestry output value (X_9); resident population (X_{10}); population density (X_{11}); and rural employment (X_{12}).

Table 8. Spatial autocorrelation tests of each ESV in Guizhou Province.

	2000	2010	2018
Moran's I	0.189071	0.187922	0.208962
Z Scores	2.817848	2.800435	3.09133
p Value	0.004835	0.005103	0.001993

We utilized the MGWR model to identify the spatial distribution of the impacts of potential independent variables selected based on their p values ($p < 0.01$) and calculated by the GD model for each ESV. Before the MGWR model was implemented, the independent variables needed to be tested for multicollinearity between them. The high value of the variance inflation factor (VIF) indicated that a significant degree of multicollinearity existed between these variables. If VIF > 7.5, a distinct multicollinearity between each factor and variable redundancy in the model existed. In contrast, if VIF $\leq$ 7.5, no variable redundancy and no multilinear relationship between each factor existed [67].

Group 1 included six potential independent variables: per capita GDP, per capita disposable income of rural residents, agricultural output value, forestry output value, population density, and rural employment (Table 9). In 2000, the VIF values of agricultural output value and rural employment were greater than 7.5 and remained greater than 7.5 in 2018. These results demonstrate that variable redundancy existed in Group 1.

Group 2 included five potential independent variables: per capita GDP, per capita disposable income of rural residents, agricultural output value, forestry output value, and population density (Table 9). The VIF values from 2000 to 2018 were less than 7.5, indicating that the variables in this group had no strong correlation, and therefore these independent variables were used in the MGWR model for subsequent analysis.

The MGWR model provided comparative performance parameters with the OLS and GWR models. Table 10 shows the performance comparison between the OLS, GWR, and MGWR models. A higher adjusted R^2 value indicates a higher explanatory power and model fitness, whereas a lower AICc value signifies model concision and a more reliable regression estimation [66,68]. Table 10 summarizes how the MGWR model defined a nonstationarity relationship for each variable, more accurately reflected the phenomena, and compared the results with OLS and GWR. The AICc values found using MGWR were smaller than those found using OLS and GWR. Moreover, the five driving factors selected in this study explained 79.9%, 77.8%, and 75.2% of the ESV in 2000, 2010, and 2018, respectively.

Table 9. Variance inflation factor (VIF) of potential independent variables in 2000, 2010, and 2018.

Group 1	X_6	X_7	X_8	X_9	X_{11}	X_{12}
2000	4.0538	3.2261	10.6942	2.9255	2.7847	11.0815
2010	3.2997	4.7933	4.9752	2.7548	3.3155	5.4463
2018	3.2598	3.6690	7.1642	4.2891	2.4023	8.5471
Group 2	X_6	X_7	X_8	X_9	X_{11}	
2000	3.4680	3.1247	1.7056	2.9123	2.7550	
2010	3.2987	3.8115	1.1606	2.6833	2.8136	
2018	2.9539	3.6688	1.5667	4.0262	2.2611	

Per capita GDP (X_6); per capita disposable income of rural residents (X_7); agricultural output value (X_8); forestry output value (X_9); population density (X_{11}); and rural employment (X_{12}).

Table 10. Model fit metrics for OLS, GWR, and MGWR.

Model	**OLS**		**GWR**		**MGWR**	
	R^2 (Adjust)	**AICc**	**R^2 (Adjust)**	**AICc**	**R^2 (Adjust)**	**AICc**
2000	0.793	130.292	0.776	128.963	0.799	120.204
2010	0.773	138.989	0.776	128.963	0.778	128.428
2018	0.753	146.473	0.750	137.060	0.752	136.653

As shown in Table 11 and Figure 3, from 2000 to 2018, significant negative correlations between ESV and per capita GDP (X_6) existed, whereas the impacts of per capita GDP (X_6) on ESV decreased significantly from 2000 to 2010 and then increased slightly after 2010. The correlation coefficient of per capita GDP (X_6) was higher in the eastern regions of the study area and lower in the western regions in 2000; however, the higher correlation coefficients of per capita GDP (X_6) in 2010 were distributed in the north, and the lower correlation coefficients were distributed in the south. In 2018, the spatial pattern of the correlation coefficient of per capita GDP (X_6) showed a negative trend, with higher values in the west and lower values in the east. The per capita disposable income of rural residents (X_7) had a significant positive correlation with ESV in 2000 but negative correlations with ESV in 2010 and 2018. The higher influence of per capita disposable income of rural residents (X_7) on ESV was higher in 2010 and lower in 2018. The spatial correlation coefficient patterns of per capita disposable income of rural residents (X_7) from 2000 to 2010 were similar, with a higher coefficient in the eastern areas and a lower coefficient in the western areas. In contrast, the spatial pattern in 2018 was reversed.

Table 11. Mean statistics of MGWR coefficients between ESV and driving factors.

	X_6	X_7	X_8	X_9	X_{11}
2000	−0.293	0.070	0.529	−0.011	−0.451
2010	−0.011	−0.124	0.462	0.164	−0.411
2018	−0.051	−0.038	0.454	0.096	−0.471

Per capita GDP (X_6); per capita disposable income of rural residents (X_7); agricultural output value (X_8); forestry output value (X_9); and population density (X_{11}).

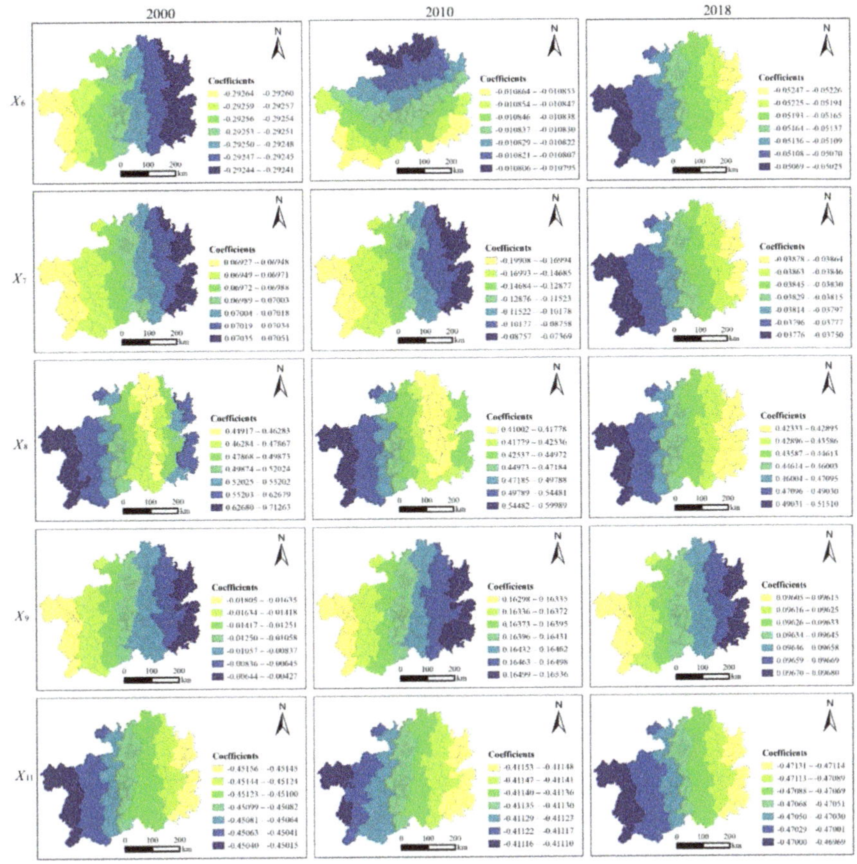

Figure 3. MGWR coefficients between ESV and driving factors in 2000, 2010, and 2018 in Guizhou Province. Per capita GDP (X_6); per capita disposable income of rural residents (X_7); agricultural output value (X_8); forestry output value (X_9); and population density (X_{11}).

Among the five driving factors, agricultural output value (X_8) primarily influenced ESV throughout the whole study period, with a significantly positive correlation, whereas the effects of agricultural output value (X_8) on ESV consistently declined. The lower correlation coefficient of the agricultural output value (X_8) was scattered in the middle of the study area in 2000 and then gradually extended in the eastern regions in 2010. Meanwhile, the higher correlation coefficient was distributed in both the east and the west of the study region in 2000, whereas it was distributed only in the west in 2010. Additionally, the spatial pattern of the correlation coefficient exhibited an increase from east to the west in 2018. ESV had a significant negative correlation with the forestry output value (X_9) in 2000 and a positive correlation in both 2010 and 2018. The influence of the forestry output value (X_9) on the ESV increased dramatically in the first decade and thereafter dropped. Furthermore, the spatial patterns of correlation coefficients from 2000 to 2018 were consistent: higher coefficients existed in the eastern areas and were lower in the western areas. Population density (X_{11}) was negatively correlated with ESV in the study period; its effects on ESV declined from 2000 to 2010 and rose from 2010 to 2018. A higher correlation coefficient appeared in the west, and a lower correlation coefficient appeared in the east of the study area from 2000 to 2018. The five selected driving factors were distinctly correlated with ESV, and there were apparent distinctions in the spatial variations in the correlation property and intensity. The results indicate that different factors dominate ESV in different regions—that is, zoning management was critical to promoting sustainable growth in them. In general, economic and social factors were the predominant drivers for regional ESV, which meant that optimizing both economic policies and human activities would most significantly improve regional ESV.

4. Discussion

4.1. Limitations of Value Estimate

In this study, ESV was calculated by the equivalent value table proposed by Costanza et al. [6] which has been applied by a number of researchers for similar studies [21,69,70] and improved by Xie et.al. [28,29]. This value technique to assess the economic value of ecosystem services was just one of many methods used for the environmental valuation [1,71], but important in studying the ESV in response to land ecosystem changes; however, based on assuming spatial homogeneity of services within ecosystems, the environmental suitability of a particular land use was not considered, and the equivalent values of ES were rough and, usually, underestimated the contributions of some land use types [21,72,73]. For instance, due to varied topography in Guizhou Province, cultivated lands located either on a steep slope or in a flat area received a similar estimation, although their ESV could greatly differ, for example, concerning the provisions of food or protection against rocky desertification and erosion, while the same is true for forests found either in a suitable or non-suitable area. Such environmental and ecological heterogeneity should be taken into account in the future development of ecosystem service value coefficients to improve the understandings of the regional characteristics of ecosystem service values.

4.2. Patterns of ESV Changes

The quantitative results of our study in three years (2000, 2010 and 2018) revealed the extent of ESV changes from 2000 to 2010 but also the decrease from 2010 to 2018 that occurred as a result of land use dynamics throughout the whole studying period. In particular, we found that various land use types affected the ESV differently; forestland had the most significant effect, followed by cultivated land and grassland; these findings are consistent with the research of Pan et al. [58]. From 2000 to 2010, the raise in the area's forest coverage led to a significant increase in ESV; however, this was accompanied by a decrease in the value of food production, owing to the decrease in cultivated land area. Previous studies have found that the process of urbanization in China significantly negatively affected the ecological environment and weakened regional ESV [74–76]. From 2010 to 2018, rapid urbanization and construction land growth in Guizhou Province exceeded the vegetation coverage growth, resulting in the decline in ESV. Severe pressure exists to implement stricter ecological protection as a result of these overall changes in land use and ESV. In general, while our study showed that conversion among cultivated lands, forestlands, grasslands and construction lands was quite intense and common in the study area, it also showed that the decline of ESV were mainly linked with the huge conversion of forests, which is identified to be the main provider of ecosystem services [77]. Findings from the literature also expressed that such changes were common in affecting the corresponding ecosystem service values in other study areas. For instance, Mekuria et al. (2021) evaluated the changes in land use over a period of 47 years (1973–2020) resulting in a total loss of USD 62,110.4 $\times$ 10^6 in ESV [14]. Kusi et al. (2022) revealed that an increase in all the land use categories caused an increase in the total ecosystem service value at USD 5.1 billion between 1992 and 2015 in Morocco [21]. On the other hand, in San Antonio, Texas, Kreuter et al. (2001) estimated that there appeared to be only a 4% net decline in the estimated annual value of ecosystem services ($5.58/ha per year) which could be attributed to the neutralizing effect of a 403% increase in the area of the woodlands between 1976 and 1991 [70]. Crespin et al. (2016) reported that ESV in El Salvador decreased by 2.6% from USD 9764.4 million per year to USD 9505.9 million per year during the 1998–2011 period and this loss was provided by tropical forests that account for 90% [78]. All these studies mirror our findings that land use dynamics have resulted in a significant change of ecosystem service values.

4.3. Evolution Mechanism of ESV

According to the spatial distribution of ESV in county-level regions, significant imbalances in the ESV per hectare existed between counties, with higher ESV located in the eastern region and lower ESV in the western region. In the west, this low ESV may be partially due to lower vegetation coverage and a more fragile ecological environment, making land use changes difficult. Residents living in this area have been forced to cultivate scattered sloping farmlands for the sake of basic food demand, giving rise to an increase in landscape fragmentation but a decrease in vegetation coverage [59]. In addition, changes in ESV per hectare indicate that ESV increases per hectare largely existed in the former period (2000–2010), mainly because urbanization efforts lagged behind ecological protection in these regions. Decreases in ESV per hectare became dominant in the latter period (2010–2018), although growth continued in a select few counties due to strict water resource protection for the Chishui River, Wujiang River, Beipanjiang River, Nanpanjiang River, and Qingshuijiang River. This transition principally arose from an expressway network integrating Guizhou with nearby economic zones, which has become an accelerator for the economy and urbanization but has reduced the regional ESV.

Efficient spatial planning of ecological protection and ecosystem management must include the identification of the dominant factors affecting ESV [49,79]. In our research, we detected significant correlations between ESV and selected driving factors using the GD model. Our results showed that socioeconomic factors (i.e., anthropogenic factors) dominated the spatial distribution of regional ESV changes; however, natural factors (including geomorphological factors and climate) had an insignificant explanatory power on the spatial differentiation of ESV changes. Regarding socioeconomic factors, population density, with a higher value in higher ESV areas, was not only the beneficiary of the ecological system but also the most significant factor in the ecological processes that altered the environment [80]. In particular, the impact of forestry output value on ESV became stronger, which may explain the benefits of implementing ecological engineering, which improves ESs and increases economic income from forests to prompt civilian involvement in the Grain for Green Program (GFGP).

4.4. Sustainable Development Implications

Despite the acknowledged limitations of rough estimations, this comprehensive framework for understanding the complex interactions between ESV and socioeconomic systems in Guizhou Province can positively contribute to policy-making, ecosystem service maintenance, territorial spatial planning, and ecological environmental managements for sustainable use of land resources. As a representative of karst landforms and China's national ecological civilization pilot zone, Guizhou is an important ecological security screen in the upper reaches of the Yangtze and Pearl Rivers, and thus, the ecological environment should be prioritized.

According to our results, targeted measures to alleviate the contradiction between ES and urbanization demands should be implemented. For instance, urban renewal planning and ecological red lining can be employed to minimize construction land and reduce pressure on cultivated land. Additionally, our findings show that anthropogenic factors had a significant effect on ESV, with spatial heterogeneity at the county-level scale. Therefore, we suggest that natural capital protection measures should be implemented according to regional variations of the primary driving factors [69]. The western region should focus on improving the agricultural output value, and the eastern regions should focus on the promotion of ecological and economic benefits from mixed forests [81–83]; however, as the only province without flatland in China, Guizhou has been confronted with a series of factors restricting the improvement of agriculture, such as a reduction in cultivated land, a weak foundation in agriculture, and a severe shortage of water. Therefore, these findings suggest that the development of agricultural technology for characteristic agriculture with comprehensive land consolidation could significantly increase the area's agricultural output value, leading to an increase in ESV in the Guizhou Province.

5. Conclusions

In this study, we analyzed the response of Guizhou Province's ESV changes from 2000 to 2018 and constructed an improved framework to identify the natural and socioeconomic factors from a geospatial perspective. In contrast to previous studies, our research provides a straightforward and flexible approach that incorporates the MGWR model with the GD model to better characterize the spatial distribution of ESV, understand the dominant factors affecting ESV, and identify the contribution of each factor to the ESV for this karst region of China.

Our results demonstrated a significant increase in ESV at first and then a subsequent decline during the later stage of the study period. In general, forestland was the dominant ecological land in Guizhou Province, followed by cultivated land and grassland, which have greatly improved due to regulation services and biodiversity protection, constituting the main ecosystem body in Guizhou Province. Meanwhile, the GD and MGWR models explicitly revealed the interaction effects and complex nexus between ESV and its driving forces. The results of the spatial recognition models show that the socioeconomic factors were robustly correlated with ESV from 2000 to 2018, with obvious diversity in the spatial variation of correlation properties and intensity, among which population density had a significantly negative effect on ESV, whereas the agricultural output and forestry output values had significantly positive effects. We expect the findings of our research and our proposed policy suggestions to provide notable references to land administrators and policy-makers for adopting suitable land resource conservation and management plans, thereby improving the overall ecological status in karst areas.

Author Contributions: L.Y. analyzed the data and wrote the manuscript. H.J. designed the experiments and provided editorial advice. All authors have read and agreed to the published version of the manuscript.

Funding: This research was funded by the National Natural Science Foundation of China "Rural spatial restructuring in poverty-stricken mountainous areas of Guizhou based on Spatial equity: A case study of Dianqiangui Rocky Desertification Area", grant number "41861038" and the National Key Research and Development Program "Research and Development of Emergency Response and Collaborative Command System with Holographic Perception of Traffic Network Disaster", grant number "2020YFC1512002".

Data Availability Statement: (1) land-use maps from 2000, 2010, and 2018 provided by the Resource and Environment Science and Data Central (http://www.resdc.cn, accessed on 19 September 2020); (2) a digital elevation model (DEM), the average annual temperature, and the average annual precipitation obtained from the Resource and Environment Science and Data Central (http://www.resdc.cn, accessed on 19 September 2020); and (3) statistical data, including grain yield, grain price, GDP data primarily collected from the Statistical Yearbooks of Guizhou Province (http://data.cnki.net, accessed on 29 September 2020), and the statistical bulletin of national economic and social development of Guizhou Province (http://stjj.guizhou.gov.cn, accessed on 29 September 2020).

Conflicts of Interest: The authors declare no conflict of interest.

Appendix A

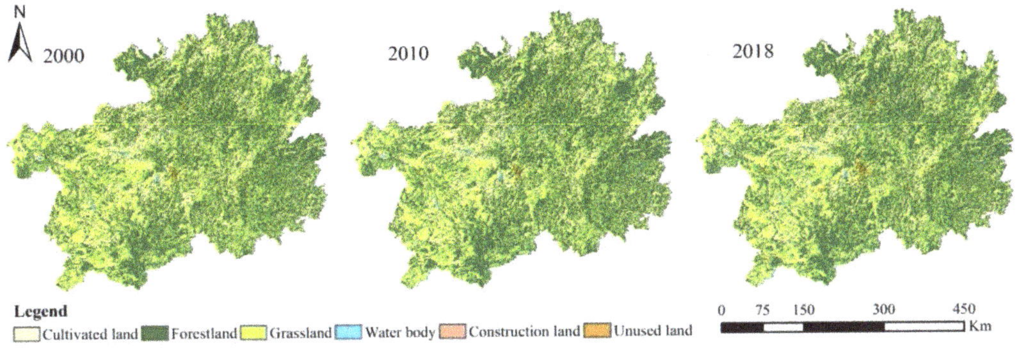

Figure A1. Spatial distribution of land use in 2000, 2010 and 2018.

Table A1. The annual ESV for each land use type per hectare in Guizhou Province (CNY/ha yr).

Ecosystem Service and Functions	Cultivated Land	Forestland	Grassland	Water Body	Construction Land	Unused Land
Food production	1590.60	530.20	681.69	833.17	0.00	25.25
Raw material	631.19	4746.55	580.69	555.45	0.00	75.74
Gas regulation	1136.14	6867.35	2398.52	807.92	0.00	100.99
Climate regulation	1540.10	6463.39	2474.26	3282.19	0.00	201.98
Water supply	1237.13	6513.88	2423.77	29,867.91	0.00	100.99
Waste treatment	2221.79	2726.74	2095.55	23,631.75	0.00	403.96
Soil formation and retention	2348.03	6387.64	3559.91	656.44	0.00	277.72
Biodiversity protection	1615.85	7170.32	2979.22	5453.48	0.00	631.19
Recreation and culture	277.72	3307.44	1388.62	7069.33	0.00	378.71
Total	12,598.55	44,713.50	18,582.23	72,157.64	0.00	2196.54

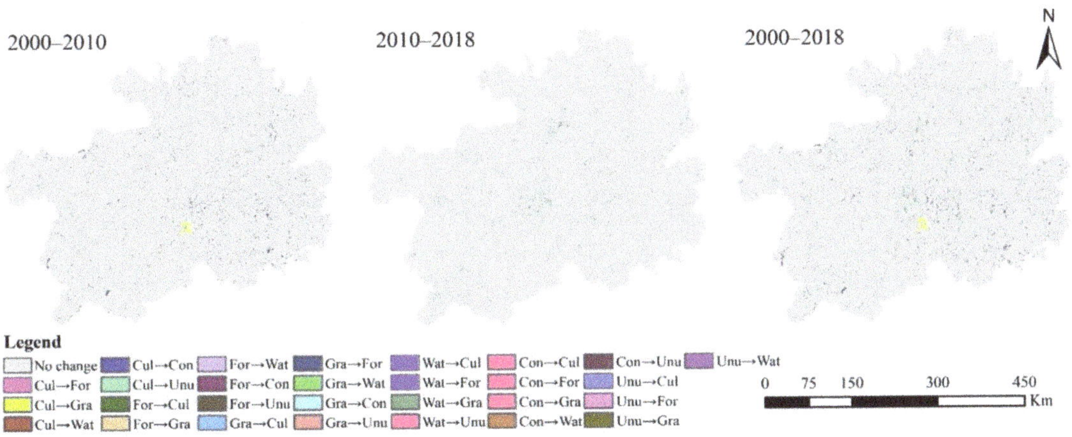

Figure A2. Spatial distribution of land use change from 2000 to 2018. Cultivated land (Cul); Forestland (For); Grassland (Gra); Water body (Wat); Construction land (Con); Unused land (Unu); → (transfer to).

Table A2. Land use and ESV changes in Guizhou Province in 2000, 2010, 2018, 2000–2010, and 2010–2018.

Land Use Types		Cultivated Land	Forestland	Grassland	Water Body	Construction Land	Unused Land	Total
2000	Land area (10^4 ha)	493.73	918.99	329.99	9.18	8.80	0.40	1761.09
	ESV (Million CNY)	62,203.13	410,912.34	61,320.15	6622.81	0.00	8.77	541,067.21
2010	Land area (10^4 ha)	491.82	933.88	315.55	9.94	9.60	0.30	1761.09
	ESV (Million CNY)	61,962.49	417,571.63	58,636.76	7169.54	0.00	6.54	545,346.97
2018	Land area (10^4 ha)	484.46	931.01	313.29	10.41	21.61	0.31	1761.09
	ESV (Million CNY)	61,034.60	416,288.40	58,216.24	7513.01	0.00	6.71	543,058.97
2000–2010	Land area (10^4 ha)	−1.91	14.89	−14.44	0.76	0.80	−0.10	0.00
	ESV (Million CNY)	−240.65	6659.29	−2683.38	546.73	0.00	−2.23	4279.76
2010–2018	Land area (10^4 ha)	−7.36	−2.87	−2.26	0.48	12.01	0.01	0.00
	ESV (Million CNY)	−927.88	−1283.24	−420.52	343.47	0.00	0.17	−2288.00
2000–2018	Land area 10^4 ha)	−9.28	12.02	−16.70	1.23	12.82	−0.09	0.00
	ESV (Million CNY)	−1168.53	5376.05	−3103.90	890.20	0.00	−2.06	1991.76

Table A3. Value changes of different ecosystem service functions in Guizhou Province in 2000, 2010, and 2018.

Ecosystem Service Functions	ESV (Million CNY)			Changes of ESV (Million CNY)		
	2000	2010	2018	2000–2010	2010–2018	2000–2018
Food production	15,051.87	15,008.30	14,864.48	−43.57	−143.82	−187.39
Raw material	48,704.21	49,319.34	49,126.14	615.13	−193.20	421.93
Gas regulation	76,709.22	77,369.95	77,038.76	660.73	−331.19	329.54
Climate regulation	75,468.79	76,069.34	75,730.07	600.56	−339.28	261.28
Water supply	76,710.04	77,532.73	77,342.00	822.70	−190.73	631.96
Waste treatment	45,113.93	45,353.63	45,176.83	239.70	−176.80	62.90
Soil formation and retention	82,103.57	82,500.67	82,067.00	397.10	−433.67	−36.57
Biodiversity protection	84,206.70	84,854.20	84,487.99	647.49	−366.20	281.29
Recreation and culture	36,998.90	37,338.83	37,225.71	339.93	−113.12	226.81
Total ESV	541,067.23	545,346.99	543,058.99	4279.76	−2288.00	1991.76

References

1. Fisher, B.; Turner, R.K.; Morling, P. Defining and classifying ecosystem services for decision making. *Ecol. Econ.* **2009**, *68*, 643–653. [CrossRef]
2. Daily, G.C.; Polasky, S.; Goldstein, J.; Kareiva, P.M.; A Mooney, H.; Pejchar, L.; Ricketts, T.H.; Salzman, J.; Shallenberger, R. Ecosystem services in decision making: Time to deliver. *Front. Ecol. Environ.* **2009**, *7*, 21–28. [CrossRef]
3. Halpern, B.S.; Walbridge, S.; Selkoe, K.A. A global map of human impact on marine ecosystems. *Science* **2008**, *319*, 948–952. [CrossRef] [PubMed]
4. Cowie, A.L.; Orr, B.J.; Sanchez, V.M.C.; Chasek, P.; Crossman, N.D.; Erlewein, A.; Louwagie, G.; Maron, M.; Metternicht, G.I.; Minelli, S.; et al. Land in balance: The scientific conceptual framework for Land Degradation Neutrality. *Environ. Sci. Policy* **2018**, *79*, 25–35. [CrossRef]
5. Shoyama, K.; Kamiyama, C.; Morimoto, J.; Ooba, M.; Okuro, T. A review of modeling approaches for ecosystem services as-sessment in the Asian region. *Ecosyst. Serv.* **2017**, *26*, 316–328. [CrossRef]
6. Costanza, R.; d'Arge, R.; Groot, R.D.; Farberk, S.; Belt, M.V.D. The value of the world's ecosystem services and natural capital. *Nature* **1997**, *387*, 253–260. [CrossRef]
7. Syrbe, R.-U.; Walz, U. Spatial indicators for the assessment of ecosystem services: Providing, benefiting and connecting areas and landscape metrics. *Ecol. Indic.* **2012**, *21*, 80–88. [CrossRef]
8. Hou, Y.; Burkhard, B.; Müller, F. Uncertainties in landscape analysis and ecosystem service assessment. *J. Environ. Manag.* **2013**, *127*, S117–S131. [CrossRef]
9. Costanza, R.; Groot, R.; Sutton, P.D.; Ploeg, S.V.D. Changes in the global value of ecosystem services. *Glob. Environ. Change-Hum. Policy Dimens.* **2014**, *26*, 152–158. [CrossRef]
10. Zhang, Z.; Xia, F.; Yang, D.; Huo, J.; Wang, G.; Chen, H. Spatiotemporal characteristics in ecosystem service value and its interaction with human activities in Xinjiang, China. *Ecol. Indic.* **2020**, *110*, 105826. [CrossRef]
11. Carpenter, S.R.; DeFries, R.; Dietz, T.; Mooney, H.A.; Polasky, S.; Reid, W.V.; Scholes, R.J. Millennium Ecosystem Assessment: Research Needs. *Science* **2006**, *314*, 257–258. [CrossRef] [PubMed]

12. Mooney, H.A.; Cropper, A.; Reid, W. The millennium ecosystem assessment: What is it all about? *Trends Ecol. Evol.* **2004**, *19*, 221–224. [CrossRef] [PubMed]
13. De Groot, R.S.; Fisher, B.; Christie, M.; Aronson, J.; Braat, L.; Haines-Young, R.; Ring, I. Integrating the ecological and economic dimensions in biodiversity and ecosystem service valuation. In *The Economics of Ecosystems and Biodiversity (TEEB): Eco-Logical and Economic Foundations*; Earthscan Routledge: London, UK, 2010; pp. 9–40.
14. Mekuria, W.; Diyasa, M.; Tengberg, A.; Haileslassie, A. Effects of Long-Term Land Use and Land Cover Changes on Ecosystem Service Values: An Example from the Central Rift Valley, Ethiopia. *Land* **2021**, *10*, 1373. [CrossRef]
15. Schmeller, D.S.; Niemelä, J.; Bridgewater, P. The Intergovernmental Science-Policy Platform on Biodiversity and Ecosystem Services (IPBES): Getting involved. *Biodivers. Conserv.* **2017**, *26*, 2271–2275. [CrossRef]
16. Haines-Young, R.; Potschin, M. *Common International Classification of Ecosystem Services (CICES): 2011 Update*; European Environment Agency: Copenhagen, Denmark, 2012; Available online: https://cices.eu/content/uploads/sites/8/2009/11/CICES_Update_Nov2011.pdf (accessed on 15 January 2022).
17. Haines-Young, R.; Potschin, M. CICES Version 4: Response to Consultation, London. 2012. Available online: https://cices.eu/resources (accessed on 15 January 2022).
18. Sinha, P.; Van, H.G. *National Ecosystem Services Classification System (NESCS): Framework Design and Policy Application*; United States Environmental Protection Agency: Washington, DC, USA, 2015. Available online: https://www.epa.gov/eco-research (accessed on 15 January 2022).
19. Costanza, R.; de Groot, R.; Braat, L.; Kubiszewski, I.; Fioramonti, L.; Sutton, P.; Farber, S.; Grasso, M. Twenty years of eco-system services: How far have we come and how far do we still need to go? *Ecosyst. Serv.* **2017**, *28*, 1–16. [CrossRef]
20. Turner, K.G.; Anderson, S.; Gonzalez-Chang, M.; Costanza, R.; Courville, S.; Dalgaard, T.; Dominati, E.; Kubiszewski, I.; Ogilvy, S.; Porfirio, L.; et al. A review of methods, data, and models to assess changes in the value of ecosystem services from land degradation and restoration. *Ecol. Model.* **2016**, *319*, 190–207. [CrossRef]
21. Kusi, K.K.; Khattabi, A.; Mhammdi, N. Analyzing the impact of land use change on ecosystem service value in the main watersheds of Morocco. *Environ. Dev. Sustain.* **2022**, *24*, 1–28. [CrossRef]
22. Kindu, M.; Schneider, T.; Teketay, D.; Knoke, T. Changes of ecosystem service values in response to land use/land cover dynamics in Munessa–Shashemene landscape of the Ethiopian highlands. *Sci. Total Environ.* **2016**, *547*, 137–147. [CrossRef]
23. Wang, Z.; Zhang, B.; Zhang, S.; Li, X.; Liu, D.; Song, K.; Li, J.; Li, F.; Duan, H. Changes of Land Use and of Ecosystem Service Values in Sanjiang Plain, Northeast China. Environ. *Monit. Assess.* **2006**, *112*, 69–91. [CrossRef]
24. Frank, S.; Fürst, C.; Koschke, L.; Makeschin, F. A contribution towards a transfer of the ecosystem service concept to landscape planning using landscape metrics. *Ecol. Indic.* **2012**, *21*, 30–38. [CrossRef]
25. Schägner, J.P.; Brander, L.; Maes, J.; Hartje, V. Mapping ecosystem services' values: Current practice and future prospects. *Ecosyst. Serv.* **2013**, *4*, 33–46. [CrossRef]
26. Iverson, L.R.; Echeverria, C.; Nahuelhual, L.; Luque, S. Ecosystem services in changing landscapes: An introduction. *Landsc. Ecol.* **2014**, *29*, 181–186. [CrossRef]
27. Fu, B.; Forsius, M. Ecosystem services modeling in contrasting landscapes. *Landsc. Ecol.* **2015**, *30*, 375–379. [CrossRef]
28. Xie, G.D.; Zhen, L.; Lu, C.X.; Xiao, Y.; Chen, C. Expert Knowledge Based Valuation Method of Ecosystem Services in China. *J. Nat. Resour.* **2008**, *23*, 911–919. [CrossRef]
29. Xie, G.D.; Xiao, Y.; Zhen, L.; Lu, C.X. Study on ecosystem services value of food production in China. *Chin. J. Eco-Agric.* **2005**, *13*, 10–13.
30. Li, S.M.; Shi, P.J.; Zhou, Q. Research on the Dynamic Changes of Ecosystem Service Value of Composite System of Economy and Environment in Xining City. In Proceedings of the International Conference on Earth and Environmental Science (ICEES), Kunming, China, 18–20 October 2013.
31. Zhou, J.; Sun, L.; Zang, S.Y.; Wang, K.; Liu, X.R. Effects of the land use change on ecosystem service value. *Glob. J. Environ. Sci. Manag. Gjesm* **2017**, *3*, 121–130. [CrossRef]
32. Xue, M.; Ma, S. Optimized Land-Use Scheme Based on Ecosystem Service Value: Case Study of Taiyuan, China. *J. Urban Plan. Dev.* **2018**, *144*, 04018016. [CrossRef]
33. Shifaw, E.; Sha, J.; Li, X.; Bao, Z.; Zhou, Z. An insight into land-cover changes and their impacts on ecosystem services before and after the implementation of a comprehensive experimental zone plan in Pingtan island, China. *Land Use Policy* **2019**, *82*, 631–642. [CrossRef]
34. Wu, C.; Chen, B.; Huang, X.; Wei, Y.D. Effect of land-use change and optimization on the ecosystem service values of Jiangsu province. *China. Ecol. Indic.* **2020**, *117*, 106507. [CrossRef]
35. Mooney, H.; Larigauderie, A.; Cesario, M.; Elmquist, T.; Hoegh-Guldberg, O.; Lavorel, S.; Yahara, T. Biodiversity, climate change, and ecosystem services. *Curr. Opin. Environ. Sustain.* **2009**, *1*, 46–54. [CrossRef]
36. Mitchell, M.G.E.; Suarez-Castro, A.F.; Martinez-Harms, M.; Maron, M.; McAlpine, C.; Gaston, K.J.; Rhodes, J.R. Reframing landscape fragmentation's effects on ecosystem services. *Trends Ecol. Evol.* **2015**, *30*, 190–198. [CrossRef] [PubMed]
37. Scholes, R.J. Climate change and ecosystem services. *Wiley Interdiscip. Rev.-Clim. Change* **2016**, *7*, 537–550. [CrossRef]
38. Li, Z.H.; Zhihui, L.; Jun, X.; Xiangzheng, D.; Haiming, Y. Multilevel modelling of impacts of human and natural factors on ecosystem services change in an oasis, Northwest China. *Resour. Conserv. Recycl.* **2021**, *169*, 105474. [CrossRef]

39. Cai, Y.-B.; Zhang, H.; Pan, W.-B.; Chen, Y.-H.; Wang, X.-R. Land use pattern, socio-economic development, and assessment of their impacts on ecosystem service value: Study on natural wetlands distribution area (NWDA) in Fuzhou city, southeastern China. *Environ. Monit. Assess.* **2013**, *185*, 5111–5123. [CrossRef] [PubMed]
40. Zhu, Z.; Li, B.; Zhao, Y.; Zhao, Z.; Chen, L. Socio-Economic Impact Mechanism of Ecosystem Services Value, a PCA-GWR Approach. Pol. *J. Environ. Stud.* **2020**, *30*, 977–986. [CrossRef]
41. Pilogallo, A.; Saganeiti, L.; Scorza, F.; Murgante, B. Assessing the Impact of Land Use Changes on Ecosystem Services Value. In Proceedings of the 20th International Conference on Computational Science and Its Applications (ICCSA), Online, 1–4 July 2020.
42. Berihun, M.L.; Tsunekawa, A.; Haregeweyn, N.; Tsubo, M.; Fenta, A.A. Changes in ecosystem service values strongly influ-enced by human activities in contrasting agro-ecological environments. *Ecol. Processes* **2021**, *10*, 1–18. [CrossRef]
43. Loomis, D.K.; Paterson, S.K. The human dimensions of coastal ecosystem services: Managing for social values. *Ecol. Indic.* **2014**, *44*, 6–10. [CrossRef]
44. Drukker, D.M.; Egger, P.; Prucha, I.R. On Two-Step Estimation of a Spatial Autoregressive Model with Autoregressive Disturbances and Endogenous Regressors. *Econom. Rev.* **2013**, *32*, 686–733. [CrossRef]
45. Zhang, Z.; Gao, J. Linking landscape structures and ecosystem service value using multivariate regression analysis: A case study of the Chaohu Lake Basin, China. *Environ. Earth Sci.* **2016**, *75*, 1–16. [CrossRef]
46. Le Cle'H, S.; Jégou, N.; Decaens, T.; Dufour, S.; Grimaldi, M.; Oszwald, J. From Field Data to Ecosystem Services Maps: Using Regressions for the Case of Deforested Areas Within the Amazon. *Ecosystems* **2018**, *21*, 216–236. [CrossRef]
47. Su, S.; Xiao, R.; Jiang, Z.; Zhang, Y. Characterizing landscape pattern and ecosystem service value changes for urbanization impacts at an eco-regional scale. *Appl. Geogr.* **2012**, *34*, 295–305. [CrossRef]
48. Yoskowitz, D.; Carollo, C.; Pollack, J.B.; Santos, C.; Welder, K. Integrated ecosystem services assessment: Valuation of changes due to sea level rise in Galveston Bay, Texas, USA. *Integr. Environ. Assess. Manag.* **2017**, *13*, 431–443. [CrossRef] [PubMed]
49. Lyu, R.; Clarke, K.C.; Zhang, J.; Feng, J.; Jia, X.; Li, J. Spatial correlations among ecosystem services and their socio-ecological driving factors: A case study in the city belt along the Yellow River in Ningxia, China. *Appl. Geogr.* **2019**, *108*, 64–73. [CrossRef]
50. Li, H.; Peng, J.; Yanxu, L.; Yi'Na, H. Urbanization impact on landscape patterns in Beijing City, China: A spatial heterogeneity perspective. *Ecol. Indic.* **2017**, *82*, 50–60. [CrossRef]
51. Brunsdon, C.; Fotheringham, A.S.; Charlton, M.E. Geographically weighted regression: A method for exploring spatial nonstationarity. *Geogr. Anal.* **1996**, *28*, 281–298. [CrossRef]
52. Brunsdon, C.; Fotheringham, A.S.; Charlton, M. Spatial Nonstationarity and Autoregressive Models. *Environ. Plan. A Econ. Space* **1998**, *30*, 957–973. [CrossRef]
53. Leung, Y.; Mei, C.-L.; Zhang, W.-X. Statistical Tests for Spatial Nonstationarity Based on the Geographically Weighted Regression Model. *Environ. Plan. A Econ. Space* **2000**, *32*, 9–32. [CrossRef]
54. Dadashpoor, H.; Azizi, P.; Moghadasi, M. Land use change, urbanization, and change in landscape pattern in a metropolitan area. *Sci. Total Environ.* **2019**, *655*, 707–719. [CrossRef]
55. Tenerelli, P.; Demšar, U.; Luque, S. Crowdsourcing indicators for cultural ecosystem services: A geographically weighted approach for mountain landscapes. *Ecol. Indic.* **2016**, *64*, 237–248. [CrossRef]
56. Su, S.; Li, D.; Hu, Y.; Xiao, R.; Zhang, Y. Spatially non-stationary response of ecosystem service value changes to urbanization in Shanghai, China. *Ecol. Indic.* **2014**, *45*, 332–339. [CrossRef]
57. Wang, X. The Ecotourism Development in Guizhou. In Proceedings of the 2nd International Conference on Economy, Management and Education Technology (ICEMET), Chongqing, China, 28–29 May 2016.
58. Pan, S.; Liang, J.; Chen, W.; Li, J.; Liu, Z. Gray Forecast of Ecosystem Services Value and Its Driving Forces in Karst Areas of China: A Case Study in Guizhou Province, China. *Int. J. Environ. Res. Public Health* **2021**, *18*, 12404. [CrossRef] [PubMed]
59. Qiu, S.; Peng, J.; Dong, J.; Wang, X.; Ding, Z.; Zhang, H.; Mao, Q.; Liu, H.; Quine, T.A.; Meersmans, J. Understanding the relationships between ecosystem services and associated social-ecological drivers in a karst region: A case study of Guizhou Province, China. *Prog. Phys. Geogr. Earth Environ.* **2021**, *45*, 98–114. [CrossRef]
60. Wang, J.; Xu, C. Geodetector: Principle and prospective. *Acta Geogr. Sin.* **2017**, *72*, 116–134. [CrossRef]
61. Wang, J.-F.; Zhang, T.-L.; Fu, B.-J. A measure of spatial stratified heterogeneity. *Ecol. Indic.* **2016**, *67*, 250–256. [CrossRef]
62. Liu, Y.; Chen, Y.; Wu, Z.; Wang, B.; Wang, S. Geographical detector-based stratified regression kriging strategy for mapping soil organic carbon with high spatial heterogeneity. *Catena* **2021**, *196*, 104953. [CrossRef]
63. Tobler, W.R. A Computer Movie Simulating Urban Growth in the Detroit Region. *Econ. Geogr.* **1970**, *46*, 234–240. [CrossRef]
64. Oshan, T.M.; Li, Z.; Kang, W.; Wolf, L.J.; Fotheringham, A.S. mgwr: A Python Implementation of Multiscale Geographically Weighted Regression for Investigating Process Spatial Heterogeneity and Scale. *ISPRS Int. J. Geo-Inf.* **2019**, *8*, 269. [CrossRef]
65. Yu, H.C.; Fotheringham, A.S.; Li, Z.; Oshan, T.; Kang, W.; Wolf, L.J. Inference in Multiscale Geographically Weighted Regression. *Geogr. Anal.* **2020**, *52*, 87–106. [CrossRef]
66. Fotheringham, A.S.; Yang, W.; Kang, W. Multiscale Geographically Weighted Regression (MGWR). *Ann. Am. Assoc. Geogr.* **2017**, *107*, 1247–1265. [CrossRef]
67. Geng, T.; Hai, C.; Hang, Z. Spatiotemporal evolution of land ecosystem service value and its influencing factors in Shaanxi province based on GWR. *J. Nat. Resour.* **2020**, *35*, 1714–1727. [CrossRef]
68. Hu, B.; Kang, F.; Han, H.; Cheng, X.; Li, Z. Exploring drivers of ecosystem services variation from a geospatial perspective: Insights from China's Shanxi Province. *Ecol. Indic.* **2021**, *131*, 108188. [CrossRef]

69. Kreuter, U.P.; Harris, H.G.; Matlock, M.D.; Lacey, R. Change in ecosystem service values in the San Antonio area, Texas. *Ecol. Econ.* **2001**, *39*, 333–346. [CrossRef]
70. Kubiszewski, I.; Costanza, R.; Dorji, L.; Thoennes, P.; Tshering, K. An initial estimate of the value of ecosystem services in Bhutan. *Ecosyst. Serv.* **2013**, *3*, e11–e21. [CrossRef]
71. Van der Ploeg, S.; de Groot, D. *The TEEB Valuation Database—A Searchable Database of 1310 Estimates of Monetary Values of Ecosystem Services*; Foundation for Sustainable Development: Wageningen, The Netherlands, 2010; Available online: https://www.researchgate.net/publication/287965940 (accessed on 15 January 2022).
72. Naidoo, R.; Balmford, A.; Costanza, R.; Fisher, B.; Green, R.E.; Lehner, B.; Malcolm, T.R.; Ricketts, T.H. Global mapping of ecosystem services and conservation priorities. *Proc. Natl. Acad. Sci. USA* **2008**, *105*, 9495–9500. [CrossRef] [PubMed]
73. Eigenbrod, F.; Armsworth, P.R.; Anderson, B.J.; Heinemeyer, A.; Gillings, S.; Roy, D.B.; Thomas, C.D.; Gaston, K.J. The impact of proxy-based methods on mapping the distribution of ecosystem services. *J. Appl. Ecol.* **2010**, *47*, 377–385. [CrossRef]
74. Li, Z.; Wang, J.; Bai, Z.; Guo, Y.; Yu, L. Land use and ecosystem service values and their grey forecast in Guizhou province. *Prog. Geogr.* **2012**, *31*, 577–583. [CrossRef]
75. Liu, Y.; Song, W.; Deng, X. Understanding the spatiotemporal variation of urban land expansion in oasis cities by integrating remote sensing and multi-dimensional DPSIR-based indicators. *Ecol. Indic.* **2019**, *96*, 23–37. [CrossRef]
76. Qiu, L.; Pan, Y.; Zhu, J.; Amable, G.S.; Xu, B. Integrated analysis of urbanization-triggered land use change trajectory and implications for ecological land management: A case study in Fuyang, China. *Sci. Total Environ.* **2019**, *660*, 209–217. [CrossRef]
77. Martínez, M.L.; Pérez-Maqueo, O.; Vázquez, G.; Castillo-Campos, G.; García-Franco, J.; Mehltreter, K.; Equihua, M.; Landgrave, R. Effects of land use change on biodiversity and ecosystem services in tropical montane cloud forests of Mexico. *For. Ecol. Manag.* **2009**, *258*, 1856–1863. [CrossRef]
78. Crespin, S.J.; Simonetti, J.A. Loss of ecosystem services and the decapitalization of nature in El Salvador. *Ecosyst. Serv.* **2016**, *17*, 5–13. [CrossRef]
79. Wang, H.; Liu, G.; Li, Z.; Zhang, L.; Wang, Z. Processes and driving forces for changing vegetation ecosystem services: Insights from the Shaanxi Province of China. *Ecol. Indic.* **2020**, *112*, 106105. [CrossRef]
80. Mace, G.M.; Norris, K.; Fitter, A.H. Biodiversity and ecosystem services: A multilayered relationship. *Trends Ecol. Evol.* **2012**, *27*, 19–26. [CrossRef] [PubMed]
81. Knoke, T. Mixed forests and finance—Methodological approaches. *Ecol. Econ.* **2008**, *65*, 590–601. [CrossRef]
82. Wang, L.; Wu, S.; Yuan, H. Analysis of Ecological Economic Advantages of Mixed Forest and Pure Forest. *Issues For. Econ.* **2018**, *38*, 18–22. [CrossRef]
83. Yang, Y.; Yao, Y.; Hao, S. Spatial-Temporal Variations and Influencing Factors of Intensified Cultivated Land Use Level in Guizhou Province Based on GWR Model. *Res. Soil Water Conserv.* **2022**, *29*, 326–332+338.

sustainability

MDPI

Article

Estimating Land-Use Change Using Machine Learning: A Case Study on Five Central Coastal Provinces of Vietnam

Nguyen Hong Giang [1,2], Yu-Ren Wang [2,*], Tran Dinh Hieu [1,*], Nguyen Huu Ngu [3] and Thanh-Tuan Dang [4,5]

1 Faculty of Architecture, Thu Dau Mot University, Thu Dau Mot 820000, Vietnam; giangnh@tdmu.edu.vn
2 Department of Civil Engineering, National Kaohsiung University of Science and Technology, Kaohsiung 807618, Taiwan
3 Faculty of Land Resources and Agricultural Environment, Hue University, Hue 49118, Vietnam; nhuungu@hueuni.edu.vn
4 Department of Industrial Engineering and Management, National Kaohsiung University of Science and Technology, Kaohsiung 807618, Taiwan; i108143106@nkust.edu.tw
5 Department of Logistics and Supply Chain Management, Hong Bang International University, Ho Chi Minh 72320, Vietnam
* Correspondence: yrwang@nkust.edu.tw (Y.-R.W.); hieutd@tdmu.edu.vn (T.D.H.)

Abstract: Population growth is one factor relevant to land-use transformation and expansion in urban areas. This creates a regular mission for local governments in evaluating land resources and proposing plans based on various scenarios. This paper discussed the future trend of three kinds of land-use in the five central coast provinces. Afterwards, the paper deployed machine learning such as Multivariate Adaptive Regression Splines (MARS), Random Forest Regression (RFR), and Lasso Linear Regression (LLR) to analyze the trend of rural land use and industrial land-use to urban land-use in the Central Coast Region of Vietnam. The input variables of land-use from 2010 to 2020 were obtained by the five provinces of the Department of Natural Resources and Environment (DONRE). The results showed that these models provided pieces of information about the relationship between urban, rural, and industrial land-use change data. Furthermore, the MARS model proved to be accurate in the Quang Binh, Quang Tri, and Quang Nam provinces, whereas RFR demonstrated efficiency in the Thua Thien-Hue province and Da Nang city in the fields of land change prediction. Furthermore, the result enables to support land-use planners and decision-makers to propose strategies for urban development.

Keywords: Multivariate Adaptive Regression Spline (MARS); Random Forest Regression (RFR); Lasso Linear Regression (LLR); rural land-use; industrial land-use; urban land-use; decision-making

Citation: Giang, N.H.; Wang, Y.-R.; Hieu, T.D.; Ngu, N.H.; Dang, T.-T. Estimating Land-Use Change Using Machine Learning: A Case Study on Five Central Coastal Provinces of Vietnam. *Sustainability* **2022**, *14*, 5194. https://doi.org/10.3390/su14095194

Academic Editors: Dianfeng Liu, Wenwu Tang and Jianxin Yang

Received: 18 February 2022
Accepted: 22 April 2022
Published: 25 April 2022

1. Introduction

Estimating land-use change provides valuable information about potential conversion that might occur over time in the earth's surface landscapes. Furthermore, the prediction enables support for land-use planners and land resource managers to propose land-use change strategies, urban planning through modeling rural development, and selecting areas for setting industrial zones [1]; therefore, simulating models to observe and examine land-use change related to landscape dynamics over time is an interesting issue at both global and local scales [2].

Three major approaches are commonly applied in land-use change prediction, which are spatial pattern, statistical analysis, and artificial intelligence [3–13]. Models based on the simulation of the spatial pattern of land-use change processes, such as the Markov model, are deployed to know and interpret regional land changes and trends of regional land-use in effective ways [14]. Using machine learning based on quarterly or annual land-use statistics in localities to analyze land-use changes is also widely applied, such as Multivariate Adaptive Regression Splines (MARS) methodology, which is a kind of nonparametric (nonparametric covers techniques that do not rely on data belonging to any particular parametric family of probability distributions) and nonlinear technique used in statistical learning [15]. The Random Forest Regression (RFR) model also belongs to nonparametric learning, and the model is used in those areas [16,17]. The Lasso Linear Regression (LLR) model is the earliest form of least-squares prediction in classification, and its properties are similar to RFR and MARS [18–20]; hence, Yilmaz et al. (2018) [21] used MARS to estimate the suspended sediment load in Coruh River Basin, Turkey. The result showed that MARS outperformed the best model with R-squared is approximately 0.9, and they summarized that the MARS might be easily applied in modeling. Bui et al. (2019) [22] applied the MARS model to analyze and predict spatial patterns of forest fire danger for tropical forest fires in Lao Cai province, Vietnam. The result of the study pointed out that the model is the ability to solve the complexity of modeling forest fire danger. Nguyen et al. (2018) [23] deployed the RFR model and Landsat data for 10 classes consisting of multiple forest classes in Vietnam. The study result indicated that the overall accuracy is estimated at 0.90. Ha et al. (2020) [24] employed RFR and Landsat data with seven land-cover classes consisting of forest land to evaluate the land-cover classification in the northeast subtropical region of Vietnam. The study showed that the overall study accuracies were higher than 0.90. Dennedy-Frank et al. (2019) [25] used LLR with cross-validation forecasting streamflow impacts of forest restoration and conservation based on simulation of the hydrology of 29 located models worldwide. The result demonstrated that the model for water yield change after the development of agriculture with R-squared is around 0.69 when using LLR model.

Although not many types of research have been announced to compare methods correlated to accuracy levels land-use changes in the five central coastal provinces of Vietnam; therefore, deploying these three proposal models to estimate urban land-use change for this region is feasible and may provide a high forecasting accuracy.

The Quang Binh, Quang Tri, Thua Thien-Hue, Da Nang, and Quang Nam provinces belong to the Central Coast Region of Vietnam that has made positive changes in the process of urbanization in recent years. As a result, there are a lot of larger urban areas in the region that are formed in the eastern coastal area; however, the process of urbanization in the region has revealed significant shortcomings, limitations, and challenges.

This study aims to present the estimation of the urban land-use change using LLR, RFR, and MARS models. The input vectors used in the models are based on the land-use change as rural land-use, industrial land-use, and urban land-use 44 quarters in five central coastal provinces in Vietnam between 2010 and 2020. With the support of the collected data, the paper discusses the role of three types of land-use in the region's urbanization process. After that, it highlights a comparison between three models based on statistical accuracy indicators' results. Furthermore, the collection of results of these three models may show the working efficiency of the models for land-use prediction, and it may develop future scenarios that can support land-use planning and decision-making.

The structure of the paper is organized as follows. Section 1 gives the paper introduction. Section 2 introduces the materials and methodology of MARS, RFR, and LLR models. Sections 3 and 4 describe the results and discussions. Finally, Section 5 presents the conclusions.

2. Materials and Methods

The process of the following experimental stages in this study is described in Figure 1. Firstly, the database of urban, rural, and industrial land use is preprocessed and tested by statistical methods in the input layer. Secondly, the database is divided into 70% for the training phase and 30% for the testing phase, and the MARS, RFR, and LLR models are used to learn the training samples and obtain the optimal network parameters during the process. Finally, the three models' implementations are showed the base functions and are compared using metrics from the accuracy measurement indicators as RMSE, MAE, MSE,

R, R^2 in the possible result stage, at the same time, looking for the most suitable prediction model for the study.

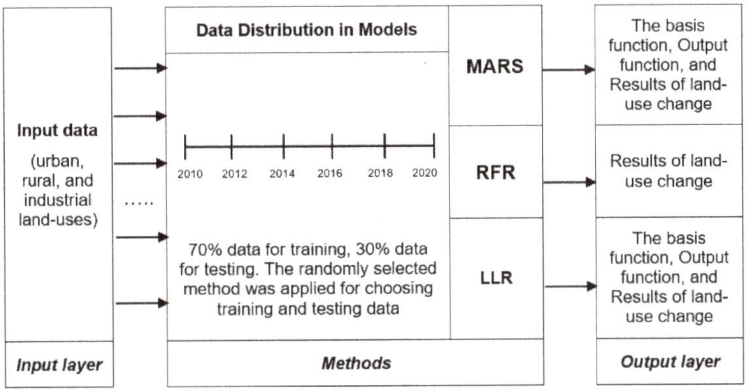

Figure 1. Diagram of the research steps used in this study.

2.1. Study Area

The Middle Central Coast region comprises five provinces and cities that serve as a link between the country's three primary socioeconomic hubs: the Red River Delta, the South Central Coast, and the Central Highlands (see Figure 2). The topography of this area is divided into two parts: the flat to gently rolling plains in the east and the most rugged forest-covered mountains in the western two-thirds. The delta area has a coastal-mountainous nature, and it is separated by mountain branches close to the sea as Hoanh Son mountain range—Ngang pass, Bach Ma mountain range—Hai Van pass [26,27]. Furthermore, this region plays a critical role in Vietnam's maritime economic development strategy in terms of tourism development, research, and technology, the Vietnamese seaport system, and vital logistics. When it comes to administrative matters, the North Central region is divided into six provinces with a total size of 2,948,430 hectares (9.8% of the country's total area) and a population of 5,366,500 people (5.70 percent of the total population); the rates of average population growth of the region is 1.1% [28]. These points show that the urbanization rate in this area is not high, and the population growth rate is only low compared to Hanoi and Ho Chi Minh City regions. Furthermore, there are two first-grade cities in the region (Da Nang and Hue), two second-grade cities (Tam Ky and Dong Hoi), two third-grade cities (Dong Ha and Hoi An), and eight fourth-grade cities in total, and 35 fifth-grade cities (see Appendix A Table A1) [29–33].

On the other hand, the region's average urbanization rate is 47.26 percent, with Da Nang having the highest rate at 87 percent and Quang Binh having the lowest at roughly 30 percent. Following that, agriculture, forestry, fisheries, and industry—construction, services, and product tax minus product subsidies accounted for 15%, 28%, 48%, and 9% of the region's GRDP in 2021, respectively. Per capita income based on GRDP is 2643 USD/person, with Da Nang having the most at 3822 USD/person and Quang Tri province having the lowest at 2087 USD/person. As this field expands, vocational training is designed to help people change occupations. Feeding a portion of the rural population and those who no longer have fertile land is a critical challenge for this region. By 2021, those with vocational training will make up approximately 66 percent of the working-age population. Their earnings will rise because they are well-educated, lowering the poverty rate to around 3.5 percent in 2021 [29–33]; therefore, vocational training demonstrates that urbanization stimulates and expands children's opportunities. People are more dynamic and innovative in their search for, and selection of, methods and forms of production, and organizations rise to become wealthy legally. The main trend and best aspect of urbanization is economic development, which improves employees' living standards. The rapid development of non-manufacturing industries is also aided by urbanization. Large cities also provide more work options, greater pay, better social services, and increased labor productivity. It is a driving force for economic transformation in both urban and rural areas, contributing to further economic development. At the same time, the metropolitan region serves as a big and diverse consumer of goods, a location to employ a skilled workforce and a hub for sophisticated technology and infrastructure facilities that draw significant domestic and foreign investment.

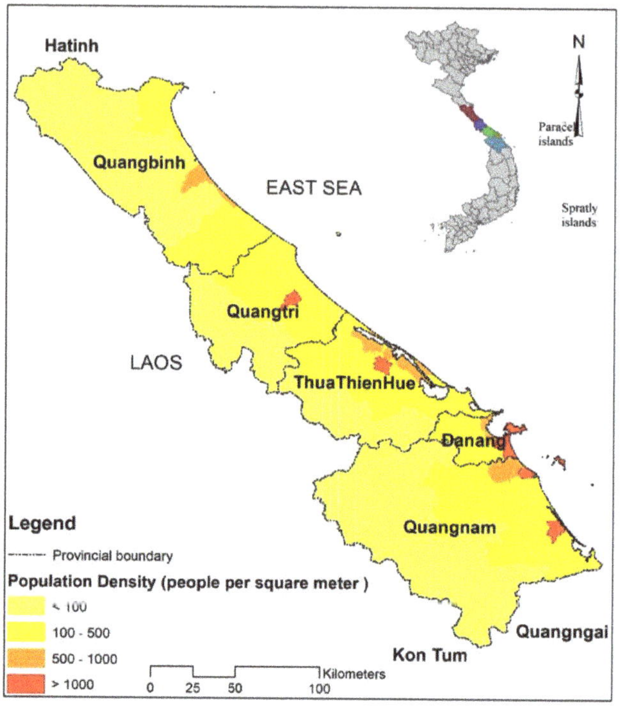

Figure 2. Location of provinces for study.

Urbanization in these provinces is established mainly in two ways, as follows. Firstly, the land-cover transforms rural areas into urban areas, then the villages and communes surrounding urban centers are gradually merged into urban areas. Alternatively, some rural areas have developed enough infrastructure, and the total population of the rural region meets the criterion of an urban population (higher than 50,000 people), that the place will be recognized as an urban area. Lastly, the developing industrial, commercial, and tourism zones promote neighboring suburb areas to develop as urban areas.

The total land-use of rural, urban, and industry areas within these provinces estimate about 71,067 ha in 2020 (see Table 1), in which Quang Binh, Quang Tri, Thua Thien-Hue, Da Nang, and Quang Nam occupy 9973 ha, 6341 ha, 14,510 ha, 11,834 ha, and 28,409 ha, respectively. In addition, Figure 3 indicates the square of land-use change in this study scope from 2010 to 2020. Figure 3a shows that the most change in urban land-use occurs in the Thua Thien Hue province, followed by Da Nang city, and the lowest variation is in the Quang Binh and Quang Tri provinces. Figure 3b indicates that the variation in rural land-use use in the Quang Nam province is the most, followed by the change in Thua Thien Hue and Quang Binh provinces, with the lowest variation occurring in the Da Nang and Quang Tri provinces. Finally, Figure 3c points out that the change in land-use of industrial zones in Quang Nam province is the highest, followed by Da Nang, then Thua Thien Hue, with the lowest being in the Quang Binh and Quang Tri provinces.

Table 1. The area of land-use categories by province in the study area in 2020 (Unit: ha).

Province	Rural Land-Use (ha)	Industrial Land-Use (ha)	Urban Land-Use (ha)	Sub-Total
Quang Binh	5632	3103	1238	9973
Quang Tri	3067	1740	1534	6341
Thua Thien-Hue	6420	4596	3494	14,510
Da Nang	2464	4694	4676	11,834
Quang Nam	17,024	6751	4634	28,409
Total	34,607	20,884	15,576	71,067

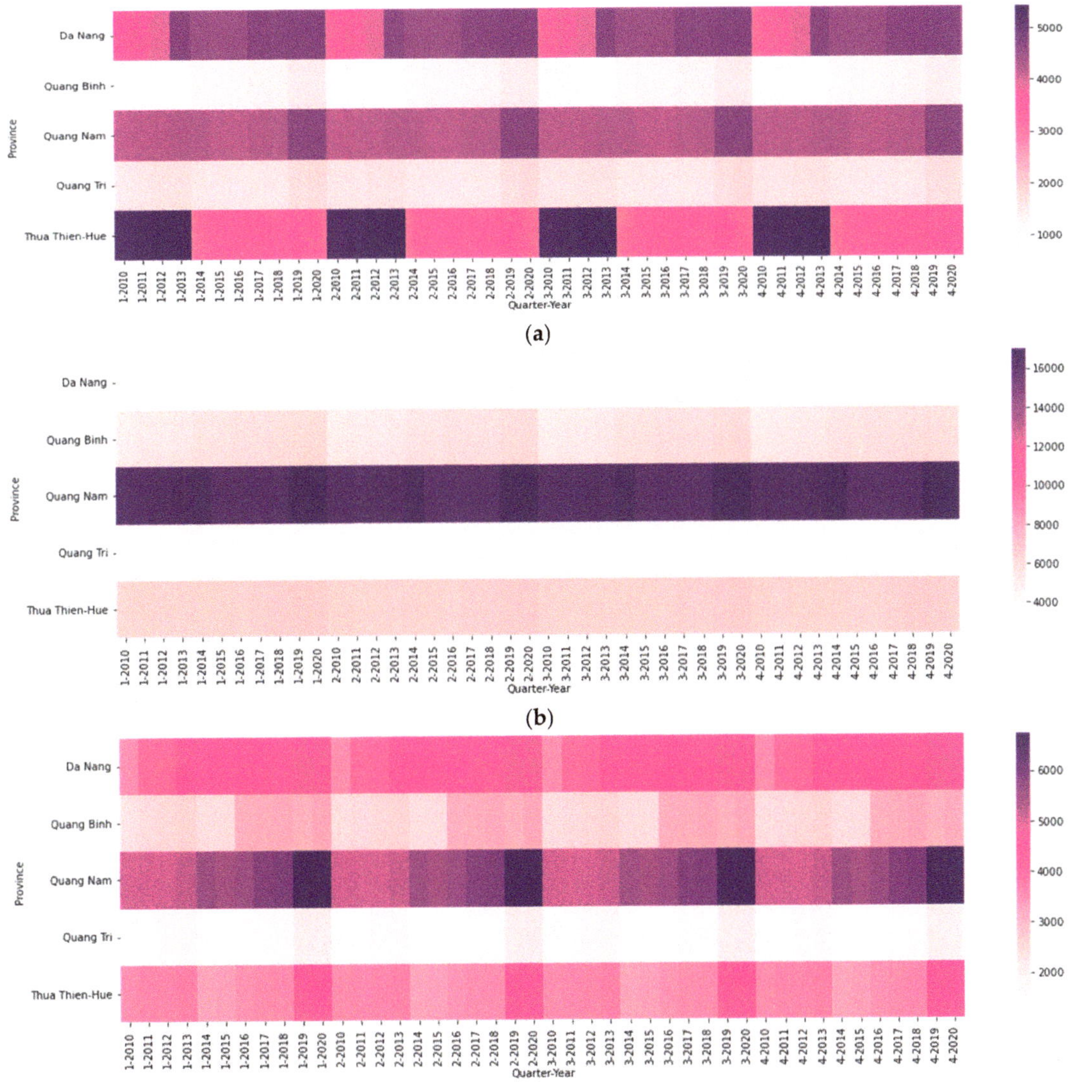

(c)

Figure 3. The area of (**a**) urban land-use (unit: ha), (**b**) rural land-use (unit: ha), and (**c**) industrial land-use (unit: ha) of five provinces from 2010 to 2020.

2.2. Database

This paper contains a database containing 44 quarters of land-use change from 2010 to 2020 (4 quarters for each year) for five provinces obtained from the Department of Natural Resources and Environment (DONRE). Three input variables include rural land-use, industrial land-use, and urban land-use, which were collected from five provinces. Furthermore, the characteristic statistical results for urban land-use in Table 2 demonstrate that the mean ranges from 878 ha in Quang Binh to 4317 ha in Da Nang, the standard deviation (St Dev) ranges from 84 ha in Quang Tri to 959 ha in Da Nang, and the minimum (Min) and maximum (Max) range from 608 ha and 1238 ha in Quang Binh to 4093 ha and 4634 ha in Quang Nam. The ranges of skewness (Skew) and kurtosis (Kurt) parameters of the five provinces fluctuate from 0.17 and −1.72 to 1.03 and 1.65. These Skew and Kurt indicators approach low values that are highly appropriate for modeling [34]. The input data patterns of five provinces were randomly selected in two parts. About 70% of the dataset was selected for the training sample, whereas 30% was used for the testing sample.

Table 2. Statistical urban land-use data from 2010 to 2020.

Province	St Dev (ha)	Mean (ha)	Min (ha)	Max (ha)	Skewness	Kurtosis
Quang Binh	214	878	608	1238	0.17	−1.17
Quang Tri	84	1369	1262	1534	0.64	−0.86
Thua Thien-Hue	959	4076	3272	5434	0.58	−1.72
Da Nang	403	4317	3514	4676	1.03	−0.67
Quang Nam	183	4219	4093	4634	0.87	1.65

2.3. Descriptions of Models

2.3.1. Multivariate Adaptive Regression Splines (MARS)

MARS is a nonparametric regression model, and it was introduced by Friedman [35]. MARS seems like a method for a fitted relationship between prediction and dependent variables. MARS is fast and based on a divide-and-conquer strategy, which divides the training dataset into distinct regions, each with its regression line [36–38]. The MARS algorithm feature is the procedure of the backward and forwards stepwise and may explain and control the complex nonlinear mapping between the inputs and output variables. This function predicts the new output y and the input variable x that uses either of the two base functions [39] and deploys a value or knot of variables that demonstrates the point of inflection along with the range of the inputs [40]. The general form of MARS forecasting is as below:

$$y = f_{(x)} = \beta_0 + \sum_{j=1}^{P} \alpha_j \beta_j \tag{1}$$

where y is the dependent variable predicted by the function $f_{(x)}$; β_0 is the constant value; P is the number of terms, each of them formed by a coefficient α_j, j $\in \{1, \ldots, P\}$; x_j is predictor variable; β_j is an individual base function. The base functions of $Max(0, x - H)$ and $Max(0, H - x)$ are univariate and do not have to each be present if their β coefficients are 0; the H values are called *"hinges"* or *"knots"*; x is an independent variable.

The function of backward stepwise relates to removing basis functions one at a time until the criterion of *"lack of fit"* is a minimum. In the deleting stage of backward stepwise, the last crucial important base functions are demolished one at a time. The lack of an applied fitting measurement is leaned on in the Generalized Cross-Validation (GCV) [41,42]:

$$GCV = A* \sum_{i=1}^{P} (y_i - \hat{f}(x))/N \tag{2}$$

where N is a number of data; $A = \left[1 - \frac{C(M)}{N}\right]^{-2}$ and $C(M) = (M+1) + dM$ are the complexity function [35]; d is a penalty for each basis function included in the model; M is the number of base functions in Equation (1). The criterion of GCV is examined for the average residual error multiplied by a penalty to modify the variability associated with more indicator prediction in the model [39,43].

2.3.2. Lasso Linear Regression (LLR)

The lasso linear regression method is widely used in domains with massive datasets, and it is also necessary to use when algorithms are efficient and quick [44]; however, the lasso is not vigorous in terms of determining the high correlation between predictors; it will randomly choose one and ignore the others, and split when all predictors are identical [44]. Moreover, the lasso penalty looks at many coefficients that are close to zero and only a small subset that is larger (and non-zero). The lasso estimator [45,46] uses the l_1 penalized least-squares criterion to get a sparse solution to the problem of optimization as below:

$$\hat{\beta}(Lasso) = \underset{\beta}{argmin} \; \|y - X\beta\|_2^2 + \gamma\|\beta\|_1 \tag{3}$$

where $\|\beta\|_1 = \sum_{j=1}^{p} |\beta_j|$ is the l_1- norm penalty on β, which is the cause of the sparse solution, and $\gamma \geq 0$ is a tuning parameter.

The l_1 penalty allows the lasso to simultaneously fit the smallest squares to and shrink some components of $\hat{\beta}(Lasso)$ to zero for a suitably chosen γ [44]. The cyclic coordinate reduction algorithm [44] efficiently computes the entire path of the Lasso solution paths for γ for the Lasso estimator and is faster than the Generalized Least Angle Regression (LARS) well-known algorithm. These properties make Lasso an attractive and popular method of variable selection.

2.3.3. Random Forest Regression (RFR)

Random forest is a regression technique that associates the performance of multiple decision tree algorithms to classify or forecast the value of a variable [47–49]. When an x input vector is received by random forest, in conjunction with the different evidential features values analyzed for a given training area, a number K of regression trees, on averages, the results are built by random forest. After K, such trees $\{T(x)\}_1^K$ are grown, and the random forest regression predictor is as follows:

$$\hat{f}_{rf}^{K}(x) = \frac{1}{K}\sum_{k=1}^{K} T(x) \tag{4}$$

To avoid the correlation of different trees, the random forest raises the tree's diversity by improving from different subsets of training data generated through a procedure called bagging [50]. Bagging is a technique used to generate training data by randomly resampling the original dataset with a replacement. As a result, some data may be used multiple times during training, whereas others may never be used. Thus, greater stability is achieved, as it makes it more robust in the face of small variations in the input data, and at the same time, it increases the accuracy of the prediction [47,51].

2.3.4. Performance Metrics

Predicting results is based on calculating and comparing the actual values to the forecasted values. These metrics of the accuracy measurement parameters include the Mean Square Error (MSE), Root Mean Square Error (RMSE), Mean Absolute Error (MAE),

Correlation Coefficient (R), and Correlation of Determination (R^2). Furthermore, the error metrics are defined as follows [52–54]:

$$\text{MSE} = \frac{\sum_{t=1}^{n}(x_t - x'_t)^2}{n} \tag{5}$$

$$\text{MAE} = \frac{\sum_{t=1}^{n}|x_t - x'_t|}{n} \tag{6}$$

$$\text{RMSE} = \sqrt{\frac{\sum_{t=1}^{n}(x_t - x'_t)^2}{n}} \tag{7}$$

$$R^2 = 1 - \frac{\sum_{t=1}^{n}(x_t - x'_t)^2}{\sum_{t=1}^{n}\left(x_t - \frac{1}{n}\sum_{t=1}^{n}x_t\right)^2} \tag{8}$$

$$R = \frac{\sum_{t=1}^{n}(x_t - \overline{x})(x'_t - \overline{x\prime})}{\sqrt{\sum_{t=1}^{n}(x_t - \overline{x})^2}\sqrt{\sum_{t=1}^{n}(x'_t - \overline{x\prime})^2}} \tag{9}$$

where x_t, x'_t are the observed and estimated values in the period time t, and n is the number of the observed values in the testing data. $\overline{x}$, $\overline{x'}$ are mean of the observed and estimated value. The R^2 and R (correlation coefficient) should be approaching 1 to indicate strong model performance, and the MSE, MAE, and RMSE should be as close to zero as possible.

3. Results Analysis

Regarding this platform, the output function of MARS and LLR of land-use in five provinces are presented as below:

Quang Binh (QB) land-use output function:

$\text{MARS}_{\text{Quang Binh}} = 990 + 1.27F_{1QB} - 0.55F_{2QB} - 0.18F_{3QB} + 0.65F_{4QB} - 0.52F_{5QB} + 0.27F_{6QB}$, where F_{1QB} = max(0, Rural-5424), F_{2QB} = max(0, 5424-Rural), F_{3QB} = max(0, Industry-2251), F_{4QB} = max(0, 2251-Industry), F_{5QB} = max(0, Industry-2995), and F_{6QB} = max(0, Industry-2366).

$$\text{LLR}_{\text{Quang Binh}} = -2563 + 0.7X_{1QB} - 0.1X_{2QB}.$$

Fi_{QB} (i = 1, 2, ... ,6) is the base function. F_{QB1} may be explained as the maximum value of 0 and Rural-5424. The minus sign ahead of the maximum value is equivalent to a minimum value. In addition, the $\text{MARS}_{\text{Quang Binh}}$ analysis indicates that the most important is in bellowing order rural land-use and industrial land-use.

Quang Tri (QT) land-use output function:

$\text{MARS}_{\text{Quang Tri}} = 1435 + 0.94F_{1QT} + 0.45F_{2QT} + 1.57F_{3QT} - 2.45F_{4QT} - 0.58F_{5QT}$, where F_{1QT} = h(Industry-1285), F_{2QT} = max(0, 1285-Industry), F_{3QT} = max(0, 2974-Rural), F_{4QT} = max(0, 3047-Rural), and F_{5QT} = max(0, Industry-1185).

$$\text{LLR}_{\text{Quang Tri}} = -308 + 0.42X_{1QT} + 0.3X_{2QT}.$$

The $\text{MARS}_{\text{Quang Tri}}$ analysis indicates that the most important is in bellowing order industrial land-use and rural land-use.

Thua Thien-Hue (TTH) land-use output function:

$\text{MARS}_{\text{Thua Thien-Hue}} = 1718 + 14.97F_{1TTH} + 12.88F_{2TTH} - 2.91F_{3TTH} - 12.79F_{4TTH}$, where F_{1TTH} = max(0, 6277-Rural), F_{2TTH} = max(0, Industry-3428), F_{3TTH} = max(0, 3428-Industry), F_{4TTH} = max(0, Industry-3558).

$$\text{LLR}_{\text{Thua Thien-Hue}} = 105{,}973 - 18.43X_{1TTH} + 3.40X_{2TTH}.$$

The $\text{MARS}_{\text{Thua Thien-Hue}}$ analysis indicates that the most important is in bellowing order rural land-use and industrial land-use.

Da Nang (DN) land-use output function:

$MARS_{Da\ Nang} = -3086 + 1.18F_{1DN} - 1.11F_{2DN} + 1.72F_{3DN}$, where F_{1DN} = max (0, Industry-4000), F_{2DN} = max(0, Industry-4408), F_{3DN} =max(0, Industry).

$$LLR_{Da\ Nang} = -1280 + 0.34X_{1DN} + 1.08X_{2DN}.$$

The $MARS_{Da\ Nang}$ analysis indicates that the most important is in bellowing order industrial land-use and rural land-use.

Quang Nam (QN) land-use output function:

$MARS_{Quang\ Nam} = 4215 - 0.5F_{1QN} - 0.11F_{2QN} - 0.17F_{3QN} + 0.06F_{4QN}$, where F_{1QN} = h(Industry-5922), F_{2QN} = max(0, 5922-Industry), F_{3QN} = max(0, 16,532-Rural), and F_{4QN} = max(0, 16,532-Rural).

$$LLR_{Quang\ Nam} = -1651 + 0.29X_{1QN} + 0.19X_{2QN}.$$

The $MARS_{Quang\ Nam}$ analysis indicates that the most important is in bellowing order industrial land-use and rural land-use.

Furthermore, the output function for RFR does not occur.

The data in Figures 4a–c, 5a–c, 6a–c, 7a–c and 8a–c present the relationship between the three types of land-use, and the area of industrial and rural land-use increases as the square of urban land-use increases. LLR makes a forecasting form that resembles a flat surface of paper. Additionally, the RFR and MARS charts are the same as the image of papers with some folds, and the folds enable a better fit to the data. In addition, Figures 4a–c, 5a–c and 8a–c demonstrate that the area of land usage in the Quang Nam province increased significantly from 2010 to 2020. Moreover, the MARS predicted algorithm shows that the red dots are evenly distributed on the surface, and the LLR and RFR forecasting algorithms demonstrate that the red dots are relatively far from the surfaces. The data in Figure 6a–c imply that the area used for the three categories of land experienced an upward tendency between 2010 and 2013; however, the urban land use area decreased dramatically and tended to be saturated, whereas the rural land use area and industrial zones grew steadily per annum from 2014 to 2020. The output in Figure 7a–c illustrates that the urban and industrial land use increased significantly from 2010 to 2020, whereas rural land use grew slowly and even reached its saturation in 2019. Additionally, the RFR forecasting model of Figures 6a and 7a shows that these red dots are moderately distributed closer to the surface compared with the two algorithms in Figures 6b,c and 7b,c; however, it is difficult to investigate the difference between the three models. Hence, this study performed an accurate metric to explore the potential models for each province. From comparing the three models, Table 3 shows that the MARS model supplies a better fit than the other models for Quang Binh, where R^2 value = 0.91 was the largest along with MSE and MAE, whereas RMSE obtained the lowest value out of the established models. According to the implementation of the other models, the hierarchical carrying out considers the order of MARS > RFR > LLR. The simultaneous determination of urban land-use change prediction of Quang Nam and Quang Tri also proves the forecasting skills of these models in urban land-use change prediction in these provinces. Using the experiment results from the three provinces, the MARS model indicated to supply the best forecasting accuracy, the hierarchical order of the models, and other models for three provinces are MARS > RFR > LLR in Quang Tri and Quang Nam. Moreover, the prediction result of urban land-use change prediction in Thua Thien-Hue and Da Nang presented in Table 3 denotes that the order of hierarchical models with performance accuracy is RFR > MARS > LLR models. Regarding the GVC parameter for MARS, it generates an equilibrium between flexibility and generalization capability of the MARS model function [55]. The data in Table 3 also indicates that the order of hierarchical models with the accuracy of the GVC value of the MARS model in five provinces are $GVC_{Thua\ Thien\text{-}Hue} > GVC_{Da\ Nang} > GVC_{Quang\ Tri} > GVC_{Quang\ Binh} > GVC_{Quang\ Nam}$. Furthermore, the scatter charts in Figure 9a,b,e proved that the line of MARS models (red plus lines) of urban land used change estimations fit better than the line of RFR models (green triangle lines) and the line of LLR models (purple multiply lines) for the Quang Binh, Quang Tri, and Quang Nam provinces. In contrast, Figure 9c,d showed that the line of RFR models of urban land used change prediction in Thua Thien-Hue which showed that Da Nang has the best fit, followed by the line of MARS models and line of LLR models, respectively. These points may explain that the distribution of land-use change data with random selection for training and testing data is suitable for the MARS model of Quang Binh, Quang Tri, and Quang Nam, and is comfortable for the RFR model in Thua Thien-Hue and Da Nang.

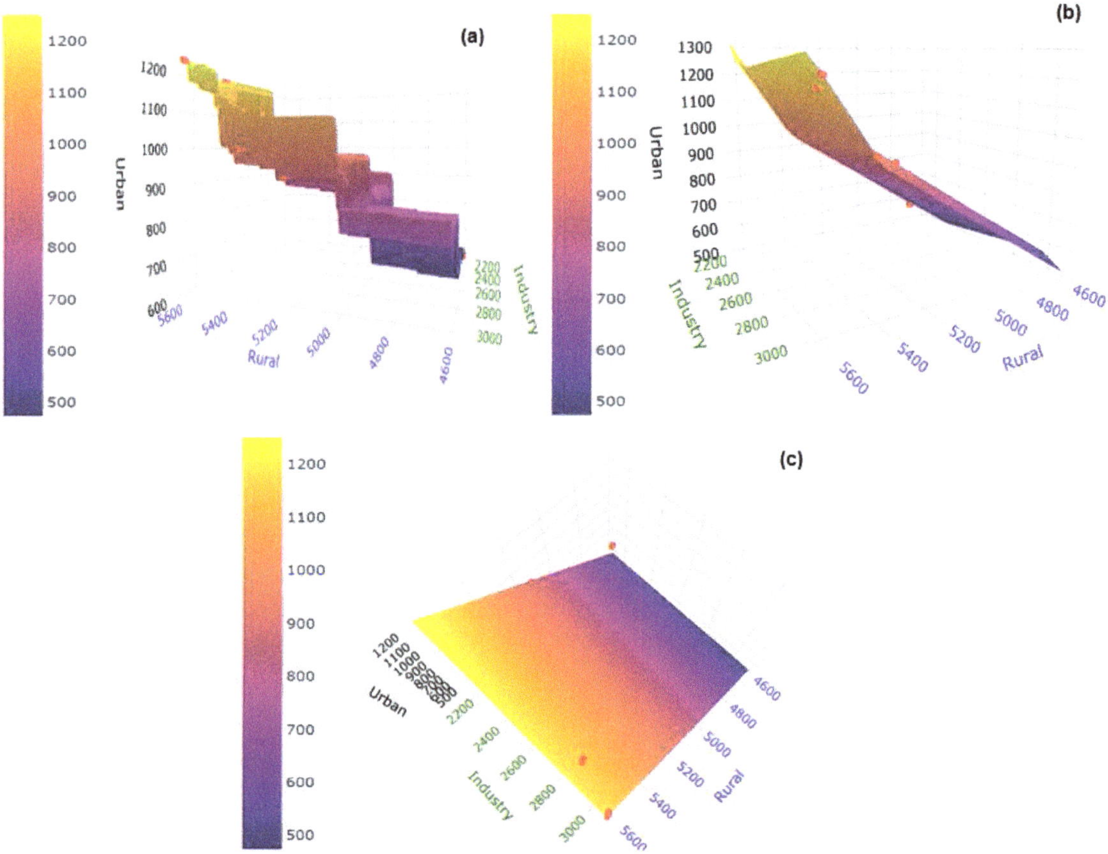

Figure 4. Land-use prediction with (**a**) RFR model, (**b**) MARS model, and (**c**) LLR model in Quang Binh (unit: ha).

Table 3. Accuracy parameters for land-use prediction.

	Quang Binh			Quang Tri			Thua Thien-Hue			Da Nang			Quang Nam			Average		
	LLR	*RFR*	*MARS*	*LLR*	*RFR*	*MARS*	*LLR*	*RFR*	*MARS*	*LLR*	*RFR*	*MARS*	*LLR*	*RFR*	*MARS*	*LLR*	*RFR*	*MARS*
MSE	143	10	5	107	10	4	535	151	162	98	58	84	58	10	8	188.2	47.8	52.6
MAE	27	4	5	33	6	6	442	70	162	78	79	75	39	8	6	123.8	33.4	50.8
RMSE	38	11	6	39	9	9	535	150	131	98	98	84	59	10	8	153.8	55.6	47.6
R	0.91	0.91	0.91	0.82	0.91	0.91	0.67	0.92	0.91	0.89	0.87	0.89	0.86	0.91	0.92	0.83	0.904	0.908
R^2	0.92	0.94	0.94	0.66	0.93	0.94	0.56	0.92	0.92	0.9	0.9	0.91	0.84	0.94	0.94	0.776	0.926	0.93
GCV			88			193			66,746			7522			77			

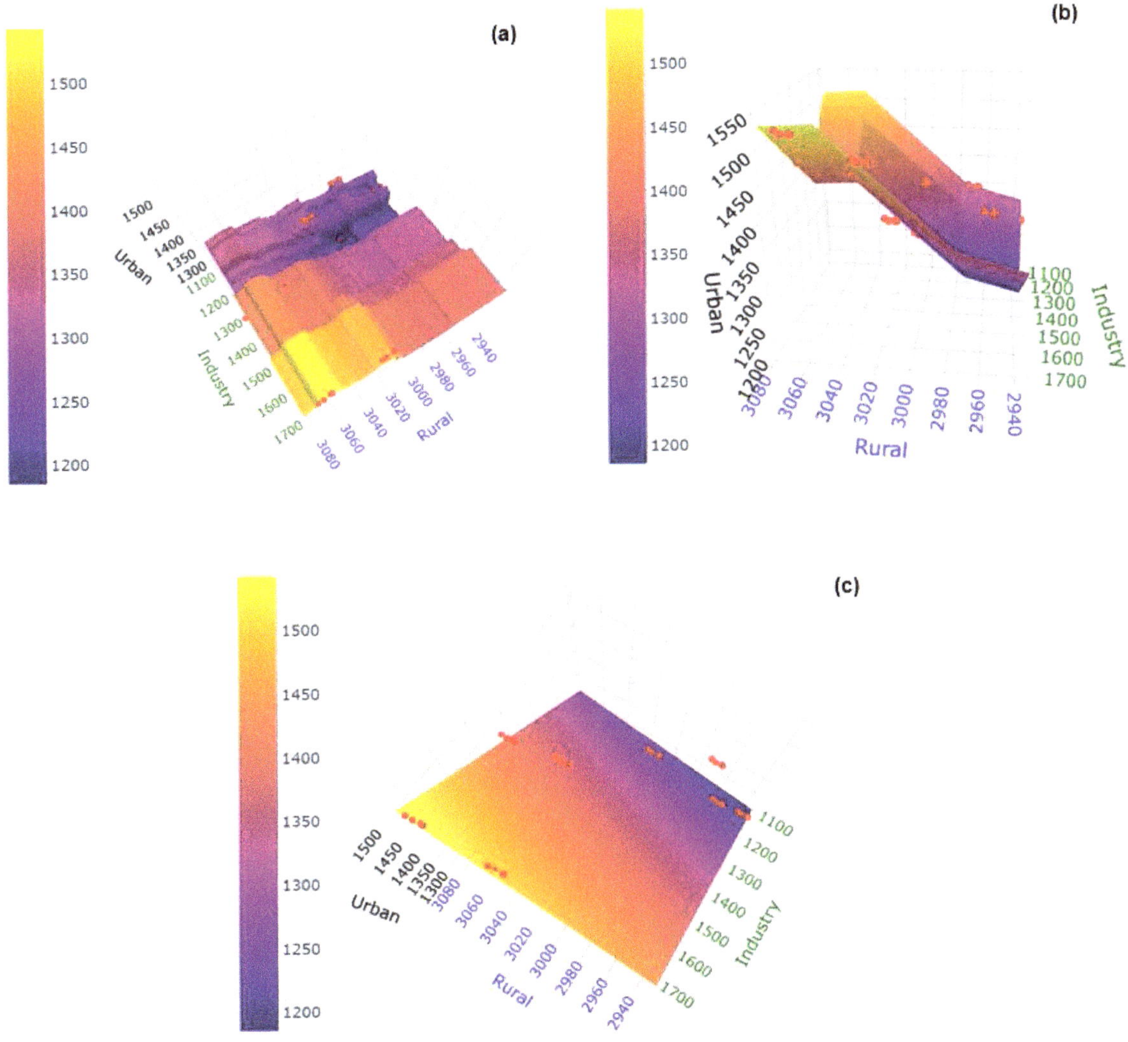

Figure 5. Land-use prediction with (**a**) RFR model, (**b**) MARS model, and (**c**) LLR model in Quang Tri (unit: ha).

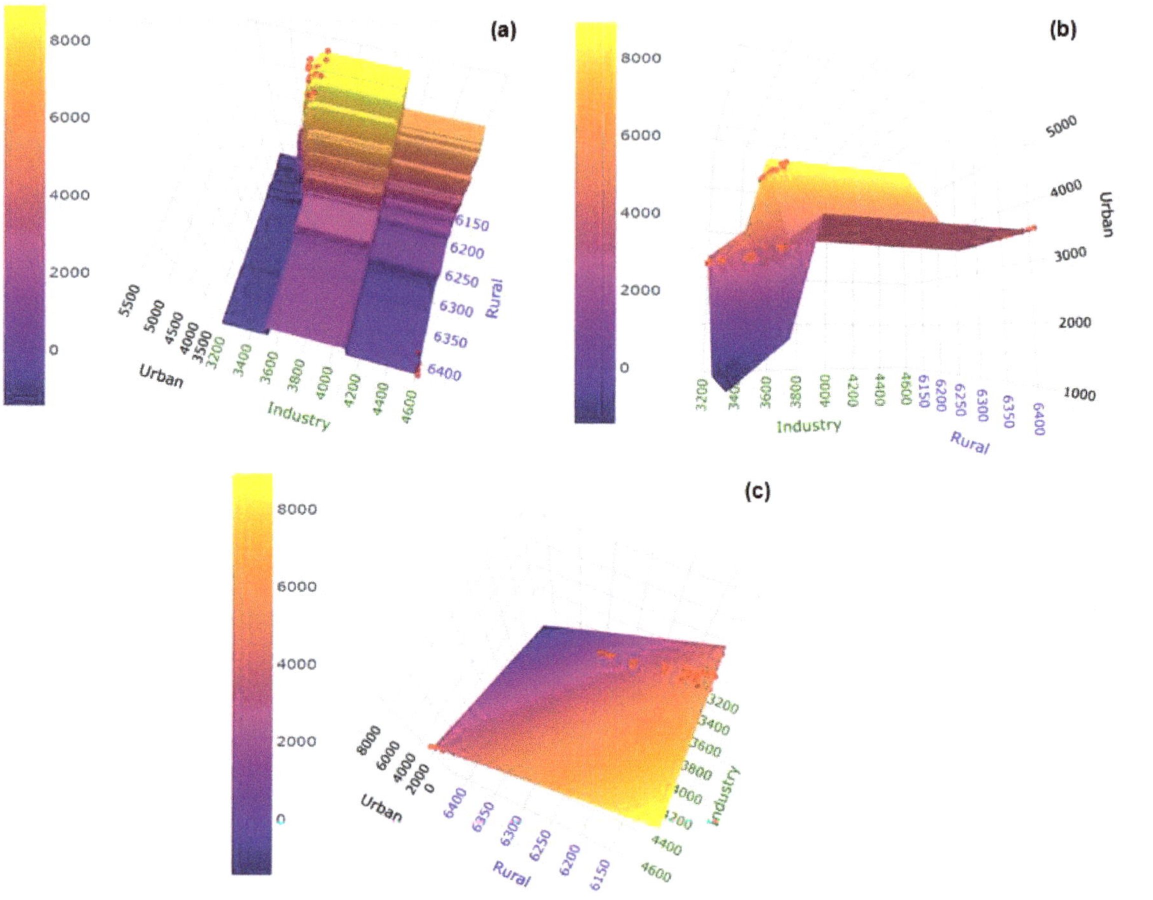

Figure 6. Land-use prediction with (**a**) RFR model, (**b**) MARS model, and (**c**) LLR model in Thua Thien-Hue (unit: ha).

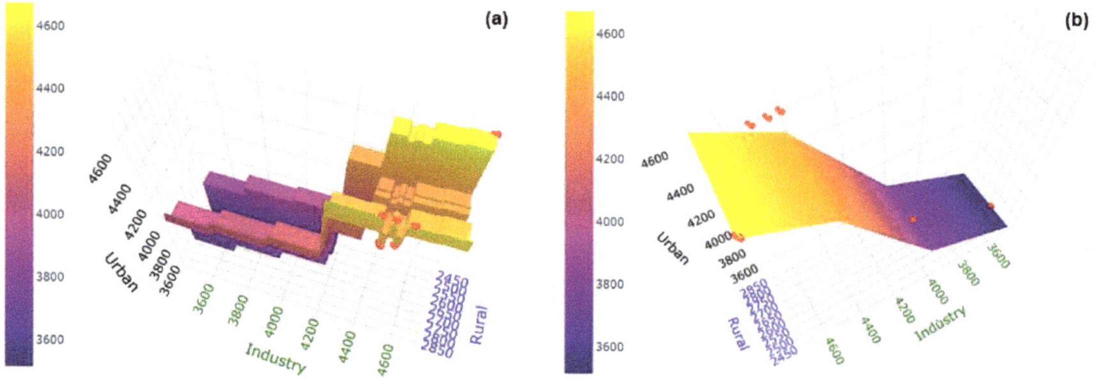

Figure 7. *Cont.*

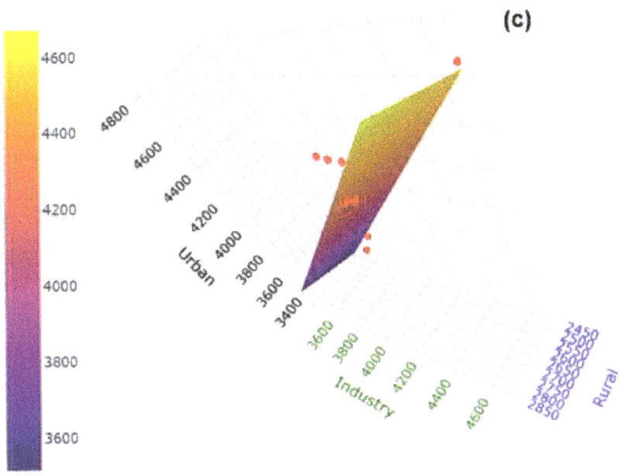

Figure 7. Land-use prediction with (**a**) RFR model, (**b**) MARS model, and (**c**) LLR model in Da Nang (unit: ha).

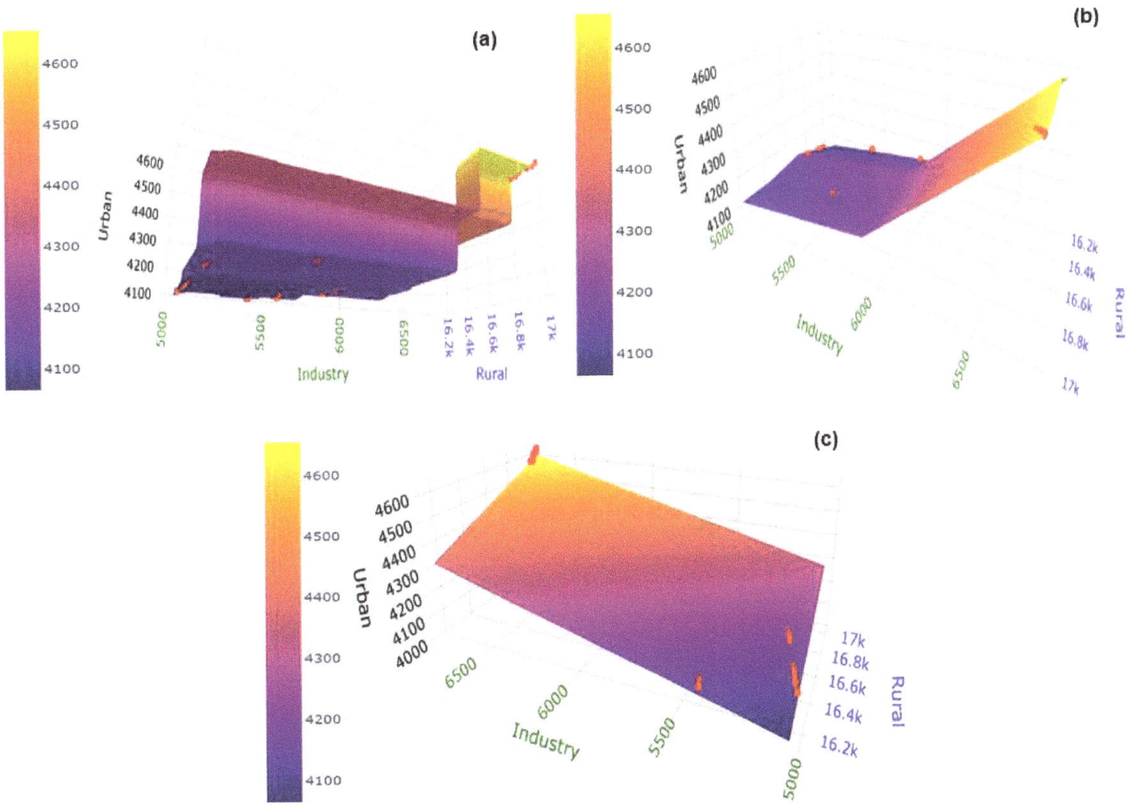

Figure 8. Land-use prediction with (**a**) RFR model, (**b**) MARS model, and (**c**) LLR model in Quang Nam (unit: ha).

Furthermore, the predicting performance of the model models is also visualized and examined by the Taylor diagram. The diagram summarizes the St Dev and correlation coefficient (CC) that is comprised concomitantly in assessing the respective model [56,57]. The St Dev, and CC between the observed and predicted datasets for all the land-use models of the provinces are described in the Taylor diagram. Figure 10a–e may be observed for LLR, RFR, and MARS in Quang Binh (CC_{LLR} = 0.91, CC_{RFR} = 0.91, CC_{MARS} = 0.91), in Quang Tri (CC_{LLR} = 0.82, CC_{RFR} = 0.91, CC_{MARS} = 0.91), in Thua Thien-Hue (CC_{LLR} = 0.67, CC_{RFR} = 0.92, CC_{MARS} = 0.91), in Da Nang (CC_{LLR} = 0.89, CC_{RFR} = 0.87, CC_{MARS} = 0.89), and in Quang Nam (CC_{LLR} = 0.86, CC_{RFR} = 0.91, CC_{MARS} = 0.92). The Taylor diagram demonstrates that these models were optimal accuracies of almost all models' outcomes and were significantly closer to 1. Moreover, the LLR model of land-use in Thua Thien-Hue with CC_{LLR} = 0.67 indicates that the level of accuracy achieved is only above medium.

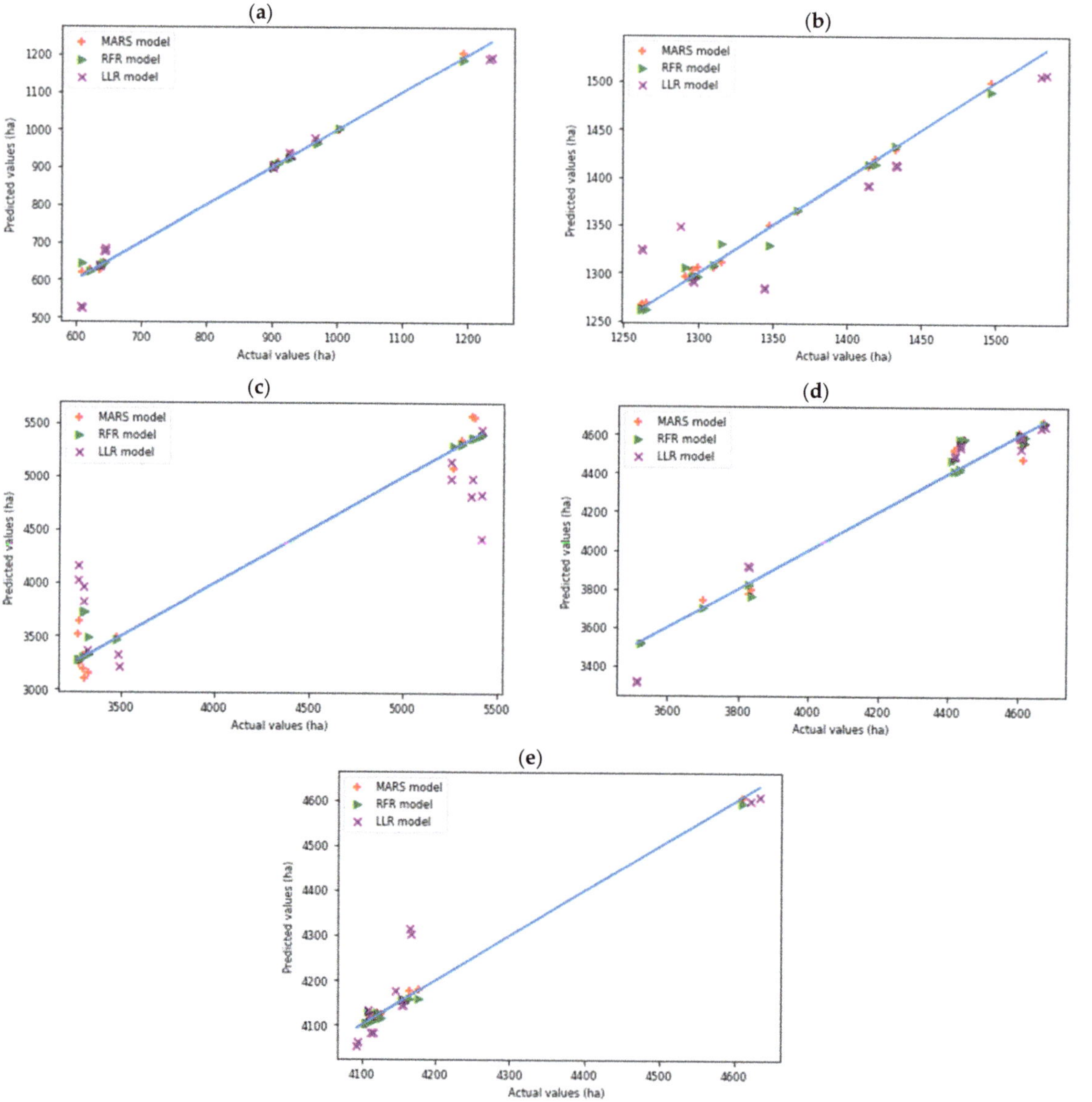

Figure 9. The best performance models for urban land use change prediction; (**a**) QuangBinh province, (**b**) QuangTri province, (**c**) ThuaThienHue province, (**d**) Da Nang City, (**e**) QuangNam province.

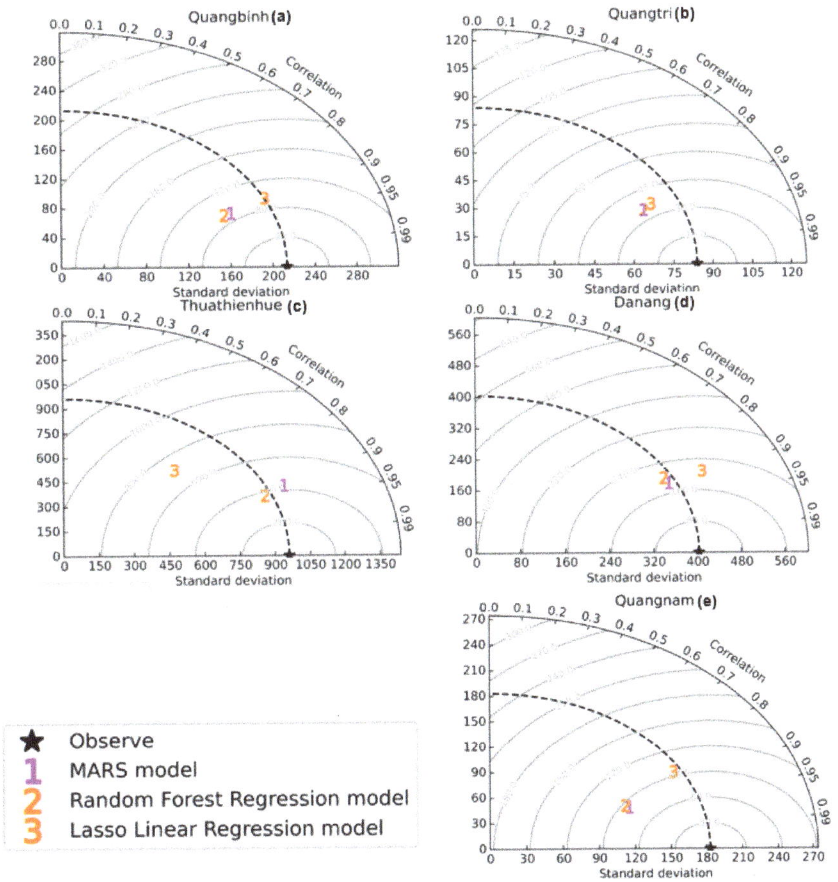

Figure 10. Taylor diagram representing the best performance of MARS, RFR, LLR models at (**a**) Quang Binh, (**b**) Quang Tri, (**c**) Thua Thien-Hue, (**d**) Da Nang, (**e**) Quang Nam.

4. Discussion

Urban areas in the five central coastal provinces are organized evenly along the coast, primarily on urban beam space. Rural areas still account for a much larger proportion of land than urban areas, and approximately 15% to 20% of the land use belongs to the urban administrative boundary, and the population accounts for over 60%. The urbanization process creates a sharp change in land-use in peri-urban and rural areas. The conversion of a large part of agricultural land to land for the construction of industrial, service, and urban residential areas. The process of expanding urban space along with the appearance of housing projects, real estate, concentrated industrial parks, large-scale commercial service works in the peri-urban communes has caused a sharp decline in production land funds, natural land, and spatial change of the rural ecological landscape, which causes the mechanical population of peri-urban communes to increase, despite insufficient infrastructure, leading to rapid overcrowding. Moreover, it also has an impact upon technical infrastructure, social infrastructure, especially traffic, education, water supply, and drainage, environmental sanitation; however, urbanization has lagged behind the growth of industrial zones, and industrial development lacks a vision for future urbanization. In industrial zones, a large number of workers leaving the agricultural production area tend to move to the industrial production area, forming areas with high population density, centralization, creating demand for services such as food, accommodation, living, studying, and commuting purposes which are the premise for the initial formation of an industrial residential area, an industrial town, and in the future, it will become an industrial city. Hence, the following are the primary types of the relationship between the growth of industrial parks and the process of urbanization. Firstly, many industrial parks are located in rural areas but not in urban areas. Secondly, several industrial parks were formerly located in rural areas, but now it is still within urban areas' boundaries. Following that, there are many industrial parks in rural areas, which are now within the proposed expansion boundaries of the neighboring urban master plan. Consequently, the above situation shows that assessing the influence of rural land-use and industrial zone land-use is significantly vital for urban land-use and urbanization in this area.

This study implemented three machine learning models for land-use change to assess the speed of urbanization taking place in the Central Coast Region in Vietnam. Three models gave high accurate results for predicting urban land-use fluctuations, in which the MARS and RFR models showed more accuracy for Quang Binh, Quang Tri, Quang Nam, and Thua Thien-Hue, Da Nang, respectively, compared with the LLR model. In addition, the estimated values of types of land-use changes made by the LLR model also provided acceptable results. The data of this study was based on the statistics of land-use types that have been measured quarterly; therefore, these estimation values supplied the total types of land-use change that have been urbanized based on the process of forming urban areas in industrial and rural areas.

To evaluate the predicted accuracy of these study models, spatial models are needed to estimate accuracy parameter values. Comparing RMSE and MAE using the MARS model of this study result with the study result of Yilmaz et al. (2018) [21] about suspended sediment load, their RMSE = 3592, and MAE = 3483 are found to be greater than the values in this study with $RMSE_{Average} = 47.9$, $MAE_{Average} = 50.8$. Jamali (2019) [58] deployed RFR to predict land-use/land-cover mapping using Landsat 8 OLI in the northern region of Iran. The RMSE and MAE for the model are 5 and 5, respectively; these points are lower than this result study. Finally, the result of the study Adab et al. (2020) [59] concerns Estimate Surface Soil Moisture in the semi-arid region of west Khorasan-Razavi province of Iran, and it shows that RMSE and MAE of LRR model (at 7 March 2017) are 6.67 and 5.55, respectively. These points also indicate that the prediction model is lower than this study model. Duong et al. (2018) [60] deployed the kernel density estimation and remotely sensed data from multiple sensors to generate the land cover maps over Central Vietnam during the period of 2007 to 2017. The result indicated that the overall accuracies of the maps for 2007 and 2017 are 90.5% (kappa coefficient of 90%) and 90.6% (kappa coefficient of 90%), respectively, in which the urban prediction was approximately 91%. This point also proved that using machine learning to show the results of this study consider equivalent to the remote sensing method for estimating the land use/ land cover for the Center of Vietnam; however, the study deploys machine learning models and statistical algorithms that majorly focus on land use transition/change. Due to sufficient published literature relating to other aspects of land use planning such as zoning, land allocation, and land restrictions, land-use mapping was not mentioned in this study; therefore, combining multiple methods for land-use information would be useful in future research.

Although classification accuracies for land-use were not particularly large, estimating urban, rural, and industrial land-use change is still useful for five central coastal provinces of Vietnam. This study result will assist the provinces' authorities and other stakeholders in decision-making and planning regarding three kinds of land-use. The usual practice is for the Ministry of Natural Resources and Environment (MONRE) to carry out urban inventory and set up urban land-use change maps every five years. Then, the DONRE provinces obtain the predictive data and update them manually. In addition, many jobs are created as a result of the development of the service, commerce, and manufacturing industries in cities. At the same time, a lot of individuals lose arable land due to urbanization to make way for industrial parks, handicrafts, or concentrated craft villages. Moreover, urbanization will affect policymakers regarding labor reorganization, changing production methods, and enhancing human resource training solutions to adapt to new employment standards. Moreover, new industries and services drive economic growth. Furthermore, sustainable urbanization development is a concern, and several criteria need to be proposed as below. Firstly, harmonious development of the economy, society, environmental protection, and ecological balance is required. Secondly, the municipality must ensure that the amount of space available for activities, the infrastructure engineering system, and social infrastructure are all up to par with high-quality standards. Thirdly, cities must have a well-organized population distribution system to close the gap between urban, rural, and industrial zones. Fourthly, urban development must balance the ecology in the inner city and suburbs. Finally, the authorities have to enforce appropriate policies related to population, land use, technical infrastructure development, environmental protection, and preservation of natural and social ecosystems.

Therefore, sustainable urban development for urban provinces can be suggested as follows:

1. Da Nang city, the most developed urban area in the region, has industrial parks equivalent to the urban land-use area. As a result, it is critical to relocate industrial zones in the ancient city, rationalize land use functions, employ high-tech equipment, and create a green environment. More importantly, to accommodate the influx of migrants from all over the country into the city's working streets, local authorities must plan to build land funding and infrastructure in industrial zones or rural areas near industrial zones, lowering stress in Da Nang's central city.
2. The Quang Nam and Thua Thien Hue provinces, two provinces with many tangible cultural heritage sites such as Hue City, Hoi An Ancient Town, and My Son Holyland, need to build satellite urban areas to relieve the pressure on infrastructure and population for urban heritage areas. In addition, because the land fund for rural use is enormous, a strategy for converting agricultural land to industrial and commercial zones in rural areas is required to support rural growth and urban areas while also creating jobs for rural residents.
3. Quang Binh and Quang Tri are two provinces with slower urban and industrial zone development than the Da Nang, Quang Nam, and Thua Thien-Hue provinces; however, plenty of rural land use and industrial land use funds are being used in these two provinces. As a result, these two provinces will need to construct satellite cities based on highly populated areas near industrial parks. In addition, it is necessary to form sub-regional centers in the district in the direction of commodity production with high technology.

5. Conclusions

Urbanization is an inevitable process for the economic and social development of the five central coastal provinces; therefore, this study has shown the role of rural land-use and industrial zone land-use in informing and expanding urban areas. However, this development has not been synchronized and has not yet ensured the infrastructure of an urban area; hence, this study used the MARS, RFR, and LLR models to estimate urban land-use change based on rural and industrial land-use from the five provinces. In addition, the projected and observed values were compared using five widely used statistical parameters (i.e., RMSE, MAE, MSE, R, and R^2). The results of the study of the models also show that the MARS model improves the accuracy of performance more than RFR, LLR in the Quang Binh, Quang Tri, Quang Nam provinces, and the RFR model gives a more accurate forecasting implementation than the MARS, LLR models in the Da Nang and Thua Thien-Hue provinces. The accuracy of the models may depend on the distribution of land-use change data with random selection for training and testing data. The prediction of land-use change may support the authorities' land-use planning and decision-making. Furthermore, the research also suggested sustainable urban development for each specific province, and the region in general. The future of the current work consists of using a hybrid of MARS and LLR in modeling land-use change and other studies to enhance the model estimating capability, or these methods may combine with the spatial pattern models to estimate land-use change.

Author Contributions: Conceptualization, N.H.G.; data curation, N.H.N.; formal analysis, Y.-R.W.; funding acquisition, N.H.G.; investigation, Y.-R.W.; methodology, N.H.G.; project administration, Y.-R.W.; software, N.H.G.; validation, Y.-R.W.; writing—original draft, N.H.G.; writing—review and editing, T.D.H. and T.-T.D. All authors have read and agreed to the published version of the manuscript.

Funding: This research received no external funding.

Institutional Review Board Statement: Not applicable.

Informed Consent Statement: Not applicable.

Data Availability Statement: Not applicable.

Acknowledgments: The authors appreciate the support from the National Kaohsiung University of Science and Technology, Taiwan; and Thu Dau Mot University, Vietnam.

Conflicts of Interest: The authors declare no conflict of interest.

Appendix A

Table A1. Urban classification of Vietnam.

Criteria/Indicators	Type I	Type II	Type III	Type IV	Type V
Population	(a) >1 million: Central government-run city (b) 500,000: Provincial city	(c) 300,000 to 1 million: If class 2 is central government-run city, poulation should be more than 800,000	(d) 100,000 to 350,000	(e) 500,000 to 350,000	(f) >4000
Nonagricultural labor	85%	80%	70%	70%	>65%
Popolation density	(a) 12,000/km^2 (b) 10,000/km^2	8000 /km^2 or 10,000 /km^2 if the city is directly uncer central government control	6000 km^2	4000 km^2	2000 km^2

References

1. Pijanowski, B.C.; Brown, D.; Shellito, B.A.; Manik, G.A. Using neural networks and GIS to forecast land use changes: A Land Transformation Model. *Comput. Environ. Urban Syst.* **2002**, *26*, 553–575. [CrossRef]
2. Sang, L.; Zhang, C.; Yang, J.; Zhu, D.; Yun, W. Simulation of land use spatial pattern of towns and villages based on CA–Markov model. *Math. Comput. Model.* **2011**, *54*, 938–943. [CrossRef]
3. Wang, S.W.; Gebru, B.M.; Lamchin, M.; Kayastha, R.B.; Lee, W.-K. Land Use and Land Cover Change Detection and Prediction in the Kathmandu District of Nepal Using Remote Sensing and GIS. *Sustainability* **2020**, *12*, 3925. [CrossRef]
4. Liping, C.; Yujun, S.; Saeed, S. Monitoring and predicting land use and land cover changes using remote sensing and GIS techniques—A case study of a hilly area, Jiangle, China. *PLoS ONE* **2018**, *13*, e0200493. [CrossRef]
5. Saputra, M.H.; Lee, H.S. Prediction of Land Use and Land Cover Changes for North Sumatra, Indonesia, Using an Artificial-Neural-Network-Based Cellular Automaton. *Sustainability* **2019**, *11*, 3024. [CrossRef]
6. Veldkamp, A.; Lambin, E.F. Predicting land-use change. *Agric. Ecosyst. Environ.* **2001**, *85*, 1–6. [CrossRef]
7. Viney, N.R.; Bormann, H.; Breuer, L.; Bronstert, A.; Croke, B.; Frede, H.; Gräff, T.; Hubrechts, L.; Huisman, J.A.; Jakeman, A.; et al. Assessing the impact of land use change on hydrology by ensemble modelling (LUCHEM) II: Ensemble combinations and predictions. *Adv. Water Resour.* **2009**, *32*, 147–158. [CrossRef]
8. Su, L.; Zhu, J.H.; Wang, W.; Liu, M. Application of ARMA Model in Prediction of Land Use Demand Take Farmland in Jin-Hu Coastal Area as Example. *Hunan Agric. Sci.* **2012**, *5*, 61–63.
9. Zhang, P.; Ke, Y.; Zhang, Z.; Wang, M.; Li, P.; Zhang, S. Urban Land Use and Land Cover Classification Using Novel Deep Learning Models Based on High Spatial Resolution Satellite Imagery. *Sensors* **2018**, *18*, 3717. [CrossRef]
10. Luus, F.P.S.; Salmon, B.P.; Bergh, F.V.D.; Maharaj, B.T. Multiview Deep Learning for Land-Use Classification. *IEEE Geosci. Remote Sens. Lett.* **2015**, *12*, 2448–2452. [CrossRef]

11. Mu, L.; Wang, L.; Wang, Y.; Chen, X.; Han, W. Urban Land Use and Land Cover Change Prediction via Self-Adaptive Cellular Based Deep Learning With Multisourced Data. *IEEE J. Sel. Top. Appl. Earth Obs. Remote Sens.* **2019**, *12*, 5233–5247. [CrossRef]
12. Azad, A.; Wang, X. Land Use Change Ontology and Traffic Prediction through Recurrent Neural Networks: A Case Study in Calgary, Canada. *ISPRS Int. J. Geo-Inf.* **2021**, *10*, 358. [CrossRef]
13. Kumar, S.; Radhakrishnan, N.; Mathew, S. Land use change modelling using a Markov model and remote sensing. *Geomat. Nat. Hazards Risk* **2013**, *5*, 145–156. [CrossRef]
14. Lu, Y.; Wu, P.; Ma, X.; Li, X. Detection and prediction of land use/land cover change using spatiotemporal data fusion and the Cellular Automata–Markov model. *Environ. Monit. Assess.* **2019**, *191*, 68. [CrossRef] [PubMed]
15. Lu, C.-J.; Lee, T.-S.; Lian, C.-M. Sales forecasting for computer wholesalers: A comparison of multivariate adaptive regression splines and artificial neural networks. *Decis. Support Syst.* **2012**, *54*, 584–596. [CrossRef]
16. Genuer, R.; Poggi, J.-M.; Tuleau-Malot, C.; Villa-Vialaneix, N. Random Forests for Big Data. *Big Data Res.* **2017**, *9*, 28–46. [CrossRef]
17. Antoniadis, A.; Lambert-Lacroix, S.; Poggi, J.-M. Random forests for global sensitivity analysis: A selective review. *Reliab. Eng. Syst. Saf.* **2020**, *206*, 107312. [CrossRef]
18. Van de Geer, S.A. High-dimensional generalized linear models and the Lasso. *Ann. Stat.* **2008**, *36*, 614–645. [CrossRef]
19. Wang, H.; Leng, C. Unified LASSO Estimation by Least Squares Approximation. *J. Am. Stat. Assoc.* **2007**, *102*, 1039–1048. [CrossRef]
20. Dyar, M.; Carmosino, M.; Breves, E.; Ozanne, M.; Clegg, S.; Wiens, R. Comparison of partial least squares and lasso regression techniques as applied to laser-induced breakdown spectroscopy of geological samples. *Spectrochim. Acta Part B At. Spectrosc.* **2012**, *70*, 51–67. [CrossRef]
21. Yilmaz, B.; Aras, E.; Nacar, S.; Kankal, M. Estimating suspended sediment load with multivariate adaptive regression spline, teaching-learning based optimization, and artificial bee colony models. *Sci. Total Environ.* **2018**, *639*, 826–840. [CrossRef] [PubMed]
22. Bui, D.T.; Hoang, N.D.; Samui, P. Spatial pattern analysis and prediction of forest fire using new machine learning approach of Multivariate Adaptive Regression Splines and Differential Flower Pollination optimization: A case study at Lao Cai province (MARS). *J. Environ. Manag.* **2019**, *237*, 476–487.
23. Nguyen, H.T.T.; Doan, T.M.; Radeloff, V. Applying random forest classification to map land use/land cover using landsat 8 oli. *ISPRS Int. Arch. Photogramm. Remote Sens. Spat. Inf. Sci.* **2018**, *42*, W4. [CrossRef]
24. Ha, T.V.; Tuohy, M.; Irwin, M.; Tuan, P.V. Monitoring and mapping rural urbanization and land use changes using Landsat data in the northeast subtropical region of Vietnam. *Egypt. J. Remote Sens. Space Sci.* **2018**, *23*, 11–19. [CrossRef]
25. Dennedy-Frank, P.J.; Gorelick, S.M. Insights from watershed simulations around the world: Watershed service-based restoration does not significantly enhance streamflow. *Glob. Environ. Chang.* **2019**, *58*, 101938. [CrossRef]
26. Trinh-Tuan, L.; Matsumoto, J.; Ngo-Duc, T.; Nodzu, M.I.; Inoue, T. Evaluation of satellite precipitation products over Central Vietnam. *Prog. Earth Planet. Sci.* **2019**, *6*, 54. [CrossRef]
27. Luu, C.; Von Meding, J.; Kanjanabootra, S.; Pham, D. A proposed flood risk assessment method for Central Vietnam. In Proceedings of the 5th International Conference on Building Resilience, Newcastle, NSW, Australia, 15–17 July 2015; pp. 336-1–336-11.
28. Van Khanh, N. Identify and Assess the Impact of Climate Change and Sea Level Rise to the System of Landfills and Solid Waste Treatment Facilities in the Central Coast Region of Vietnam. In *Waste Management and Resource Efficiency*; Springer: Singapore, 2018; pp. 195–208. [CrossRef]
29. Location and Natural Conditions. Available online: https://www.danang.gov.vn/web/en/detail?id=26029&_c=16407111 (accessed on 24 February 2022).
30. Overview of Quang Binh Province. Available online: https://www.quangbinh.gov.vn/3cms/gioi-thieu-chung-14532.htm (accessed on 24 February 2022).
31. Quang Nam Portal. Available online: https://quangnam.gov.vn/webcenter/portal/ubnd_en (accessed on 24 February 2022).
32. Overview of Quang Tri Province. Available online: https://www.quangtri.gov.vn/xem-chi-tiet-gioi-thieu-tong-quan/-/view-article/1/3500113539863336577/1573630224087QuangTri (accessed on 24 February 2022).
33. Population of Thua Thien Hue Province. Available online: https://thuathienhue.gov.vn/en-us/Home/Detail/tid/Population/newsid/65F39533-85E0-4C1B-BC34-A8B600A82A8E/cid/AEBA5AE7-F4B9-4D9B-A507-DE8802BF1D14TTH (accessed on 24 February 2022).
34. Olyaie, E.; Abyaneh, H.Z.; Mehr, A.D. A comparative analysis among computational intelligence techniques for dissolved oxygen prediction in Delaware River. *Geosci. Front.* **2017**, *8*, 517–527. [CrossRef]
35. Friedman, J.H. Multivariate adaptive regression splines. *Ann. Stat.* **1991**, *19*, 1–67. [CrossRef]
36. Sekulic, S.; Kowalski, B.R. MARS: A tutorial. *J. Chemom.* **1992**, *6*, 199–216. [CrossRef]
37. Steinberg, D. An alternative to neural nets: Multivariate adaptive regression splines (MRAS). *PC AI* **2001**, *15*, 38–41.
38. Fan, J.; Wu, L.; Ma, X.; Zhou, H.; Zhang, F. Hybrid support vector machines with heuristic algorithms for prediction of daily diffuse solar radiation in air-polluted regions. *Renew. Energy* **2020**, *145*, 2034–2045. [CrossRef]
39. LeBlanc, M.; Tibshirani, R. Adaptive principal surfaces. *J. Am. Stat. Assoc.* **1994**, *89*, 53–64. [CrossRef]
40. Sharda, V.N.; Prasher, S.O.; Patel, R.M.; Ojasvi, P.R.; Prakash, C. Performance of Multivariate Adaptive Regression Splines (MARS) in predicting runoff in mid-Himalayan micro-watersheds with limited data/Performances de régressions par splines multiples et adaptives (MARS) pour la prévision d'écoulement au sein de micro-bassins versants Himalayens d'altitudes intermédiaires avec peu de données. *Hydrol. Sci. J.* **2008**, *53*, 1165–1175.

41. Craven, P.; Wahba, G. Smoothing noisy data with spline functions. *Numer. Math.* **1978**, *31*, 377–403. [CrossRef]
42. Mohsen, S. Computational Estimation of Biliary Excretion of Compounds and the Role of Transporters. Ph.D. Thesis, University of Kent, Canterbury, UK, 2014.
43. Aydin, D.; Yilmaz, E. Modified Spline Regression based on Randomly Right-Censored Data: A Comparative Study. *Commun. Stat. Simul. Comput.* **2017**, *47*, 2587–2611. [CrossRef]
44. Friedman, J.H.; Hastie, T.; Tibshirani, R. Regularization Paths for Generalized Linear Models via Coordinate Descent. *J. Stat. Softw.* **2010**, *33*, 1–22. [CrossRef]
45. Tibshirani, R. Regression shrinkage and selection via the Lasso. *J. R. Stat. Soc. Ser. B (Methodol.)* **1996**, *58*, 267–288. [CrossRef]
46. Chouzenoux, E.; Pesquet, J.-C. A Stochastic Majorize-Minimize Subspace Algorithm for Online Penalized Least Squares Estimation. *IEEE Trans. Signal Process.* **2017**, *65*, 4770–4783. [CrossRef]
47. Breiman, L. Random forests. *Mach. Learn.* **2001**, *45*, 5–32. [CrossRef]
48. Guo, L.; Chehata, N.; Mallet, C.; Boukir, S. Relevance of airborne lidar and multispectral image data for urban scene classification using Random Forests. *ISPRS J. Photogramm. Remote Sens.* **2011**, *66*, 56–66. [CrossRef]
49. Rodriguez-Galiano, V.F.; Ghimire, B.; Rogan, J.; Chica-Olmo, M.; Rigol-Sanchez, J.P. An assessment of the effectiveness of a random forest classifier for land-cover classification. *ISPRS J. Photogramm. Remote Sens.* **2012**, *67*, 93–104. [CrossRef]
50. Rodriguez-Galiano, V.; Sanchez-Castillo, M.; Chica-Olmo, M.; Chica-Rivas, M.J.O.G.R. Machine learning predictive models for mineral prospectivity: An evaluation of neural networks, random forest, regression trees and support vector machines. *Ore Geol. Rev.* **2015**, *71*, 804–818. [CrossRef]
51. Seni, G.; Elder, J.F. Ensemble Methods in Data Mining: Improving Accuracy Through Combining Predictions. *Synth. Lect. Data Min. Knowl. Discov.* **2010**, *2*, 1–126. [CrossRef]
52. Kardani, N.; Zhou, A.; Nazem, M.; Shen, S.-L. Estimation of Bearing Capacity of Piles in Cohesionless Soil Using Optimised Machine Learning Approaches. *Geotech. Geol. Eng.* **2019**, *38*, 2271–2291. [CrossRef]
53. Touzani, S.; Granderson, J.; Fernandes, S. Gradient boosting machine for modeling the energy consumption of commercial buildings. *Energy Build.* **2018**, *158*, 1533–1543. [CrossRef]
54. Yang, J.-H.; Yang, M.-S. A control chart pattern recognition system using a statistical correlation coefficient method. *Comput. Ind. Eng.* **2005**, *48*, 205–221. [CrossRef]
55. Çevik, A.; Weber, G.-W.; Eyüboglu, B.M.; Oğuz, K.K. Voxel-MARS: A method for early detection of Alzheimer's disease by classification of structural brain MRI. *Ann. Oper. Res.* **2017**, *258*, 31–57. [CrossRef]
56. Taylor, K.E. Summarizing multiple aspects of model performance in a single diagram. *J. Geophys. Res. Atmos.* **2001**, *106*, 7183–7192. [CrossRef]
57. Ghorbani, M.A.; Deo, R.C.; Yaseen, Z.M.; Kashani, M.H.; Mohammadi, B. Pan evaporation prediction using a hybrid multilayer perceptron-firefly algorithm (MLP-FFA) model: Case study in North Iran. *Arch. Meteorol. Geophys. Bioclimatol. Ser. B* **2017**, *133*, 1119–1131. [CrossRef]
58. Jamali, A. Evaluation and comparison of eight machine learning models in land use/land cover mapping using Landsat 8 OLI: A case study of the northern region of Iran. *SN Appl. Sci.* **2019**, *1*, 1448. [CrossRef]
59. Adab, H.; Morbidelli, R.; Saltalippi, C.; Moradian, M.; Ghalhari, G.A.F. Machine Learning to Estimate Surface Soil Moisture from Remote Sensing Data. *Water* **2020**, *12*, 3223. [CrossRef]
60. Duong, P.C.; Trung, T.H.; Nasahara, K.N.; Tadono, T. JAXA High-Resolution Land Use/Land Cover Map for Central Vietnam in 2007 and 2017. *Remote Sens.* **2018**, *10*, 1406. [CrossRef]

MDPI
St. Alban-Anlage 66
4052 Basel
Switzerland
Tel. +41 61 683 77 34
Fax +41 61 302 89 18
www.mdpi.com

Sustainability Editorial Office
E-mail: sustainability@mdpi.com
www.mdpi.com/journal/sustainability

www.ingramcontent.com/pod-product-compliance
Lightning Source LLC
Chambersburg PA
CBHW061436210926
51850CB00008B/1527
* 9 7 8 3 0 3 6 5 8 3 4 4 0 *